W0107632

EXS 74

Epithelial–Mesenchymal Interactions in Cancer

Edited by I.D. Goldberg
E.M. Rosen

Birkhäuser Verlag
Basel · Boston · Berlin

Editors
Dr. I.D. Goldberg
Dr. E.M. Rosen
Long Island Jewish
Medical Center
Institute of Oncology
New Hyde Park
N.Y. 11042
USA

Library of Congress Cataloging-in-Publication Data
Epithelial-mesenchymal interactions in cancer / edited by I.D.
  Goldberg, E.M. Rosen.
     p. cm. — (Exs; 74)
  Includes bibliographical references and index.

  1. Epithelium—Tumors.   2. Mesenchyme—Pathophysiology.   3. Cancer
invasiveness.   4. Carcinogenesis.   5. Fibroblasts.   6. Transforming
growth factors.   I. Goldberg, I.D. (Itzhak David), 1948–
II. Rosen, E.M. (Eliot M.)   III. Series: Experientia. Supplementum; v. 74.
  [DNLM: 1. Epithelium—physiology.   2. Mesoderm—physiology.
3. Stromal Cells—physiology.   4. Cell Transformation, Neoplastic.
5. Neoplasm Metastasis.   6. Neoplasm Invasiveness.   EX23 v. 74 1995
/ QZ 365 E64 1995]
RC280.E66E65   1995
616.99'4071—dc20
DNLM/DLC
for Library of Congress

Deutsche Bibliothek Cataloging-in-Publication Data
EXS. – Basel; Boston; Berlin: Birkhäuser.
    Früher Schriftenreihe
    Fortlaufende Beil. zu: Experientia
74. Epithelial–mesenchymal interactions in cancer. – 1995
Epithelial–mesenchymal interactions in cancer / ed. by I.D. Goldberg; E.M. Rosen. – Basel;
Boston; Berlin: Birkhäuser, 1995
    (EXS; 74)

NE: Goldberg, Itzak D. [Hrsg.]

The publisher and editor can give no guarantee for the information on drug dosage and administration contained in this publication. The respective user must check its accuracy by consulting other sources of reference in each individual case.
The use registered names, trademarks etc. in this publication, even if noot identified as such, does not imply that they are exempt from the relevant protective laws and regulations or free for general use.

This work is subject to copyright. All rights are reserved, whether the whole or part of the material is concerned, specifically the rights of translation, reprinting, re-use of illustrations, recitation, broadcasting, reproduction on microfilms or in other ways, and storage in data banks. For any kind of use permission of the copyright owner must be obtained.

© 1995 Birkhäuser Verlag, P.O. Box 133, CH–4010 Basel, Switzerland

Softcover reprint of the hardcover 1st edition 1995

Printed on acid-free paper produced from chlorine-free pulp. TCF ∞

ISBN-13: 978-3-0348-9893-5   e-ISBN-13: 978-3-0348-9070-0
DOI: 10.1007/978-3-0348-9070-0
9 8 7 6 5 4 3 2 1

Dedicated to the memory of Gertrude Kurz,
whose character, devotion and courage
were an inspiration to all who knew her.

Itzhak D. Goldberg

Contents

Epithelial–Mesenchymal Interactions in Cancer
ed. by I.D. Goldberg & E.M. Rosen
© 1995 Birkhäuser Verlag Basel/Switzerland

Introduction

The idea that interactions between malignant epithelia and surrounding stromal cells might contribute to neoplasia is not new. The contribution of the stroma to tumor development and progression has been recognized in the past. Stromal cells actively interact with tumor epithelia and actively support malignant growth. In addition, tumor stromal cells (eg., fibroblasts) are phenotypically distinct from their non-tumor counterparts; these tumor fibroblasts may phenotypically resemble embryonic fibroblasts more than normal adult fibroblasts (see van den Hooff, A. (1988) Stromal involvement in malignant growth. *Adv. Cancer Res.* 50: 159–196). The implication is that tumor-associated fibroblasts, while not 'malignant' in the usual sense of the word, are somehow 'activated' or 'stimulated'. Within the last six years, considerable progress has been made in identifying some of the molecular mechanisms through which tumor cells interact with stroma. Some of the recent advances in this field and their potential relevance to cancer invasion and metastasis are highlighted in this book.

The importance of epithelial-mesenchyme interaction in embryogenesis and tissue formation was suspected even before its significance in malignancy came to be appreciated. Indeed, recent progress in deciphering the molecular signals that mediate the exchange between epithelia and mesenchyme suggests intriguing parallels between the processes of normal development and malignant growth. Some of the same cytokines, growth factors, proteolytic enzymes, cell adhesion molecules, and structural proteins of the extracellular matrix (ECM) have been implicated in both processes. Epithelia-mesenchyme signalling relevant to development is discussed in the chapter by Birchmeier and in some of the other chapters.

Scatter factor (SF, *aka.* hepatocyte growth factor) is a cytokine that is synthesized and secreted predominantly by mesenchymal cells; its canonical receptor, the protein product of the c-*met* proto-oncogene, is expressed predominantly on epithelial cells. SF is over-produced within tumors; and an increasing body of circumstantial evidence suggests that SF may mediate tumor-stroma interactions that contribute to tumor progression. Different aspects of these interactions are discussed in this volume. Rosen and Goldberg discuss potential mechanisms that may account for the accumulation of large quantities of SF within tumors. Evidence is presented that malignant tumor cells secrete proteins (SF-inducing factors) that stimulate stromal cells to produce more SF. Zarnegar discusses recent studies of the SF gene promoter that may be relevant to understanding the detailed molecular mechanism(s) by which soluble factors regulate SF production.

Polverini and Nickoloff discuss another mechanism by which SF may enhance tumor growth, ie., stimulation of angiogenesis, the formation of new blood vessels from pre-existing microvessels. Angiogenesis is required for continued growth of most solid tumors, and provides a mechanism by which the stroma may continue to grow along with the tumor cells. Although endothelial cells are stromal cells, they express a number of epithelial characteristics including (i) epithelial-like tight junctions and junctional proteins; (ii) the ability to organize into flattened tubular structures; (iii) the c-*met* receptor protein; and (iv) biologic responsiveness to SF. It is, perhaps, not surprising that vascular endothelial cells may both produce and respond to SF in different situations.

'Epithelialness' may be defined in two ways: (i) expression of generic epithelial structures and proteins (eg., specialized junctions, junctional proteins [eg., cadherins, Z01], cytokeratins); and (ii) production of specific differentiated products (eg., milk proteins by mammary epithelia, renin by renal tubular epithelia of the juxtaglomerular apparatus). Recent studies suggest that SF/c-*met* signalling may mediate epithelia-mesenchyme interconversion, in part by modifying some of the generic epithelial characteristics. Nusrat discusses the effects of SF on the epithelial junctional apparatus. Relatively little is known about whether and how SF regulates cell-specific differentiation.

The contributions of paracrine and autocrine SF/c-*met* signalling to the tumorigenic phenotype of NIH/3T3-derived tumors in an *in vivo* animal model is discussed by Cortner and colleagues. These studies may be relevant to mechanisms of human sarcomagenesis, the process by which tumors of mesenchymal origin are formed. The authors have shown that over-expression of both ligand and receptor are required for maximal tumorigenesis. Thus, mechanisms that normally repress c-*met* receptor expression in fibroblasts may be important in the prevention of spontaneous development of sarcomas. The authors suggest that the p53 anti-oncoprotein is one possible regulator of c-*met* expression, providing possible linkage between p53 mutation and c-*met* over-expression.

A variety of other growth factors, cytokines, and motility factors that accumulate within the tumor microenvironment contribute to tumor growth, invasion, and metastasis. Some of these factors are described in the chapters by Drs. Nicolson, Stracke, Kacinski, and Rubin. Stracke describes a group of factors originally thought to primarily mediate tumor cell motility, although in some cases, various other activities have been documented. Some of these factors are autocrine factors, ie., act upon the same cells that produce them. The designation of a factor as autocrine may be somewhat artificial, since a soluble factor produced by one tumor cell may diffuse to and stimulate a neighboring tumor cell. Autotaxin, a tumor cell-derived motility factor related to certain ectoenzymes (ie., phosphodiesterases) is described in detail in the chapter by Stracke. Nicolson describes paracrine interactions between microvascular endothelial cells and tumor cells that may explain why certain tumor types metastasize exclusively or selectively to specific organs. Some of the molecular mediators of organ specific colonization and growth of tumor cells which have been isolated and characterized are discussed in this chapter.

Kacinski reviews some new data relating to tumor-stromal interactions that may promote the growth of ovarian carcinomas, including studies from his own laboratory. Colony-stimulating factor-1 (CSF-1) and its receptor (the c-*fms* proto-oncogene product) may be of particular interest. CSF-1 simulates proliferation and invasion of ovarian carcinoma cells *in vitro*, and its expression *in vivo* is correlated with prognosis. Moreover, CSF-1 expression may be regulated in part by a paracrine mechanism involving production of inducing factors by tumor cells leading to production of CSF-1 by stromal cells. The recently discovered keratinocyte growth factor (KGF), a member of the heparin-binding family of fibroblast growth factors, stimulates epithelial cell proliferation through its receptor, a tyrosine kinase isoform encoded by the *fgfr-2* gene. Studies of the distribution and activity of KGF and its receptor suggest that KGF may act as a paracrine mediator of mesenchymal-epithelial communication. These findings are described, and their possible contribution to malignancy are discussed in the chapter by Rubin.

In addition to growth-stimulating and invasogenic cytokines, other factors may promote the invasive tumor phenotype, including proteases that degrade basement membrane and other forms of ECM. Accumulating evidence indicates that members of the family of matrix metalloproteinases (MMPs) contribute to the invasive behavior of tumor cells. Enzymes of particular interest in this respect are MMP-2 and MMP-9, two distinct forms of type IV collagenase. Molecular interactions between invading tumor cells and adjacent stroma contributory to the expression and activation of these enzymes have recently been identified. As described by Shih, it now appears that type IV procollagenase is produced by stromal fibroblasts in response to factors secreted by tumor cells at the leading edge of an invading tumor. Moreover, procollagenase IV may be activated by proteins located on the tumor cell membrane. Some of these same mechanisms may also be relevant to angiogenesis, during which endothelial cells transiently become proteolytically active and invade through basement membrane of the parent vessel. Endothelium-stroma interactions that may regulate the angiogenic phenotype are discussed in the chapter by Grant.

Studies of potential differences between tumor-associated versus normal adult fibroblasts have been pursued at the molecular level. Two chapters on this subject have been provided by Drs. Hornby and Schor.

Dr. Hornby reviews the roles of growth factors, ECM, and proteolytic enzymes as mediators of mesenchyme-epithelia interactions in tumors. She goes on to describe some specific molecular differences between fibroblasts derived from benign *versus* malignant human breast tissue. Dr. Schor describes differences between breast cancer/fetal-derived fibroblasts and normal adult fibroblasts with respect to their ability to produce and respond to a soluble migration-stimulating factor (MSF). MSF is an 'autocrine' factor in that it is produced by and acts upon fibroblasts. However, studies by Schor and his colleagues suggest that the expression of MSF in adults may reflect an abnormal mesenchyme-epithelia interaction that may predispose to cancer.

The subject of mesenchyme-epithelial interaction, even those interactions limited to neoplasia, is too broad to cover exhaustively. We hope that the material contained in this volume will provide the reader with an overview of this exciting field. We further hope that the ideas and hypothesis expressed by the authors will stimulate new avenues of research.

Epithelial–Mesenchymal Interactions in Cancer
ed. by I.D. Goldberg & E.M. Rosen
© 1995 Birkhäuser Verlag Basel/Switzerland

Epithelial–mesenchymal transitions in development and tumor progression

W. Birchmeier and C. Birchmeier

Max-Delbrueck-Centrum for Molecular Medicine, Robert-Roessle-Str. 10, D-13122 Berlin, Germany

Summary. Epithelial–mesenchymal transitions play important roles in development and malignancy. Here we discuss molecular events in the control of such transitions: changes in cellular adhesion components, action of oncogenes and tyrosine kinase receptors, as well as activation of transcription factors. In development, epithelial–mesenchymal transitions take place in a temporally and spatially controlled manner, whereas in tumors these changes are highly uncontrolled. Loss of epithelial character is typically observed late in progression of human carcinomas, and correlates there with the acquisition of invasive and metastatic potential.

Introduction

Epithelia and mesenchyme are two distinct types of tissues found virtually in every organ: epithelia are composed of closely associated, largely immobile cells; in contrast, mesenchyme contains more mobile, fibroblast-like cells that form loosely associated cell agglomerations. During normal development, transitions of epithelia to mesenchyme occur (for example in gastrulation), and mesenchyme can differentiate into new epithelia (for example in kidney development). Such transitions are not confined to development. In particular, the loss of epithelial character in malignant carcinomas, which results in the appearance of invasive, motile cells, is of major importance in tumor progression. In the past, the biology of epithelial–mesenchymal and mesenchymal–epithelial transitions have been well characterized. However, these studies have remained essentially descriptive and only recently have the molecular control events become amenable to experimentation. Here we will discuss various mechanisms that regulate transitions between epithelia and mesenchyme during development and in tumor progression.

Mesenchymal and epithelial characteristics

Mesenchymal cells are generally non-polarized, loosely associated, and surrounded by extracellular matrix. Fibroblasts in culture can be considered representative of such mesenchymal cells: Their adhesive contacts are primarily to the extracellular matrix. Fibroblasts express various integrin receptors that recognize components of the matrix (Hynes, 1992). Fibroblasts also produce various ligands for epithelial-specific receptor tyrosine kinases, eg. scatter factor/hepatocyte growth factor (SF/HGF), keratinocyte growth factor (KGF), and neu differentiation factor (NDF), also called neuregulin, that are important for signal exchange between mesenchymal and epithelial cell compartments (Birchmeier and Birchmeier, 1993).

Epithelial cells are morphologically entirely different; they form continuous cell layers, and they are generally polar and less mobile. In single-layered epithelia (eg. the mature intestine), apical and basolateral cell surfaces are separated by tight junctions. An example of a multilayered epithelium is the skin, where basal cells (stem cells) are covered by layers of differentiating cells. Epithelial cells are interconnected by adherens junctions and desmosomes, organelles responsible for strong intercellular adhesion, and they form basement membranes at their basal surface. Hemidesmosomes are responsible for cell contacts to the basement membrane (Timpl, 1989; Sonnenberg et al., 1991). In addition, tyrosine kinase receptors for a variety of mesenchymal ligands are expressed on epithelial cells, for instance c-met, KGF receptor, and c-erbBs (Kraus et al., 1989; Press et al., 1990; Plowman et al., 1993; Sonnenberg et al., 1993; Orr-Urteger et al., 1993). Interactions of specific ligands with the tyrosine kinase receptors is thus one of several possible mechanisms used for signal exchange between the mesenchymal and epithelial cell compartments.

Epithelial cells are the first identifiable embryonic cell type and appear during compaction of the morula. During gastrulation, the three germ layers, ectoderm, mesoderm and endoderm are formed. Whereas ectoderm and endoderm remain essentially epithelial in character, the mesoderm consists of the first mesenchymal cells in development and expresses a set of proteins typical for the mesenchymal lineage. Mesoderm formation is thus an example of a controlled epithelial–mesenchymal transition. During further development, the various parenchymal organs are formed and, in general, mesenchyme and epithelia participate in this process. New epithelia usually derive from the ectoderm or endoderm, but can also be formed from the mesoderm by a mesenchymal–epithelial transition, for example during development of the kidney (Saxen, 1987).

Tumors of epithelial origin are of major medical importance, since more than 90% of all fatal malignancies are carcinomas (DeVita et al., 1993). In human carcinomas, the loss of epithelial differentiation is a prognostic marker and correlates with a poor outcome of the disease (Gabbert et al., 1985). Loss of epithelial differentiation is accompanied by greater motility and invasiveness of the tumor cells and by reduced intercellular adhesion. The ability to form metastases is critically dependent on a permanent or transient loss of epithelial characteristics (Birchmeier and Behrens, 1994).

Experimentally-induced transitions

Epithelial–mesenchymal transitions can be induced *in vitro* by several experimental approaches: Epithelial cells can be forced to assume fibroblast-like properties by the disturbance of their adherens junctions, by the expression of various oncogenes, or by several soluble factors. For example, antibodies against the epithelial cell adhesion molecule E-cadherin reversibly induce a fibroblast-like morphology in epithelial cells (Imhof et al., 1983; Behrens et al., 1989). Expression of a temperature-sensitive v-*src* (Behrens et al., 1993; Hamaguchi et al., 1993) or an inducible c-*fos* protein (Reichmann et al., 1992) converts epithelial cells to a mesenchymal cell shape and induces the synthesis of various mesenchymal markers. Motility factors such as scatter factor/hepatocyte growth factor (SF/HGF) or acidic fibroblast growth factor (aFGF) dissociate epithelial cells, induce high motility and a fibroblast-like morphology (Stoker et al., 1987; Weidner et al., 1990; Jouanneau et al., 1991). Finally, over-expression of c-*erbB*2 in breast epithelial cells downregulates E-cadherin expression and induces fibroblastoid phenotype. Treatment of the cells with antibodies against c-*erbB*2 that inhibit tyrosine phosphorylation of the receptor, allows a reversion to an epithelial morphology and upregulates E-cadherin (D'souza et al., 1994).

Conversely, transitions of mesenchymal cells to epithelia have been accomplished by several methods: ectopic expression of E-cadherin in fibroblasts results in cells which become adhesive to each other and show polar characteristics (Nagafuchi et al., 1987; Ringwald et al., 1987; Gallin et al., 1987). Similarly, poorly differentiated, fibroblast-like carcinoma cells can be converted to epitheloid cells by the expression of E-cadherin and then lose their invasive potential (Vlemickx et al., 1991; Frixen et al., 1991). During normal development, kidney mesenchyme is induced to form new epithelia by signals derived from the ureter bud; the newly induced cells then express *int-4* (Stark et al., 1994) and the transcription factors N-*myc* and *pax-2* (Mugrauer et al., 1988; Dressler et al., 1993). Recent experimental evidence indicates that the closely related factor, *int-1*, is a signal that can convert kidney mesenchyme to epithelium (Herzlinger et al., personal communication). Also, a retrovirus that transduces the transcription factor *pax-2* is sufficient to convert kidney mesenchyme to epithelium (Dressler et al., 1993). Co-expression of c-*met* and SF/HGF in NIH3T3 fibroblast induced epithelial characteristics (Tsarfaty et al., 1994). Additionally, the adenovirus E1a gene causes diverse tumor cells of mesenchymal origin to adapt epithelial morphology and induces expression of epithelial-specific genes (Frisch, 1994).

The structure of intercellular adhesion junctions in epithelia

At the lateral sides of epithelial cells are intercellular junctions such as tight and adherens junctions as well as desmosomes. A consequence of the laterally located junctional complexes is the cell polarity characteristic for epithelial cells. Distinct proteins are expressed on the basolateral or apical surface, and free diffusion of membrane proteins to all surfaces of the cells is inhibited. Polar epithelial cells have evolved special mechanisms that allow the transport of membrane proteins to either the apical and basolateral surface (Simons and Wandinger-Ness, 1990). Adhesion molecules or receptors for signals received by epithelia are predominantly located on the basolateral surface.

Adherens junctions are specialized structures containing the transmembrane cell adhesion molecule E-cadherin, that recognises and binds E-cadherin present on the neighbouring cells in a Ca^{2+}-dependent manner. E-cadherin is a 120 kDa transmembrane glycoprotein, from which an extracellular 80 kDa tryptic fragment can be released in the presence of Ca^{2+}. The cDNAs of mouse, chicken and human E-cadherin have been characterized; they code for a signal peptide at the amino terminus, a large extracellular domain with four repeated domains important in Ca^{2+}-binding, a single transmembrane sequence, and a 15 kDa cytoplasmic domain (Takeichi, 1991). E-cadherin is the prototype of a family of Ca^{2+}-dependent cell adhesion molecules; close relatives are N-cadherin (neurons) and P-cadherin (placenta and epithelia). Further members are M-cadherin (muscle), OB-cadherin (osteoblasts), LI-cadherin (liver, intestine), as well as the desmosomal proteins desmoglein and desmocollin, and others (Takeichi, 1991; Donalies et al., 1991; Buxton and Magee, 1992; Kemler, 1992; Sano et al., 1993; Okazaki et al., 1994; Berndorff et al., 1994).

E-cadherin is not only important for the maintenance of morphology and adhesion in differentiated epithelia, but also for the acquisition of epithelial characteristics early in embryogenesis. This was originally suggested by experiments with anti-E-cadherin antibodies (Hyafil et al., 1981), and recently also demonstrated genetically (Larue et al., 1994; Riethmacher et al., 1995). We introduced a targeted mutation into the E-cadherin gene of the mouse via homologous recombination and embryonic stem cell technology. This mutation removes sequences essential for Ca^{2+}-binding and thus for the adhesive function of the molecule (Riethmacher et al., 1995). Animals that carry this mutation in a heterozygous state appear normal and are fertile. However, homozygous mutant embryos do not develop normally. They can reach the morula stage and also compact properly, but the compacted state is not sustained. The individual morula cells in the embryo lose their morphological polarization, become rounded and continue to divide. As a consequence, the embryos appear totally distorted at a time when wild type or heterozygous mutant embryos form a well-organized blastocyst with a well-formed blastocoel. Since the mutant embryos never emerge from the zona pellucida, further development, particularly implantation into the uterus, cannot take place. It has previously been reported that removal of Ca^{2+}-ions or treatment with anti-E-cadherin antibodies interferes with the compaction of the mouse morula, indicating that the initial compaction and polarization of the epithelial-like cells is dependent on E-cadherin (Hyafil et al., 1981; Vestweber and Kemler, 1984). Nevertheless, embryos which lack a functional E-cadherin gene can undergo normal compaction. This is due to maternally derived E-cadherin and not to functional compensation by other cell adhesion molecules: Anti-E-cadherin antibodies interfere with compaction of the embryo even in homozygous mutant embryos. The presence of the maternal E-cadherin in mutant embryos at early stages was demonstrated by immunohistological techniques; it disappears when the embryos develop beyond the morula stage. The maternal E-cadherin suffices thus for the initial compaction and for the formation of the first polarized, epithelial-like cells, but not for further development beyond the morula stage.

The integrity of adherens junctions also requires catenin molecules. The cytoplasmic portion of E-cadherin interacts with α-, β- and γ-catenins which mediate the interaction to the cytoskeleton, and this interaction is essential for the adhesive function of E-cadherin. α-Catenin has sequence similarities with vinculin, β-catenin is the mammalian homologue of *armadillo* from *Drosophila*, and γ-catenin is identical to plakoglobin also found at the desmosome (Takeichi, 1991; Kemler, 1992; Tsukita et al., 1993; Troyanovsky et al., 1994). Transient expression of cDNA encoding variant β-catenin which lacks sequences for different subdomains has elucidated the interactions of the molecules in the adherens junction complex. E-cadherin binds directly to β-catenin; the binding site in β-catenin consists of *armadillo* repeat sequences (named after the *Drosophila armadillo* gene where this sequence motive was first identified) located in the central part of the molecule. The N-terminus of β-catenin interacts with α-catenin, and via α-catenin the cytoskeletal proteins bind to the complex. γ-Catenin (plakoglobin) is similar in structure to β-catenin, and it interacts in a similar manner with E-cadherin and α-catenin (Huelsken et al., 1994; Naethke et al., 1994). Also the APC tumor suppressor gene product can bind to β-catenin (Rubinfeld et al., 1993; Su et al, 1993). Mutations of APC are frequently observed in colon carcinomas (Fearon et al., 1990). APC and E-cadherin have largely indistinguishable binding-sites on β-catenin, ie. they can compete for β-catenin binding (Huelsken et al., 1994).

Mechanisms that interfere with intercellular adhesion of epithelial cells

Changes in structure or expression of any of the essential components of the adherens junctions can lead to junctional disassembly and, consequently, loss of epithelial and acquisition of mesenchymal character. Various molecular alterations that result in defective adhesion of epithelial cells *in vitro* have been identified: expression of E-cadherin without a cytoplasmic domain in fibroblasts does not result in functional cell–cell adhesion, in contrast to the expression of intact E-cadherin. Apparently, the truncated E-cadherin is unable to interact with the catenins and therefore, proper junctions cannot be formed (Nagafuchi and Takeichi, 1988; Ozawa et al., 1990a). Cadherins without the entire extracellular sequence, but an intact transmembrane and cytoplasmic domain, act in a dominant–negative fashion ie. distributed epithelial cell adhesion. These truncated molecules might deplete the cellular pool of catenins, which are consequently not available to form complexes with the intact E-cadherin (Kintner, 1992; Fujimori and Takeichi, 1993). A point mutation in the first Ca^{2+}-binding motive of E-cadherin results in a non-functional cell adhesion molecule (Ozawa et al., 1990b). Similar mutations have been identified in diffuse-type gastric carcinomas (Becker et al., 1994; see also below). The adhesive properties of PC9 lung cancer cells could be restored by transfection with α-catenin cDNA; these cells harbour a null mutation in the α-catenin gene (Hirano et al., 1992; Shimoyama et al., 1992). Transformation of epithelial cells with a temperature-sensitive v-*src* gene induces a reversible loss of differentiation and tyrosine phosphorylation of β-catenin at the permissive temperature (Matsuyoshi et al., 1992; Behrens et al., 1993; Hamaguchi et al., 1993). Various receptor tyrosine kinases have also been found to phosphorylate β-catenin; in parallel, these receptors can induce the dissociation of epithelial cells (Shibamoto et al., 1994; see also Hoschuetzky et al., 1994). It has therefore been postulated that tyrosine phosphorylation of β-catenin might interfere with an intact adhesion complex and induce a non-adhesive state of epithelial cells.

In addition to mutation, adherens junction integrity can also be affected by changes in the regulation of gene expression of specific junctional constituents. In particular, the expression of E-cadherin and α-catenin is known to be down-regulated in tumors (see also below). We and others have therefore characterized the E-cadherin promoter in order to analyze the mechanisms that regulate epithelial-specific expression and down-regulation of expression during de-differentiation of tumors (Behrens et al., 1991; Ringwald et al., 1991). We found that a -178 to $+92$ bp upstream fragment from the E-cadherin promoter induced epithelial-specific expression of a reporter gene. In non-epithelial cells, this promoter fragment is not active, demonstrating the presence of cell type-specific regulatory elements. Mutations of a proximal GC-rich sequence motive (at -25 to -58) and a CCAAT-box (at -65) resulted in loss of transcriptional activity in epithelial cells. Mutations or deletions of a palindromic element termed E-Pal (at -86) reduced promoter activity in epithelial cells but, surprisingly, resulted in increased activity in non-epithelial and dedifferentiated carcinoma cells. The E-cadherin promoter is therefore composed of positive and negative regulatory cis-elements, the combination of which generates epithelial specific activity (Behrens, J., G. Hennig and W. Birchmeier, unpublished results).

Alterations of cell adhesion components in carcinomas

Since the importance of intact E-cadherin-mediated cell adhesion in the maintenance of the epithelial phenotype and the prevention of invasion and metastasis was recognized (Behrens et al., 1989), various types of human carcinomas were examined for defects in one of the various constituents of intercellular junctions. A correlation between dysfunction of E-cadherin-mediated cellular adhesion (by down-regulation of E-cadherin and α-catenin expression or mutation in the E-cadherin gene) and invasive potential of many types of carcinomas has emerged from these studies. We will discuss selected examples, with the object of reviewing all the different mechanisms of alterations observed in tumors.

Diffuse-type gastric carcinomas contain poorly adhesive cells that often express E-cadherin protein and mRNA. When the mRNA was analyzed by reverse PCR, a high frequency of altered E-cadherin transcripts were identified. Fifty percent of all the tumors tested express shortened transcripts that lack exon 9 or 10 (Becker et al., 1994); these exons encode part of the first Ca^{2+}-binding site of E-cadherin and thus sequences known to be essential for adhesive function. Analysis of the genomic E-cadherin DNA demonstrated that mutations are present in the corresponding splice junctions, and that the second, intact copy of the gene was frequently lost (Becker et al., 1994). Mutations in one copy followed by mutation or loss of the second intact allele are typical for genes that play a causative role in the development or progression of human tumors, ie. for tumor suppressor genes.

A frequently observed alteration in malignant carcinoma cells of various tissues of origin is down-regulation or even complete absence of E-cadherin expression. In head and neck cancers, for example, down-regulation of E-cadherin expression correlates with loss of epithelial differentiation of the tumors and with the formation of lymph node metastases (Schipper et al., 1991). In breast carcinomas, a correlation between down-regulation of E-cadherin expression and malignancy is observed for lobular, but not for ductal carcinomas (Moll et al., 1993; Gamallo et al., 1993). In colorectal carcinomas, down-regulation of E-cadherin was similarly observed to correlate with late stages of tumor progression (Dorudi et al., 1993; Kinsella et al., 1993). In bladder carcinomas, down-regulation of E-cadherin correlates with poor survival rates (Bringuier et al., 1994; Otto et al., 1994).

In summary, the molecular aspects of junctional deficiencies and the loss of differentiation of epithelial cells in carcinomas has turned out to be quite complex; many junctional proteins (E-cadherin, catenins) but also proteins of various signal cascades (src or tyrosine kinase receptors, see below) can be involved. The critical proteins are affected by down-regulation of expression, by mutation, or by phosphorylation. Often, down-regulation of E-cadherin expression correlates with loss of tumor differentiation and progression to lymph node metastasis. The prognostic value of the measurement of E-cadherin expression will certainly be the subject of future clinical research. Often, several components of the epithelial differentiation program appear to be deregulated in concert in carcinomas, which indicates that epithelial master genes are affected during tumorigenesis. Obviously, developmental studies aimed at the elucidation of mechanisms that regulate the epithelial phenotype will also contribute to our understanding of malignant abberations of epithelial cells during carcinogenesis.

Epithelial–mesenchymal transitions induced by tyrosine kinase receptors

Various soluble factors that function as ligands for tyrosine kinase receptors can induce transient epithelial–mesenchymal conversion and motility in cell culture, eg. SF/HGF, aFGF, EGF and others (Stoker et al., 1987; Jouanneau et al., 1993; Hoschuetzky et al., 1994). The best characterized motility factor is the ligand for the c-*met* tyrosine kinase, that has been named scatter factor (SF) because of its activity on epithelial cells in culture (Stoker et al., 1987; Weidner et al., 1991). In addition, the factor can induce growth of hepatocytes (hepatocyte growth factor, HGF) and other cells (Miyazawa et al., 1989; Nakamura et al., 1989). A third activity of SF/HGF is its morphogenic action, first discovered as the ability to induce epithelial cells to form tubules *in vitro* (Montesano et al., 1991a,b; see also below).

SF/HGF has a unique structure since it closely resembles proteases like plasminogen (40% sequence identity) but does not resemble other ligands for tyrosine kinase receptors (Miyazawa et al., 1989; Nakamura et al., 1989; Weidner et al., 1991). The protein is produced as an inactive precursor molecule (90 kD) that is cleaved outside the producing cells into a heavy (60 kD) and a light (30 kD) chain (Hartmann et al., 1992; Lokker et al., 1992; Naldini et al., 1992). The heavy chain contains an

N-terminal hairpin loop and four kringle domains; the light chain shows extensive homologies to serine proteases. However, two of the three amino acids which form the catalytic triad of serine proteases are altered in SF/HGF, and therefore the factor has no catalytic activity.

Like many other tyrosine kinase receptors, c-*met* was initially identified because of its transforming activity when mutated. The *met* oncogene was derived from an N-methyl-N-nitro-N-nitrosoguanidine-treated osteosarcoma cell line that was used in a transfection/tumorigenicity assay (Cooper et al., 1984). The gene transfered from the osterosarcoma cells was the product of a rearrangement that fused *tpr* (translocated promoter region) on chromosome 1 to c-*met* on chromosome 7 (Park et al., 1986). The oncogenic variant of *met* encodes a cytoplasmically located tyrosine kinase. In contrast, the protooncogene product is a transmembrane glycoprotein of 190 kD that is cleaved postranslationally into an a- and a b-chain (Gonzatti-Haces et al., 1988; Giordano et al., 1989). A major breakthrough in the understanding of the c-*met* receptor was the identification of its ligand, which is scatter factor/hepatocyte growth factor (Bottaro et al., 1991; Naldini et al., 1991). All the known biological activities of SF/HGF are mediated by the c-*met* receptor (Weidner et al., 1993).

We have recently analyzed the normal physiological role of the SF/HGF gene by the introduction of a targeted mutation in the mouse via homologous recombination and embryonic stem cell technology (Schmidt et al., 1995). Whereas animals with a heterozygous mutation in SF/HGF are normal and fertile, a homozygous mutation is not compatible with normal development. SF/HGF $-/-$ embryos die between day 13 and 16.5 (E13–E16.5) of development. Externally, mutant embryos appear normal; however starting on E12 their liver is considerably reduced in size. Histological examination reveals damage to the embryo on E14.5 that varies in severity and is not observed on E12.5. The essential function of SF/HGF in the development of the liver is also supported by the analysis of embryonic stem cells that carry two c-*met* mutant alleles. Such cells, when injected into blastocysts, cannot contribute to the liver, but participate in the development of a variety of other organs and cell types. In contrast, embryonic stem (ES) cells with only one mutant c-*met* allele participate appropriately in the development of the hepatic lineage (Bladt, F. and Birchmeier, C., manuscript in preparation). In addition, placental development is impaired in the SF/HGF $-/-$ embryos. The trophoblast cells are particularly affected, and consequently, a disorganisation of the labyrinth layer is observed. (In this layer, exchange of nutrients and oxygen between the maternal and fetal blood occurs.) Thus, two epithelial cell types in the embryo, hepatocytes and trophoblast cells of the placenta, are affected by the SF/HGF mutation. The c-*met* receptor is widely expressed on epithelial cells, both during development and in the adult, whereas SF/HGF is usually produced by mesenchymal cells in the vicinity (Sonnenberg et al., 1993). SF/HGF and c-*met* form a paracrine signalling system, a concept originally suggested by Stoker and colleagues, which is essential for normal development of the liver and placenta.

Among the activities of SF/HGF that were described in cell culture, the morphogeneic activity of SF/HGF is unique and no other factor with similar properties has been described (Montesano et al., 1991a,b). MDCK cells, when grown in a three-dimensional collagen matrix for several days, form hollow cysts. When SF/HGF is added, individual cells dissociate and move away from the cysts. Consequently, the cells reassociate and form continuous tubules. These tubules have a lumen surrounded by well polarized epithelial cells with a smooth basal surface in contact with the collagen matrix, and an apical surface rich in microvilli that faces the lumen. The structures formed *in vitro* resemble thus the tubular epithelia present in many organs (Montesano et al., 1991a). In the development of the mammary gland, SF/HGF plays a similar role, since it can induce the branching and growth of the tubular epithelia in organ cultures (Yang et al., 1995). Recent experimental evidence indicates, however, that not only the formation of tubular epithelia can be induced by SF/HGF, but also other morphogenic programs, such as the formation of crypt-like structures from colon and pancreas epithelial cells (Brinkmann et al., 1995). Thus, SF/HGF seems to be able to activate the intrinsic morphogenic potential of various epithelial cell types, but does not instruct the cells to form one particular structure.

In vitro data indicate that SF/HGF not only induces morphogenesis or motility of epithelial cells, but also invasiveness into collagen matrices (Weidner et al., 1990). Furthermore, it has recently been shown that SF/HGF is expressed in breast carcinomas, and that the level of expression correlates with the state of tumor progression (Yamashita et al., 1994). The question therefore arises: how can the same factor be responsible for epithelial morphogenesis on the one hand and invasiveness of epithelial cells on the other? Interestingly, the different morphogenic programs can only be induced by SF/HGF in cells with intact epithelial characteristics, including expression of E-cadherin and functional adherens junctions (Brinkmann et al., 1995). On the other hand, SF/HGF was shown to induce metastasis of the lymph nodes and lungs when expressed by transfection of cDNA is dedifferentiated breast carcinoma cells (S. Meiners, H. Naundorf and W. Birchmeier, manuscript in preparation; see also Rosen et al., 1994). Thus, the differentiation state of the epithelial cells seems to determine the response towards SF/HGF: morphogenesis requires an intact epithelial programme, whereas metastatic behaviour is observed in cells that have already undergone epithelial–mesenchymal transitions.

References

Becker, K.F., Atkinson, M.J., Reich, U., Becker, I., Nekarda, H., Siewert, J.R. and Hoefler, H. (1994) E-Cadherin gene mutations provide clues to diffuse type gastric carcinomas. *Cancer Res.* 54: 3845–3851.

Behrens, J., Mareel, M.M., Van Roy, F.M. and Birchmeier, W. (1989) Dissecting tumor cell invasion: epithelial cells acquire invasive properties after the loss of uvomorulin-mediated cell–cell adhesion. *J. Cell Biol.* 108: 2435–3447.

Behrens, J., Löwrick, O., Klein-Hitpass, L. and Birchmeier, W. (1991) The E-cadherin promoter: functional analysis of a GC-rich region and an epithelial cell-specific palindromic regulatory element. *Proc. Natl. Acad. Sci. USA* 88: 11495–11499.

Behrens, J., Vakaet, L., Friis, R., Winterhager, E., Van Roy, F., Mareel, M.M. and Birchmeier, W. (1993) Loss of epithelial differentiation and gain of invasiveness correlates with tyrosine phosphorylation of the E-cadherin/β-catenin complex in cells transformed with a temperature-sensitive v-SRC gene. *J. Cell Biol.* 120: 757–766.

Berndorff, D., Gessner, R., Kreft, B., Schnoy, N., Lajous-Petter, A.M., Loch, N., Reutter, W., Hortsch, M. and Tauber, R. (1994) Liver–intestine cadherin: molecular cloning and characterization of a novel Ca^{2+}-dependent cell adhesion molecule expressed in liver and intestine. *J. Cell Biol.* 125: 1353–1369.

Birchmeier, C. and Birchmeier, W. (1993) Molecular aspects of mesenchymal–epithelial interactions. *Ann. Rev. Cell Biol.* 9: 511–540.

Birchmeier, W. and Behrens, J. (1994) Cadherin expression in carcinomas: role in the formation of cell junctions and the prevention of invasiveness. *Biochim. Biophys. Acta* 1198: 11–26.

Bottaro, D.P., Rubin, J.S., Faletto, D.L., Chan, A.M-L., Kmiecik, T.E., Vande Woude, G.F. and Aaronson, S.A. (1991) Identification of the hepatocyte growth factor receptor as the c-*met* proto-oncogene product. *Science* 251: 802–804.

Bringuier, P.P., Umbas, R., Schaafsma, H.E., Karthans, H.F.M., Debruyne, F.M.J. and Schalken, J.A. (1993) Decreased E-cadherin immunoreactivity correlates with poor revival in patients with bladder tumors. *Cancer Res.* 53: 3241–3245.

Brinkmann, V., Sachs, M., Weidner, K.M. and Birchmeier, W. (1995) Scatter factor/hepatocyte growth factor induces tissue-specific morphogenic programs in epithelial cells. *J. Cell Biol.*; *in revision*.

Buxton, R.S. and Magee, A.I. (1992) Structure and interactions of desmosomal and other cadherins. *Sem. Cell Biol.* 3: 157–167.

Cooper, C.S., Park, M., Blair, D.G., Tainsky, M.A., Huebner, K., Croce, C.M. and Vande Woude, G.F. (1984) Molecular cloning of a new transforming gene from a chemically transformed human cell line. *Nature* 311: 29–34.

D'souza, B. and Taylor Papadimitriou, J. (1994) Overexpression of ERBB2 in human mammary epithelial cells signals inhibition of transcription of the E-cadherin gene. *Proc. Natl. Acad. Sci. USA* 91: 7202–7206.

DeVita, V.T., Hellman, S. and Rosenberg, S.A. (1993) *Cancer: Principles and Practice of Oncology*, Fourth Edition, J.B. Lippincott Co., Philadelphia.

Donalies, M., Cramer, M., Ringwald, M. and Starzinsky-Powitz, A. (1991) Expression of M-cadherin, a member of the cadherin multigene family, correlates with differentiation of skeletal-muscle cells. *Proc. Natl. Acad. Sci. USA* 88: 8024–8028.

Dorudi, S., Sheffield, J.P., Poulsom, R., Northover, J.M.A. and Hart, I.R. (1993) E-cadherin expression in colorectal cancer. An immunochemical and *in situ* hybridization study. *Am. J. Pathol.* 142: 981–986.

Dressler, G.R., Wilkinson, J.E., Rothenpieler, U.W., Patterson, L.T., Williams Simons, L. and Westphal, H. (1993) Deregulation of Pax-2 expression in transgenic mice generates severe kidney abnormalities. *Nature* 362: 65–67.

Fearon, E.R. and Vogelstein, B. (1990) A genetic model for colorectal tumorigenesis. *Cell* 61: 759–767.

Frisch, S.M. (1994) E1a induces the expression of epithelial characteristics. *J. Cell Biol.* 127: 1085–1096.

Frixen, U.H., Behrens, J., Sachs, M., Eberle, G., Voss, B., Warda, A., Lochner, D. and Birchmeier, W. (1991) E-cadherin mediated cell–cell adhesion prevents invasiveness of human carcinoma cell lines. *J. Cell Biol.* 11: 173–185.

Fujimori, T. and Takeichi, M. (1993) Disruption of epitheloid cell–cell adhesion by exoge-
nous expression of a mutated nonfunctional N-cadherin. *Mol. Biol. Cell* 4: 37–47.
Gabbert, H., Wagner, R., Moll, R. and Gerharz, C.D. (1985) Tumor dedifferentiation: an
important step in tumor invasion. *Exp. Metastasis* 3: 257–279.
Gallin, W.J., Sorkin, B.C., Edelman, G.M. and Cunningham, B.A. (1987) Sequence analysis
of a cDNA clone encoding the liver cell adhesion molecule, L-CAM. *Proc. Natl. Acad. Sci.
USA* 84: 2808–2812.
Gamallo, C., Palacios, J., Suarez, A., Pizarro, A., Navarro, P., Quintanilla, M. and Cano, A.
(1993) Correlation of E-cadherin expression with differentiation grade and histological type
in breast carcinoma. *Am. J. Pathol.* 142: 987–993.
Giordano, S., Ponzetto, C., Di Renzo, M.F., Cooper, C.S. and Comoglio, P.M. (1989)
Tyrosine kinase receptor indistinguishable from the c-*met* protein. *Nature* 339: 155–156.
Gonzatti-Haces, M., Seth, A., Park, M., Copeland, T., Oroszlan, S. and Vande Woude, G.F.
(1988) Characterization of the *tpr-met* oncogene p65 and the *met* proto-oncogene cell
surface p140 tyrosine kinases. *Proc. Natl. Acad. Sci. USA* 85: 21–25.
Hamaguchi, M., Matsuyoshi, N., Ohnishi, Y., Gotoh, B., Takeichi, M. and Nagai, Y. (1993)
p60v-*src* causes tyrosine phosphorylation and inactivation of the N-cadherin-catenin cell
adhesion system. *EMBO J.* 12: 307–314.
Hartmann, G., Naldini, L., Weidner, K.M., Sachs, M., Vigna, E., Comoglio, P.M. and
Birchmeier, W. (1992) A functional domain in the heavy chain of scatter factor/hepatocyte
growth factor binds and activates the c-*met* receptor, induces cell dissociation but not
mitogenesis. *Proc. Natl. Acad. Sci. USA* 89: 11574–11578.
Hirano, S., Kimoto, N., Shimoyama, Y., Hirohashi, S. and Takeichi, M. (1992) Identification
of a neural α-catenin as a key regulator of cadherin function and multicellular organization.
Cell 70: 293–301.
Hoschuetzky, H., Aberle, H. and Kemler, R. (1994) β-Catenin mediates the interaction of the
cadherin–catenin complex with epidermal growth factor receptor. *J. Cell Biol.* 127: 1375–
1380.
Huelsken, J., Birchmeier, W. and Behrens, J. (1994) E-Cadherin and APC compete for the
interaction with β-catenin and the cytoskeleton. *J. Cell Biol.* 127: 2061–2069.
Hyafil, F., Babinet, C. and Jacob, F. (1981) Cell–cell interactions in early embryogenesis: A
molecular approach to the role of calcium. *Cell* 26: 447–454.
Hynes, R.O. (1992) Integrins: versatility, modulation, and signaling in cell adhesion. *Cell* 69:
11–25.
Imhof, B.A., Vollmers, H.P., Goodman, S.L. and Birchmeier, W. (1983) Cell–cell interaction
and polarity of epithelial cells. Specific perturbation using a monoclonal antibody. *Cell* 35:
667–675.
Jounneau, J., Gavrilowic, J., Caruelle, D., Jaye, M., Moens, G., Caruelle, J.P. and Thiery, J.P.
(1991) Secreted or nonsecreted forms of acidic fibroblast growth factor produced by
transfected epithelial cells influence cell morphology, motility, and invasive potential. *Proc.
Natl. Acad. Sci. USA* 88: 2893–2897.
Kemler, R. (1992) Classical cadherins. *Sem. Cell Biol.* 3: 149–155.
Kinsella, A.R., Green, B., Lepts, G.C., Hill, C.L., Bowie, G. and Taylor, B.A. (1993) The role
of the cell adhesion molecule E-cadherin in large bowel tumor cell invasion and metastasis.
Br. J. Cancer 67: 904–909.
Kintner, C. (1992) Regulation of embryonic cell adhesin by the cadherin cytoplasmic domain.
Cell 69: 225–236.
Kraus, M.H., Issing, W., Miki, T., Popescu, N.C. and Aaronson, S.A. (1989) Isolation and
characterization of ERBB3, a third member of the ERBB/epidermal growth factor receptor
family: evidence for overexpression in a subset of human mammary tumors. *Proc. Natl.
Acad. Sci. USA* 86: 9193–9197.
Larue, L., Ohsugi, M., Hirchenhain, J. and Kemler, R. (1994) E-cadherin null mutant
embryos fail to form a trophectoderm epithelium. *Proc. Natl. Acad. Sci. USA* 91: 8263–
8267.
Lokker, N.A., Mark, M.R., Luis, E.A., Bennett, G.L., Robbins, K.A., Baker, J.B. and
Godowski, P.J. (1992) Structure–function analysis of hepatocyte growth factor: identifica-
tion of variants that lack mitogenic activity yet retain high affinity receptor binding. *EMBO
J.* 11: 2503–2510.

Matsuyoshi, N., Hamaguchi, M., Taniguchi, S., Nagafuchi, A., Tsukita, S. and Takeichi, M. (1992) Cadherin mediated cell-cell adhesion is perturbed by v-*src* tyrosine phosphorylation in metastatic fibroblasts. *J. Cell Biol.* 118: 703-714.

Miyazawa, K., Tsubouchi, H., Naka, D., Takahashi, K., Okigaki, M., Arakaki, N., Nakayama, H., Hirono, S., Sakiyama, O. and Takahashi, K. (1989) Molecular cloning and sequence analysis of cDNA for human hepatocyte growth factor. *Biochem. Biophys. Res. Commun.* 163: 967-973.

Moll, R., Mitze, M., Frixen, U.H. and Birchmeier, W. (1993) Differential loss of E-cadherin expression in infiltrating ductal and lobular breast carcinomas. *Am. J. Pathol.* 1433/6: 1731-1742.

Montesano, R., Schaller, G. and Orci, L. (1991a) Induction of epithelial tubular morphogenesis *in vitro* by fibroblast-derived soluble factors. *Cell* 66: 697-711.

Montesano, R., Matsumoto, K., Nakamura, T. and Orci, L. (1991b) Identification of a fibroblast-derived epithelial morphogen as hepatocyte growth factor. *Cell* 67: 901-908.

Mugrauer, G., Alt, F.W. and Ekblom, P. (1988) N-*myc* proto-oncogene expression during organogenesis in the developing mouse as revealed *in situ* hybridization. *J. Cell Biol.* 107: 1325-1335.

Naethke, I., Hinck, L., Swedlow, J.R., Papkoff, J. and Nelson, W.J. (1994) Defining interactions and distributions of cadherin and catenin complexes in polarized epithelial cells. *J. Cell Biol.* 125: 1341-1352.

Nagafuchi, A., Shirayoshi, Y., Okazaki, K., Yasuda, K. and Takeichi, M. (1987) Transformation of cell adhesion properties by exogenously introduced E-cadherin cDNA. *Nature* 329: 341-343.

Nagafuchi, A. and Takeichi, M. (1988) Cell binding function of E-cadherin is regulated by the cytoplasmic domain. *EMBO J.* 7: 3679-3684.

Nakamura, T., Nishizawa, T., Hagiya, M., Seki, T., Shimonishi, M., Sugimura, A., Tashiro, K. and Shimizu, S. (1989) Molecular cloning and expression of human hepatocyte growth factor. *Nature* 342: 440-443.

Naldini, L., Weidner, K.M., Vigna, E., Gaudino, G., Bardelli, A., Ponzetto, C., Narsimhan, P.R., Hartmann, G., Zarnegar, R., Michalopoulos, G.K., Birchmeier, W. and Comoglio, P.M. (1991) Scatter factor and hepatocyte growth factor are indistinguishable ligands for the MET receptor. *EMBO J.* 10: 2867-2878.

Naldini, L., Tamagnone, L., Vigna, E., Sachs, M., Hartmann, G., Birchmeier, W., Daikuhara, Y., Tsubouchi, H., Blasi, F. and Comoglio, P. (1992) Extracellular proteolytic cleavage by urokinase is required for activation of hepatocyte growth factor. *EMBO J.* 11: 4825-4833.

Okazaki, M., Takeshita, S., Kawai, S., Kikuno, R., Tsujimura, A., Kudo, A. and Amann, E. (1994) Molecular cloning and characterization of OB-cadherin, a new member of cadherin family expressed in osteoblasts. *J. Biol. Chem.* 269: 12092-12098.

Orr-Urtreger, A., Bedford, M., Burakova, T., Arman, E., Zimmer, Y., Yayon, A., Givol, D. and Lonai, P. (1993) Development localization of the splicing alternatives of fibroblast growth factor receptor-2 (FGFR2). *Dev. Biol.* 158: 475-486.

Otto, T., Birchmeier, W., Schmidt, U., Rembrink, K., Schipper, J., Rübben, H. and Raz, A. (1994) Inverse relation of E-cadherin and autocrine motility factor receptor expression as prognostic factors in patients with bladder carcinoma. *Cancer Res.* 54: 3120-3123.

Ozawa, M., Ringwald, M. and Kemler, R. (1990a) Uvomorulin-catenin complex formation is regulated by a specific domain in the cytoplasmic region of the cell adhesion molecule. *Proc. Natl. Acad. Sci. USA* 87: 4246-4250.

Ozawa, M., Engel, J. and Kemler, R. (1990b) Single amino acid substitutions in one Ca^{2+}-binding site of uvomorulin abolish the adhesive function. *Cell* 63: 1033-1038.

Park, M., Dean, M., Cooper, C., Schmidt, M., O'Brian, S., Blair, D. and Vande Woude, G. (1986) Mechanism of *met* oncogene activation. *Cell* 25: 895-904.

Plowman, G.D., Culouscou, J.M., Whitney, G.S., Green, J.M., Carlton, G.W., Foy, L., Neubauer, M.G. and Shoyab, M. (1993) Ligand-specific activation of HER4/p180*erb*B4, a fourth member of the epidermal growth factor receptor family. *Proc. Natl. Acad. Sci USA* 90: 1746-1750.

Press, M.F., Cordon Cardo, C. and Slamon, D.J. (1990) Expression of the HER-2/neu proto-oncogene in normal human adult and fetal tissues. *Oncogene* 5: 953-962.

Reichmann, E., Schwarz, H., Deiner, E.M., Leitner, L., Eilers, M., Berger, J., Busslinger, M. and Beug, H. (1992) Activation of an inducible c-fos ER fusion protein causes loss of epithelial polarity and triggers epithelial-fibroblastoid cell conversion. *Cell* 71: 1-20.

Riethmacher, D., Brinkmann, V. and Birchmeier, C. (1995) A targeted mutation in the mouse E-cadherin gene results in defective preimlantation development. *Proc. Natl. Acad. Sci. USA* 92: 855–859.

Ringwald, M., Schuh, R., Vestweber, D., Eistetter, H., Lottspeich, F., Engel, J., Doelz, R., Jaehnig, F., Epplen, J., Mayer, S., Mueller, C. and Kemler, R. (1987) The structure of the cell adhesion molecule uvomorulin. Insights into the molecular mechanism of Ca^{2+}-dependent cell adhesion. *EMBO J.* 6: 3647–3653.

Ringwald, M., Baribault, H., Schmidt, C. and Kemler, R. (1991) The structure of the gene coding for the mouse cell adhesion molecule uvomorulin. *Nucleic Acid Res.* 19: 6533–6539.

Rosen, E.M., Knesel, J., Goldberg, I.D., Jin, L., Bhargava, M., Joseph, A., Zitnik, R., Wines, J., Kelley, M. and Rockwell, S. (1994) Scatter factor modulates the metastatic phenotype of the EMT6 mouse mammary tumor. *Int. J. Cancer* 57: 706–714.

Rubinfeld, B., Souza, B., Albert, I., Müller, O. Chamberlain, S.H., Masiarz, F.R., Munemitsu, S. and Polakis, P. (1993) Association of the APC gene product with β-catenin. *Science* 262: 1731–1734.

Sano, K., Tamihara, H., Heimark, R.L., Obata, S., Davidson, M., St. John, T., Taketani, S. and Suzuki, S. (1993) Protocadherins: a large family of cadherin-related molecules in central nervous system. *EMBO J.* 12: 2249–2256.

Saxen, L. (1987) *Organogenesis of the kidney.* Cambridge University Press, Cambridge.

Schipper, J.H., Frixen, U.H., Behrens, J., Unger, A., Jahnke, K. and Birchmeier, W. (1991) E-cadherin expression in squamous cell carcinomas of head and neck: inverse correlation with tumor dedifferentiation and lymph node metastasis. *Cancer Res.* 51: 6328–6337.

Schipper, J.H., Unger, A. and Jahnke, J. (1994) E-cadherin as a functional marker of the differentiation and invasiveness of squamous cell carcinoma of the head and neck. *Clin. Otolaryngol.* 19: 381–384.

Schmidt, C., Bladt, F., Goedecke, S., Brinkmann, V., Zschiesche, W., Sharpe, M., Gherardi, E. and Birchmeier, C. (1995) Scatter factor/hepatocyte growth factor SF/HGF is essential for liver development. *Nature* 373: 699–702.

Shibamoto, S., Hayakawa, M., Takeuchi, K., Hori, T., Oku, N., Miyazawa, K., Kitamura, N., Takeichi, M. and Ito, F. (1994) Tyrosine phosphorylation of β-catenin and plakoglobin enhanced by hepatocyte growth factor and epidermal growth factor in human carcinoma cells. *Cell Adhes. Commun.* 1: 295–305.

Shimoyama, Y., Nagafuchi, A., Fujita, S., Gotoh, M., Takeichi, M., Tsukita, S. and Hirohashi, S. (1992) Cadherin dysfunction in a human cancer cell line: possible involvement of loss of α-catenin expression in reduced cell–cell adhesiveness. *Cancer Res.* 52: 1–5.

Simons, K. and Wandinger-Ness, A. (1990) Polarized sorting in epithelia. *Cell* 62: 207–210.

Sonnenberg, E., Meyer, D., Weidner, K.M. and Birchmeier, C. (1993) Scatter factor/hepatocyte growth factor and its receptor, the c-*met* tyrosine kinase, can mediate a signal exchange between mesenchyme and epithelia during mouse development. *J. Cell Biol.* 123: 223–235.

Sonnenberg, A., Calafat, J., Janssen, H., Daams, H., van der Raaij-Helmer, L., Falcioni, R., Kennel, S., Aplin, J., Baker, J., Loizidou, M. and Garrod, D. (1991) Integrin alpha6/beta4 complex is located in hemidesmosomes, suggesting a major role in epidermal cell–basement membrane adhesion. *J. Cell Biol.* 113: 907–917.

Stark, K., Vainio, S., Vassileva, G. and McMahon, P. (1994) Epithelial transformation of metanephric mesenchyme in the developing kidney regulated by Wnt-4. *Nature* 372: 679–681.

Stoker, M., Gherardi, E., Perryman, M. and Gray, J. (1987) Scatter factor is a fibroblast-derived modulator of epithelial motility. *Nature* 327: 239–242.

Su, L.K., Vogelstein, B. and Kinzler, K.W. (1993) Association of the APC tumor suppressor protein with catenins. *Science* 262: 1734–1737.

Takeichi, M. (1991) Cadherin cell adhesion receptors as morphology regulators. *Science* 251: 1451–1455.

Timpel, R. (1989) Structure and biological activity of basement membrane proteins. *Eur. J. Biochem.* 180: 487–502.

Troyanovsky, S.M., Troyanovsky, R.B., Eshking, L.G., Leube, R.E. and Franke, W.W. (1994) Identification of amino acid sequence motifs in desmocollin, a desmosomal glycoprotein, that are required for plakoglobin binding and plaque formation. *Proc. Natl. Acad. Sci. USA* 91: 10790–10794.

Tsarfaty, I., Rong, S., Resau, J.H., Rulong, S., da Silva, P.P. and Vande Woude, G.F. (1994) The *met* proto-oncogene mesenchymal to epithelial cell conversion. *Science* 263: 98–101.

Tsukita, S., Itoh, M., Nagafuchi, A. and Yonemura, S. (1993) Submembranous junctional plaque proteins include potential tumor suppressor molecules. *J. Cell Biol.* 123: 1049–1053.

Vestweber, D. and Kemler, R. (1984) Rabbit antiserum against a purified surface glycoprotein decompacts mouse preimplantation embryos and reacts with specific adult tissues. *Exp. Cell Res.* 152: 169–178.

Vleminckx, K., Vakaet, L., Mareel, M., Fiers, W. and Van Roy, F. (1991) Genetic manipulation of E-cadherin expression by epithelial tumor cells reveals an invasion suppressor role. *Cell* 66: 107–119.

Weidner, K.M., Behrens, J., Vandekerckhove, J. and Birchmeier, W. (1990) Scatter factor: molecular characteristics and effect on the invasiveness of epithelial cells. *J. Cell Biol.* 111: 2097–2108.

Weidner, K.M., Arakaki, N., Hartmann, G., Vandekerckhove, J., Weingart, S., Rieder, H., Fonatsch, C., Tsubouchi, H., Hishida, T., Daikuhara, Y. and Birchmeier, W. (1991) Evidence for the identity of human scatter factor and hepatocyte growth factor. *Proc. Natl. Acad. Sci. USA* 88: 7001–7005.

Weidner, K.M., Sachs, M. and Birchmeier, W. (1993) The *met* receptor tyrosine kinase transduces motility, proliferation, and morphogenic signals of scatter factor/hepatocyte growth factor in epithelia cells. *J. Cell Biol.* 121: 145–154.

Yamashita, J., Ogawa, M., Yamashita, S., Nomura, K., Kuramoto, M., Saishoji, T. and Shin S. (1994) Immunoreactive hepatocyte growth factor is a strong and independent predictor of recurrence and survival in human breast cancer. *Cancer Res.* 54: 1630–1633.

Yang, Y., Spitzer, E., Meyer, D., Sachs, M., Niemann, N., Hartmann, G., Weidner, K. Michael, Birchmeier, C. and Birchmeier, W. (1995) Sequential requirement of scatter factor/hepatocyte growth factor (SF/HGF) and neu differentiation factor/heregulin (NDF/HRG) in the morphogenesis and differentiation of the mammary gland. *J. Cell Biol.* 131; *in press.*

Epithelial–Mesenchymal Interactions in Cancer
ed. by I.D. Goldberg & E.M. Rosen
© 1995 Birkhäuser Verlag Basel/Switzerland

Regulation of scatter factor (hepatocyte growth factor) production by tumor-stroma interaction

E.M. Rosen and I.D. Goldberg

Long Island Jewish Medical Center, The Long Island Campus for Albert Einstein College of Medicine, Department of Radiation Oncology, 270-05 76th Avenue, New Hyde Park, New York 11042, USA

Introduction

Scatter factors (SF) [also known as hepatocyte growth factor (HGF)] is a cytokine produced by stromal cell types that induces dissociation (scattering) of contiguous sheets of epithelial cells into individual motile fibroblast-like cells (Stoker et al., 1987; Rosen et al., 1989). In addition to scattering, SF stimulates the following processes in multiple epithelial and carcinoma cell types: (1) random motility of isolated cells (Li et al., 1994); (2) gradient-directed migration (chemotaxis) (Rosen et al., 1990b, 1991a); (3) migration of cells out of multicellular spheroids onto flat culture surfaces (Rosen et al., 1990a); (4) migration off carrier beads onto flat surfaces (Rosen et al., 1990b, 1990c, 1991a); (5) cell surface expression of urokinase (Pepper et al., 1992; Rosen et al., 1994a); and (6) invasion through extracellular matrix proteins (Weidner et al., 1990; Rosen et al., 1991a, 1994a; Bhargava et al., 1992). Thus, SF appears to activate an entire program of cellular activities required for the invasive phenotype. SF stimulates proliferation of many normal epithelial cell types, including hepatocytes, mammary epithelium, keratinocytes, melanocytes, and renal tubular epithelium (Miyazawa et al., 1989; Nakamura et al., 1989; Rubin et al., 1991; Kan et al., 1991; Halaban et al., 1992). However, SF has little or no effect on the proliferation of most carcinoma cell types. For these cell types, the major *in vitro* action of SF appears to be stimulation of cell motility.

Several observations suggest a potential role for SF as a mediator of the invasive and metastatic tumor phenotype. First, as discussed above, SF induces motility and invasiveness of a variety of cultured carcinoma cell types. Secondly, the SF receptor, a transmembrane tyrosine kinase, is encoded by a proto-oncogene (*c-met*) (Bottaro et al., 1991). The *tpr-met* gene, the oncogenic counterpart of c-*met*, encodes a truncated protein that lacks a ligand-binding domain and is constitutively activated (Gonzatti-Haces et al., 1988). Thirdly, SF is a potent inducer of angiogenesis (Rosen et al., 1991b; Grant et al., 1993), a process that is required for the continued growth of solid tumors (Folkman, 1992). Various experimental and clinical studies provide evidence suggesting that SF may be involved in the progression of tumors toward a more biologically aggressive phenotype. For example, transfection and over-expression of human c-*met* in mouse 3T3 cells results in a tumorigenic phenotype (Rong et al., 1992). Over-expression of both human SF and human c-*met* induces a further increase in tumorigenicity of the cells. Serial passage of the SF^{hu} + c-met^{hu} transfected cells as tumors *in vivo* selects for cells with increasing ability to produce SF (Rong et al., 1993). These findings suggest that an SF→c-*met* autocrine loop is highly tumorigenic, more so than the paracrine interaction. An SF→c-*met* autocrine loop may also enhance the proliferation of some human lung cacinoma cell lines (Tsao et al., 1993). *In vitro* treatment of EMT6 mouse mammary tumor cells confers these cells with enhanced lung colonizing ability on subsequent intravenous injection into isogeneic mice (Rosen et al., 1994a).

High titers of SF are found in various types of mouse and human carcinomas (Rosen et al., 1994a; Yamashita et al., 1994; Joseph et al., in press). In a study of 258 invasive human breast carcinomas, the SF content was found to be an extremely powerful predictor of relapse and death, even after correction for the influence of conventional prognostic indicators (lymph node status, hormone receptors, tumor size, etc.) (Yamashita et al., 1994). Among a small number of transitional cell bladder cancers, high grade invasive tumors tended to show higher SF content than low grade, superficial tumors (Joseph et al., in press). Thus, high SF content appears to be correlated with aggressive tumor biologic phenotype.

During normal physiologic and reparative processes, including embryonic development and organ regeneration, the expression of SF is up-regulated and subsequently down-regulated in precise temporally and spatially controlled patterns (Sonnenberg et al., 1993; Matsumoto and Nakamura, 1993; Joannadis et al., 1994). On the other hand, SF accumulates and persists within tumors, indicating a failure of the normal regulatory mechanisms. In this chapter, we will examine some of the mechanisms that may be responsible for the accumulation of SF in tumors. Based on the foregoing considerations, these mechanisms may be relevant to the processes of tumor development and progression. The essential hypothesis to be discussed is that SF is chronically over-expressed in tumors primarily because of an abnormal tumor:stroma interaction in which carcinoma cells produce factors that stimulate production of SF by the associated non-malignant stromal cells.

SF-Producing cell types

In vitro, the major SF-producing cell types are mesenchymal cells, including fibroblasts, vascular smooth muscle cells, glial cells, macrophages, and T lymphocytes (Stoker et al., 1987; Rosen et al., 1989, 1990a, 1994b; Inaba et al., 1993; Shiota et al., 1992; Naidu et al., 1994). Most epithelial and carcinoma cell lines produce little or no SF (Stoker et al., 1987; Rosen et al., 1989), with the exception of a few such cell types that produce modest titers of the factor. These include a line of non-differentiating human keratinocytes (NDK) and several lines of human lung carcinoma cells (Adams et al., 1991; Tsao et al., 1993). For these cell types, SF appears to be an autocrine inducer of motility and proliferation. Like epithelial cells, most lines of cultured vascular endothelial cells produce little or no SF (Stoker et al., 1987; Rosen et al., 1989). Similarly, basal expression of SF mRNA *in vivo* in capillary endothelial cells is low. However, in the setting of injury to the rat liver, expression of SF mRNA in hepatic sinusoidal and pulmonary alveolar endothelia is greatly up-regulated (Noji et al., 1990; Yanagita et al., 1992). These findings suggest that suitably stimulated endothelial cells may be a source of SF production. The potential role of cell activation in the regulation of SF production by endothelial cells and other cell types is discussed below.

Regulation of SF production

Multifunctional cytokines

The promoter region of the murine SF gene is very large (at least 2.8 kb) and complex. This region contains four IL-6 response elements, two potential binding sites for NF-IL6, a TGFβ inhibitory element (TIE), a cAMP response element (CRE), estrogen response elements, cell type-specific transcription factor binding sites, and multiple positive and negative regulatory elements for which transcription factors have not yet been identified (Liu et al., 1994). Some human fibroblast lines, including lung, gingival, and skin fibroblasts, are stimulated to produce SF by interleukin-1 (IL-1α and IL-1β) and by tumor necrosis factor (TNFα) (Matsumoto et al., 1992a; Tamura et al., 1993), consistent with the findings of NF-IL6 binding sites in the SF promoter. IL-1 and TNFα are 18 kDa cytokines that modulate tissue response to injury and up-regulate the inflammatory response. These cytokines are principally produced by monocytes and macrophages, but may be produced in smaller quantities by other stromal cell types, including fibroblasts and lymphocytes (Leek et al., 1994). Thus, macrophages may not only produce SF but also induce fibroblasts to produce SF by secreting soluble modulators of SF production. On the other hand, transforming growth factor-β (TGFβ), which generally down-modulates the inflammatory response, inhibits the production of SF by fibroblasts (Gohda et al., 1992). Interestingly, although the SF promoter contains multiple IL-6 response elements, it has not been demonstrated that IL-6 modulates SF production. IL-1, TNFα, and TGFβ are each multifunctional cytokines that mediate a wide array of activities. Moreover, we have observed that not all SF-producing fibroblast or non-fibroblast cell types are responsive to IL-1 and TNFα (Rosen et al., 1994b). Thus, the existence of a more specific class of SF regulatory molecules seems likely.

Other modulators

In addition to cytokines, other mediators may regulate the expression of SF. For example, tumor promoting phorbol esters (eg., phorbol myristate acetate [PMA]) may enhance the ability of IL-1 to up-regulate SF production by fibroblasts (Matsumoto et al., 1992a). 1,25-Dihydroxyvitamin D3 inhibits PMA- and dibutyryl cAMP-stimulated SF production by HL-60 human promyelocytic leukemia cells (Inaba et al., 1993). A factor partially purified from porcine liver stimulates SF protein synthesis by MRC5 human lung fibroblasts but does not affect MRC5 cell SF mRNA expression (Okazaki et al., 1994). Chemical characterization studies indicate that this factor is heat and acid stable, nonproteinaceous, and may be an 8–15 kDa glycosaminoglycan related to heparin or heparan sulfate.

Cell:Cell contact

Co-culture of fibroblasts and epithelial cells, either by adding epithelial cells to fibroblast monolayers or vice versa, induces down-regulation of fibroblast SF expression at the mRNA and protein levels (Kamalati et al., 1992; Seslar et al., 1993). This phenomenon of contact inhibition of SF production has been observed both with normal tissue-derived epithelial cells (Tab. 1) and carcinoma cells. Therefore, it may reflect a homeostatic mechanism to allow the stable co-existence of epithelia and mesenchyme in adult tissues. In one of two studies, SF production was also down-regulated by culturing fibroblasts and carcinoma cells in a Millipore well chamber so that these two cell types could co-condition the same medium in the absence of direct cell contact (Seslar et al., 1993). This study suggests that a soluble inhibitor may be released as a result of interaction between the two cell types. The same carcinoma cells by themselves produce factors that *stimulate* fibroblast production

Table 1. Down-regulation of SF expression in co-cultures of normal human mammary epithelial cells (HMECs) with MRC5 human lung fibroblasts[1]

HMEC cell line	SF concentration (ng/ml)			
	0.0	2×10^4	1×10^5	2×10^5 cells added
184	15.3	16.7	7.3	8.3
48R	15.3	18.2	9.8	6.9
161	15.3	19.2	8.3	4.4

[1]HMECs were added to confluent MRC5 fibroblasts in 2 cm^2 wells and allowed to attach overnight in their normal growth medium. Cultures were then washed and incubated in 0.5 ml of fresh serum-free Dulbecco Modified Eagle's Medium (DMEM) for 48 h. The medium was spun and assayed for SF by ELISA. Values are averages of duplicate assays, with ranges of $\pm 15\%$.

of SF (see below). Thus, the proximity of fibroblasts may determine the nature of the fibroblast:epithelial interaction and its effect on SF production.

Cell activation

As described above, organ injury induces expression of SF by endothelial cells in both local and distant capillary beds (Noji et al., 1990; Yanagita et al., 1992). Thus, we may hypothesize that following injury, paracrine and endocrine signals stimulate or activate endothelial cell SF production. These signals may include pro-inflammatory cytokines, which are produced in response to tissue injury (see above) or a group of partially characterized soluble proteins which we call scatter factor-inducing factors (SF-IFs) (see below). In the same *in vivo* studies, Kupffer cells (a liver macrophage-like cell type) and pulmonary alveolar macrophages were found to express SF mRNA after liver injury (Noji et al., 1990; Yanagita et al., 1992). Production of SF by HL-60 promyelocytic leukemia cells was significantly stimulated by the tumor-promoting phorbol ester PMA, which induces activation of protein kinase C and acquisition of monocyte/macrophage characteristics (Inaba et al., 1993). We have hypothesized that conversion of monocytes into macrophages and further activation of macrophages result in increased SF production (Rosen and Goldberg, in press).

Resting human T lymphocytes and human T cell line HUT 78 normally produce very little SF. However, HUT 78 cells infected and transformed with a retrovirus (human T cell lymphotrophic virus type II [HTLV-II]) produce high titers if SF, comparable to those produced by human lung embryo fibroblasts (Naidu et al., 1994). Retroviruses, which include human and feline leukemia viruses as well as the AIDS virus (HIV-1) are known to activate T cells (Fulton et al., 1987). Stoker and colleagues (1987) observed that different fibroblast lines produce widely varying SF titers from 0 to very high titers. The reason for these variations is not clear, but may involve a requirement for cellular activation for SF production. Activating factors may include inflammatory cytokines, viral infection, phorbol esters, and SF-IFs.

SF-Inducing factors (SF–IFs)

Several studies indicate that carcinoma cells, which do not produce SF, produce soluble factors distinct from IL-1 and TNF that stimulate SF mRNA and protein expression by fibroblast and other SF-producing cell types (Seslar et al., 1993; Rosen et al., 1994a,b; Joseph et al., 1995). We reported that EMT6 mouse mammary tumor cells produce a < 30 kDa heat-stable protein factor that stimulates SF production by human lung fibroblasts up to 5-fold (Rosen et al., 1994a). EMT6 solid tumor nodules were found to contain high titers of SF, despite the fact that cultured EMT6 cells do not produce any detectable SF. These tumor cells are maintained by alternative passage *in vitro* as cell cultures and *in vivo* as tumors at two month intervals. Thus, it is unlikely that the cells lose the ability to produce SF during adaptation to culture. A clue to the mechanism by which SF accumulates within solid EMT6 tumors is provided by the observation that these tumors also contain extremely high titers of < 30 kDa SF-inducing activity. Thus, tumoral SF may be derived from stromal cell types (eg., fibroblasts, macrophages, endothelial cells) that are induced to produce SF by SF–IFs secreted by the tumor cells. If SF is required for the growth of EMT6 solid tumor nodules, then SF–IF may function as a tumor survival factor in the EMT6 model system.

Seslar and colleagues (1993) described a high molecular weight (> 30 kDa) heat-labile SF–IF activity in conditioned medium from human MCF7 breast carcinoma cells. We found that multiple lines of human breast carcinoma cells produce mixtures of < 30 and > 30 kDa SF–IF protein activities (Rosen et al., 1994b). A 10–30 kDa SF–IF protein activity called injurin appears in the serum of rats within several hours after liver injury (Matsumoto et al., 1992b). Injurin induces SF mRNA expression *in vitro* in fibroblast cultures and *in vivo* in various organs of rats after intraperitoneal injection of the factor. Injurin has physicochemical properties similar to those of the < 30 kDa tumor cell-derived SF-inducing factor. On the other hand, bladder carcinoma cell lines produce a > 30 kDa heat-sensitive SF–IF protein activity; and similar SF–IF activities are present in both primary bladder cancer extracts and urine from bladder cancer patients (Joseph et al., 1995). These findings suggest the existence of at least two distinct SF–IF species, a > 30 kDa heat-sensitive protein and a < 30 kDa heat-stable protein.

Table 2. Effects of various agents on SF production by human mammary fibroblasts[1]

A. Partially purified *ras*-3T3 SF–IF

SF–IF (%, v/v)	SF prod. rate (ng/10^6 cells/48 h)			
	1136TF	1136PF	1136SKF	G94EF
0.0	4.4	8.3	5.9	5.9
2.0	17.8	8.5	12.4	5.6
5.0	25.2	37.0	13.0	8.9
10.0	26.4	56.3	13.9	21.2

B. Growth factors and cytokines[2]

Cytokine (ng/ml)	1136TF	1136PF	1136SKF	G94EF	MRC5
None (0)	6.2	7.3	5.9	5.9	10.2
IL-1α (5)	8.7	8.2	20.1	2.4	28.1
IL-1β (5)	2.1	ND	20.3	4.1	21.9
TNFα (20)	4.7	7.8	8.6	3.2	29.3
EGF (20)	6.1	4.5	7.7	1.6	12.1
Onco-M (50)	4.7	10.4	1.4	ND	14.2
IL-2 (20)	4.8	6.5	6.8	ND	13.8
IL-8 (20)	5.1	6.6	5.9	ND	14.7

[1]Mammary tumor-associated fibroblasts (1136TF), fibroblasts from outside of the tumor (1136PF), mammary skin fibroblasts from the same patient (1136SKF), and normal mammary fibroblasts from a patient who had undergone reduction mammoplasty (G94EF) were cultured in DME:Ham's F12 plus 10% fetal calf serum. Confluent fibroblasts in 2 cm² wells were incubated in 0.5 ml of DMEM containing the indicated addition for 48 h. (Note: SF–IF was partially purified by ultrafiltration and chromatography on BioRex-70). The medium was assayed for SF by ELISA, and the cells were counted. Values of SF production rates are averages of two assays, with ranges of ± 15%. ND = No Data.
[2]*Abbreviations:* EGF, epidermal growth factor; IL, interleukin; Onco-M, oncostatin-M; SF–IF, scatter factor-inducing factor; TNFα, tumor necrosis factor-type α.

We have purified and characterized a new and unique murine 12 kDa SF–IF protein with properties similar to the < 30 kDa factor secreted by breast carcinoma cells (Rosen et al., 1994b). This protein was isolated from conditioned medium from *ras*-transformed NIH2/3T3 cells (clone D4), a cell type that produces extremely high titers of both SF and SF–IF. The *ras*-3T3 SF–IF protein is stable to boiling, binds tightly to cation exchange resins (including BioRex-70 and heparin sepharose), and has a unique N-terminal amino acid sequence. It induces up to (4–6)-fold increases in SF mRNA and protein production and is biologically active at physiologic concentrations (20–400 picomolar). Interestingly, *ras*-3T3 SF–IF induces SF production by several SF-producing lines of fibroblasts that are not responsive to IL-1α, IL-1β, or TNFα; these include human mammary tumor-derived fibroblasts (Tab. 2). Unlike IL-1, *ras*-3T3 SF–IF does *not* induce the expression of IL-6 mRNA by human lung fibroblasts (Fig. 1). TGFβ, which inhibits basal and IL-1 stimulated SF expression, also inhibits *ras*-3T3 SF–IF stimulated SF expression. We suggest that SF–IF may

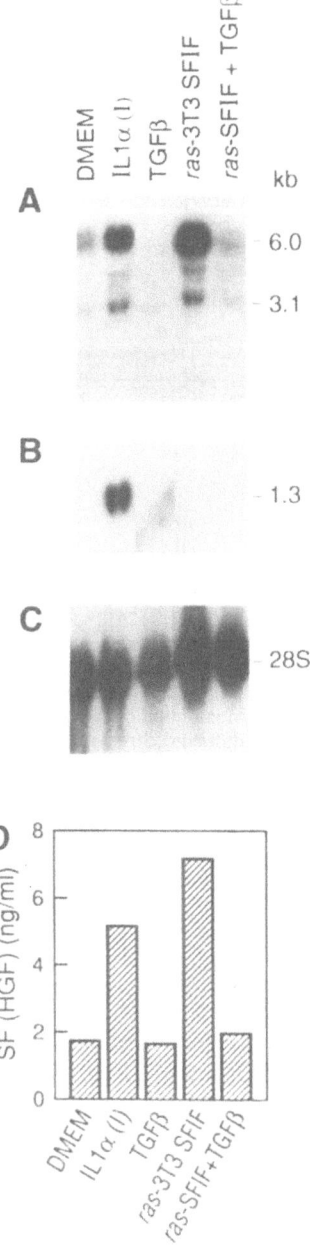

Figure 1. Effects of *ras*-3T3 SF–IF and cytokines on expression of SF and IL-6 mRNA by CCD11Lu human lung fibroblasts. Confluent CCD11Lu cells in 100 mm dishes were incubated with DMEM alone (control); recombinant human IL-1α (5 ng/ml); recombinant human TGFβ (10 ng/ml); *ras*-3T3 SF–IF [10% (v/v) of *ras*-3T3 cell CM < 30 kDa filtrate]; and *ras*-3T3 SF–IF plus TGFβ for 24 h. Total cell RNA was extracted, electrophoresed (20 μg per lane), and Northern blotted. The same blot was serially probed using cDNAs for SF (panel A), IL-6 (panel B), and then 28S ribosomal RNA (panel C). Samples of 24 h CM from the same experiment were assayed for immunoreactive SF by ELISA (panel D).

be a prototype of a new class of specific, physiologic SF regulators (Rosen et al., 1994b).

Production of and responsiveness to SF–IFs

Multiple lines of mammary carcinoma cells that do not produce SF produce low and/or high molecular weight SF–IFs (Seslar et al., 1993; Rosen et al., 1994a,b; Joseph et al., 1995). On the other hand, we found that seven out of seven SF-producing lines of fibroblasts also produce easily detectable SF–IF activity and respond to exogenous SF–IF (Rosen et al., 1994b). Three of three lines of fibroblasts that produce little or no SF also produce little or no SF–IF activity (Rosen et al., 1994b). None of these non-producing fibroblast lines are responsive to exogenous SF–IF. These findings suggest that the inability to produce SF may be due to the inability to produce and/or respond to SF–IFs. Thus, carcinoma cells that produce high titers of SF–IFs but do not produce any detectable SF may lack the mechanism(s) required to respond to SF–IFs. The variability in production of SF among different fibroblast lines previously reported (Stoker et al., 1987) may also be related to the ability of these cells both to produce and respond to SF–IFs. Since we have not found any SF-producing fibroblast lines that do not produce SF–IF activity and respond to addition of exogenous SF–IF, we also postulate that SF production *requires* both SF–IFs and the ability to respond to these factors.

The identification of an SF–IF protein that functions at physiologic (picomolar) concentrations suggests that SF–IFs may be cytokines that act through cell surface receptors (SF–IFRs). Carcinoma cells may generally lack these receptors and/or another component of the SF–IF signalling mechanism. The most common phenotypes of carcinoma and fibroblastic cells may be summarized as follows:

carcinoma cells   SF − /SFIF + or − /SFIFR − /c-*met* +

fibroblasts   SF + /SFIF + /SFIFR + /c-*met* −

However, other combinations of these properties are possible, giving rise to a variety of autocrine and paracrine loops that may influence SF production and responsiveness in tumors (illustrated in Fig. 2).

Hypotheses and models

If SF–IFs play a major role in the accumulation of SF in tumors, it is logical to next ask what regulates the production of these factors: ie., what induces the inducers? SF–IFs induce elevated levels of SF-specific

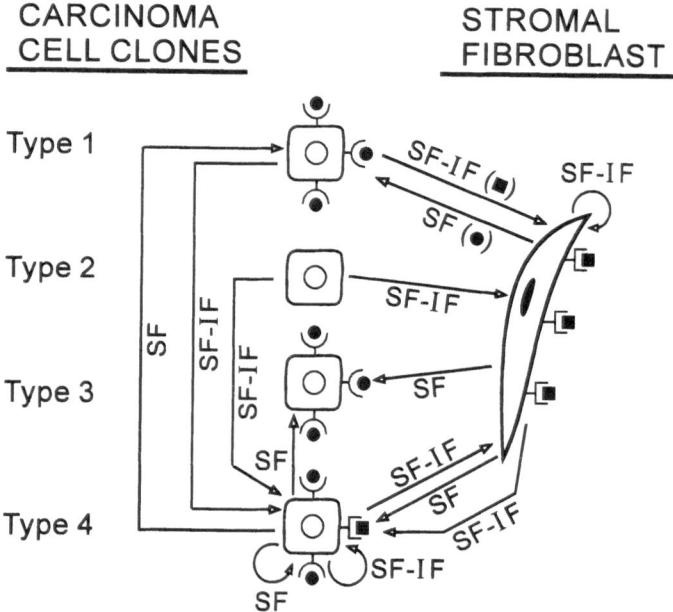

CARCINOMA
CELL CLONES

STROMAL
FIBROBLAST

Figure 2. Possible modes of tumor cell:fibroblast interaction that regulate tumoral SF production. The carcinoma cell clones differ in their expression of scatter factor (SF), SF-inducing factor (SFIF), the putative SF-inducing factor receptor (SFIFR), and the SF receptor (c-*met* protein): type 1, SF − /SFIF + /SFIFR − /c-*met* + ; type 2, SF − /SFIF + / SFIFR − /c-*met* −; type 3, SF − /SFIF − /SFIFR − /c-*met* + ; and type 4, SF + /SFIF + / SFIFR + /c-*met* + .

mRNA transcripts (Seslar et al., 1993; Rosen et al., 1994b; Joseph et al., 1995). This is due, at least in part, to activation of the SF gene promoter, since preliminary studies suggest that murine *ras*-3T3 SF−IF activates transcription of SF promoter-reporter constructs (unpublished results). If, in fact, both carcinoma and normal epithelial cell types are capable of producing SF−IFs, over-expression of SF in tumors may result from: (1) quantitatively increased SF−IF production by carcinoma cells as compared with normal cells; (2) increased ability of proliferating or migrating cells to produce SF−IFs as compared with quiescent cells; (3) increased access of stromal target cells to carcinoma cell products (ie., due to invasion of carcinoma cells into the stromal compartment); (4) enhanced responsiveness of "activated" tumor stromal cells to SF−IFs; and (5) a combination of the above.

Possibility (1) implies intrinsically increased production of SF−IFs by carcinoma cells, as might result from loss of function of an anti-oncogene (tumor suppressor gene) that encodes a transcriptional repressor of SF−IF or activation of an oncogene that transactivates the SF−IF gene. The p53 gene, which encodes a nuclear phosphoprotein that regulates cell growth by transactivating or repressing gene transcription (Tomi-

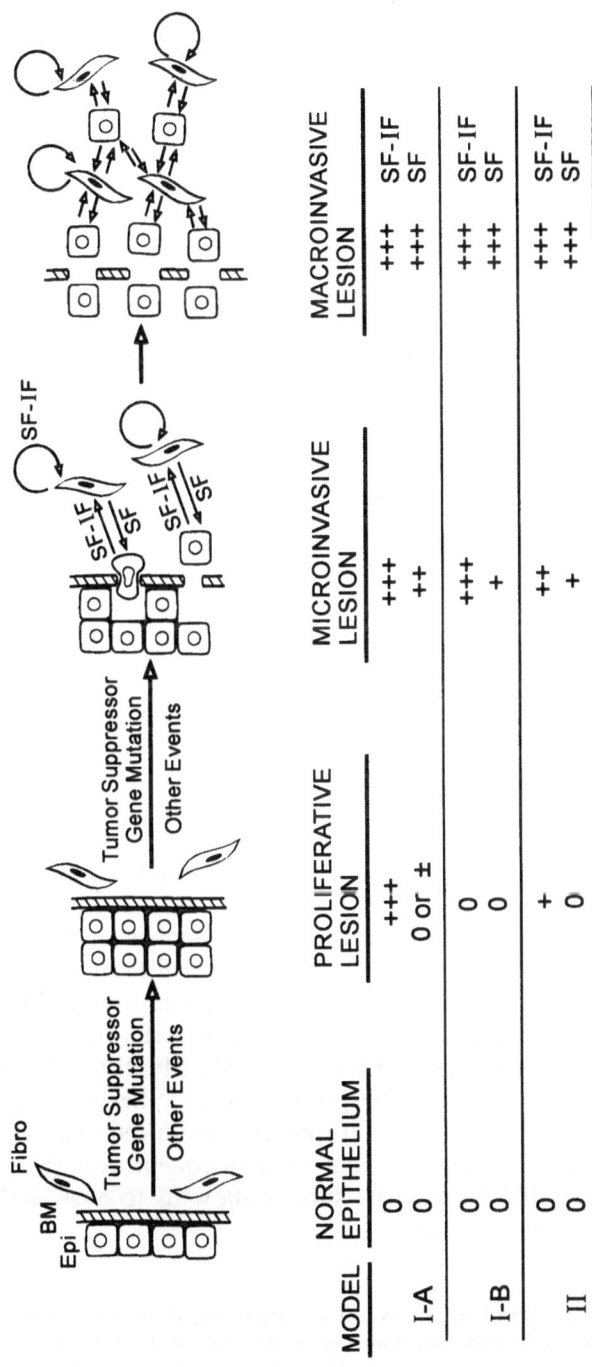

MODEL	NORMAL EPITHELIUM	PROLIFERATIVE LESION	MICROINVASIVE LESION	MACROINVASIVE LESION	
I-A	0	+++	+++	+++	SF-IF
	0	0 or ±	++	+++	SF
I-B	0	0	+++	+++	SF-IF
	0	0	+	+++	SF
II	0	+	++	+++	SF-IF
	0	0	+	+++	SF

Figure 3. Putative events leading to over-expression of SF–IFs and SF during the evolution of carcinomas. These models differ in the events which cause SF–IF expression. In model I-A, an early mutation(s) (leftmost arrow) leads to tumor cell proliferation and high expression of SF–IF. A secondary mutation(s) and/or accumulation of SF then induce early invasion (second arrow from left). In Model I-B, a secondary gene mutation(s) induces SF–IF production; whereas in Model II, dynamic cell changes related to proliferation and invasion are the primary stimuli for SF–IF production. In all cases, accumulation of SF is enhanced by increased access of tumor cells to fibroblasts during stromal invasion.

naga et al., 1992; Mack et al., 1993), is a potential SF–IF regulatory anti-oncogene. Loss of p53 function results in down-regulation of thrombospondin-1, an angiogenesis inhibitor (Dameron et al., 1994), and up-regulation of *bcl*-2, a proto-oncogene whose protein product inhibits apoptosis (programmed cell death) (Miyashita et al., 1994). Moreover, p53 mutations which result in loss of normal p53 function are the most common genetic alterations observed in human cancers.

Possibility (2) implies that SF–IF production is associated with a state of dynamic change – as occurs during normal development, tissue repair, and tumor invasion – rather than to cancer-specific alterations. These two possibilities are not mutually exclusive. The accumulation of SF–IFs and SF in tumors, as compared with normal tissues, may also result from lack of normal homeostatic down-regulation of SF–IF production in the tumor. Thus, the lack of signals that inhibit SF–IF production or the inability of tumor cells to respond to such signals may lead to steady and unrelenting SF–IF expression. Several models illustrating potential molecular events leading to activation of the SF–IF→SF paracrine interaction in tumors are illustrated in Figure 3.

Conclusions

SF appears to be overproduced within tumors, at least in part, due to an abnormal paracrine interaction in which carcinoma cells secrete factors (SF IFs) that stimulate stromal cells (eg., fibroblasts, smooth muscle, endothelium) to produce more SF. Tumor stromal cells may also produce their own autocrine SF–IFs that further enhance SF production. One such SF–IF, a 12 kDa heat stable protein with a unique aminoterminal sequence, has been purified and characterized. A second, high molecular weight (> 30 kDa) SF–IF protein(s) is produced by human breast and bladder carcinoma cells. A number of important questions remain to be answered: (1) When, during the evolution of a tumor, are SF–IFs first produced?; (2) What factors and conditions lead to excess production and accumulation of SF–IFs in tumors as opposed to normal physiologic processes?; (3) Do SF–IFs have other biologic functions (eg., inducing the production of other cytokines)?; and (4) Why are more than one SF–IF proteins produced? The molecular identification of these proteins and the consequent availability of cDNA probes and antibody reagents should allow us to answer these questions over the next few years.

Acknowledgement
Supported in part by the USPHS (CA50516 and CA64869). We thank Dr. Ralph Zitnik (Department of Medicine, Yale University School of Medicine) for performing the Northern blots shown in Figure 1. We are grateful to Drs Helene B. Smith (Geraldine Brush Cancer Research Institute) and Martha Stampfer (Lawrence Berkeley Laboratory) for providing

mammary cell types for the studies in Tables 1 and 2. Dr. Rosen is an Established Investigator of the American Heart Association (AHA 90-195).

References

Adams, J.C., Furlong, R.A. and Watt, F.M. (1991) Production of scatter factor by ndk, a strain of epithelial cells, and inhibition of scatter factor activity by suramin. *J. Cell Sci.* 98: 385–394.

Bhargava, M., Joseph, A., Knesel, J., Halaban, R., Li, Y., Pang, S., Goldberg, I., Setter, E., Donovan, M.A., Zarnegar, R., Michalopoulos, G.A., Nakamura, T., Faletto, D. and Rosen, E.M. (1992) Scatter factor and hepatocyte growth factor: Activities, properties, and mechanism. *Cell Growth & Diff.* 3: 11–20.

Bottaro, D.P., Rubin, J.S., Faletto, D.L., Chan, A.M.-L., Kmiecik, T.E., Vande Woude, G.F. and Aaronson, S.A. (1991) Identification of the hepatocyte growth factor receptor as the c-*met* proto-oncogene product. *Science* 251: 802–804.

Dameron, K.M., Volpert, O.V., Tainsky, M.A. and Bouck, N. (1994) Control of angiogenesis in fibroblasts by *p53* regulation of thrombospondin-1. *Science* 265: 1582–1590.

Folkman, J. (1992) The role of angiogenesis in tumor growth. *Semin. Cancer Biol.* 3: 65–71.

Fulton, R., Forrest, D., McFarlane, R., Onions, D. and Neil, J.C. (1987) Retroviral transduction of T-cell antigen receptor beta-chain and *myc* genes. *Nature* 326: 190–194.

Gohda, E., Matsunaga, T., Kataoka, H. and Yamamoto, I. (1992) TGF-β is a potent inhibitor of hepatocyte growth factor secretion by human fibroblasts. *Cell Biol. Int. Reports* 16: 917–926.

Gonzatti-Haces, M., Seth, A., Park, M., Copeland, T., Oroszlan, S. and Vande Woude, G.F. (1988) Characterization of the TPR-MET oncogene p65 and the MET protooncogene p140 protein tyrosine kinases. *PNAS USA* 85: 21–25.

Grant, D.S., Kleinman, H.K., Goldberg, I.D., Bhargava, M., Nickoloff, B.J., Polverini, P. and Rosen, E.M. (1993) Scatter factor induces blood vessel formation *in vivo*. *PNAS USA* 90: 1937–1941.

Halaban, R., Rubin, J., Funasaka, Y., Cobb, M., Boulton, T., Faletto, D., Rosen, E., Chan, A., Yoko, K., White, W., Cook, C. and Moellmann, G. (1992) Met and hepatocyte growth factor/scatter factor signal transduction in normal melanocytes and melanoma cells. *Oncogene* 7: 2195–2206.

Inaba, M., Koyama, H., Hino, M., Okuno, S., Terada, M., Nishizawa, Y., Nishino, T. and Morii, H. (1993) Regulation of release of hepatocyte growth factor from human promyelocytic leukemia cells, HL-60, by 1,25-dihydroxyvitamin D3, 12-O-tetradecanoylphorbol 13-acetate, and dibutyryl cyclic adenosine monophosphate. *Blood* 82: 53–59.

Joannidis, M., Spokes, K., Nakamura, T., Faletto, D. and Cantley, L.G. (1994) Regional expression of hepatocyte growth factor/c-*met* in experimental renal hypertrophy and hyperplasia. *Am. J. Physiol.* 267: F231–F236.

Joseph, A., Weiss, G.H., Jin, L., Fuchs, A., Chowdhury, S., O'Shaughnessy, P., Goldberg, I.D. and Rosen, E.M. (1995) Expression of scatter factor in human bladder carcinoma. *J. Natl. Cancer Inst.* 87: 372–377.

Kamalati, T., Thirunavukarasu, B., Wallace, A., Holder, N., Brooks, R.F., Nakamura, T., Stoker, M., Gherardi, E. and Buluwela, L. (1992) Down-regulation of scatter factor in MRC5 fibroblasts by epithelial derived cells: A model for scatter factor modulation. *J. Cell Sci.* 101: 323–332.

Kan, M., Zhang, G.H., Zarnegar, R., Michalapoulos, G., Myoken, Y., McKeehan, W.L. and Stevens, J.L. (1991) Hepatocyte growth factor-hepatopoietin A stimulates the growth of rat proximal tubule epithelial cells (rpte), rat non-parenchymal liver cells, human melanoma cells, mouse keratinocytes, and stimulates anchorage-independent growth of SV40-transformed rpte. *Biochem. Biophys. Res. Commun.* 174: 331–337.

Leek, R.D., Harris, A.L. and Lewis, C.E. (1994) Cytokine networks in solid human tumors: regulation of angiogenesis. *J. Leuk. Biol.* 56: 423–435.

Li, Y., Bhargava, M.M., Joseph, A., Jin, L., Rosen, E.M. and Goldberg, I.D. (1994) Effect of hepatocyte growth factor/scatter factor and other growth factors on motility and morphology of non-tumorigenic and tumor cells. *In Vitro Cell Dev. Biol.* 30A: 105–110.

Liu, Y., Michalopoulos, G.K. and Zarnegar, R. (1994) Structural and functional characterization of the mouse hepatocyte growth factor gene promoter. *J. Biol. Chem.* 269: 4152–4160.

Mack, D.H., Vartikar, J., Pipas, J.M. and Laimins, L.A. (1993) Specific repression of
TATA-mediated but not initiator-mediated transcription. *Nature* 363: 281–283.

Matsumoto, K., Okazaki, H. and Nakamura, T. (1992a) Up-regulation of hepatocyte growth
factor gene expression by interleukin-1 in human skin fibroblasts. *Biochem. Biophys. Res.
Commun.* 188: 235–243.

Matsumoto, K., Tajima, H., Hamanoue, M., Kohno, S., Kinoshita, T. and Nakamura, T.
(1992b) Identification and characterization of injurin, an inducer of expression of the gene
for hepatocyte growth factor. *PNAS USA* 89: 3800–3804.

Matsumoto, K. and Nakamura, T. (1993) Roles of HGF as a pleiotropic factor in organ
regeneration. *In:* I.D. Goldberg and E.M. Rosen (eds): *Hepatocyte Growth Factor–Scatter
Factor and the C-Met Receptor.* Birkhäuser Verlag, Basel, pp 225–250.

Miyashita, T., Krajewski, S., Krajewska, M., Wang, H.G., Lin, H.K., Liebermann, D.A.,
Hoffman, B. and Reed, J.C. (1994) Tumor suppressor p53 is a regulator of *bcl-2* and *bax*
gene expression *in vitro* and *in vivo*. *Oncogene* 9: 1799–1805.

Miyazawa, K., Tsubouchi, H., Naka, D., Takahashi, K., Okigaki, M., Arakaki, N.,
Nakayama, S., Hirono, S., Sakiyama, O., Gohda, E., Daikuhara, Y. and Kitamura, N.
(1989) Molecular cloning and sequence analysis of cDNA for human hepatocyte growth
factor. *Biochem. Biophys. Res. Commun.* 163: 967–973.

Naidu, Y.M., Rosen, E.M., Zitnik, R., Goldberg, I., Park, M., Naujokas, M., Polverini, P.J.
and Nickoloff, B.J. (1994) Role of scatter factor in the pathogenesis of AIDS-related
Kaposi's sarcoma. *PNAS USA* 91: 5281–5285.

Nakamura, T., Nishizawa, T., Hagiya, M., Seki, T., Shimonishi, M., Sugimura, A. and
Shimizu, S. (1989) Molecular cloning and expression of human hepatocyte growth factor.
Nature 342: 440–443.

Noji, S., Tashiro, K., Koyama, E., Nohno, T., Ohyama, K., Taniguchi, S. and Nakamura, T.
(1990) Expression of hepatocyte growth factor gene in endothelial and Kupffer's cells of
damaged rat livers as revealed by *in situ* hybridization. *Biochem. Biophys. Res. Commun.*
173: 42–47.

Okazaki, H., Matsumoto, K. and Nakamura, T. (1994) Partial purification and characterization of "injurin-like" factor which stimulates production of hepatocyte growth factor.
Biochem. Biophys. Acta 1220: 291–298.

Pepper, M.S., Matsumoto, K., Nakamura, T., Orci, L. and Montesano, R. (1992) Hepatocyte
growth factor increases urokinase-type plasminogen activator (u-PA) and u-PA receptor
expression in Madin-Darby canine kidney epithelial cells. *J. Biol. Chem.* 267: 20493–20496.

Rong, S., Bodescot, M., Blair, D., Nakamura, T., Mizuno, K., Park, M., Chan, A., Aaronson,
S. and Vande Woude, G.F. (1992) Tumorigenicity of the *met* protooncogene and the gene
for hepatocyte growth factor. *Mol. Cell Biol.* 12: 5152–5158.

Rong, S., Oskarsson, M. Faletto, D., Tsarfaty, I., Resau, J.H., Nakamura, T., Rosen, E.,
Hopkins, R.F. and Vande Woude, G.F. (1993) Tumorigenesis induced by coexpression of
human hepatocyte growth factor and the human *met* protooncogene leads to high levels of
expression of the ligand and receptor. *Cell Growth & Diff.* 4: 563–569.

Rosen, E.M., Goldberg, I.D., Kacinski, B.M., Buckholz, T. and Vinter, D.W. (1989) Smooth
muscle releases an epithelial cell scatter factor which binds to heparin. *In Vitro Cell Dev.
Biol.* 25: 163–173.

Rosen, E.M., Meromsky, L., Setter, E., Vinter, D.W. and Goldberg, I.D. (1990a) Smooth
muscle-derived factor stimulates mobility of human tumor cells. *Invasion. Metast.* 10:
49–64.

Rosen, E.M., Meromsky, L., Setter, E., Vinter, D.W. and Goldberg, I.D. (1990b) Purification
and migration-stimulating activities of scatter factor. *Proc. Soc. Exp. Biol. Med.* 195: 34–43.

Rosen, E.M., Meromsky, L., Setter, E., Vinter, D.W. and Goldberg, I.D. (1990c) Quantita-
tion of cytokine-stimulated migration of endothelium and epithelium by a new assay using
microcarrier beads. *Exp. Cell Res.* 186: 22–31.

Rosen, E.M., Goldberg, I.D., Liu, D., Setter, E., Donovan, M.A., Bhargava, M., Reiss, M.
and Kacinski, B.M. (1991a) Tumor necrosis factor stimulates epithelial tumor cell motility.
Cancer Res. 57: 5315–5321.

Rosen, E.M., Grant, D., Kleinman, H., Jaken, S., Donovan, M.A., Setter, E., Luckett, P.M.
and Carley, W. (1991b) Scatter factor stimulates migration of vascular endothelium and

capillary-like tube formation. *In:* I.D. Goldberg and E.M. Rosen (eds): *Cell Motility Factors*. Birkhäuser Verlag, Basel, pp 76–88.

Rosen, E.M., Knesel, J., Goldberg, I.D., Bhargava, M., Joseph, A., Zitnik, R., Wines, J., Kelley, M. and Rockwell, S. (1994a) Scatter factor modulates the metastatic phenotype of the EMT6 mouse mammary tumor. *Int. J. Cancer* 57: 706–714.

Rosen, E.M., Joseph, A., Jin, L., Rockwell, S., Elias, J.A., Knesel, J., Wines, J., McClellan, J., Kluger, M.J., Goldberg, I.D. and Zitnik, R. (1994b) Regulation of scatter factor production via a soluble inducing factor. *J. Cell Biol.* 127: 225–234.

Rosen, E.M. and Goldberg, I.D. (1995) Scatter factor and angiogenesis. *Adv. Cancer Res.* 67; *in press.*

Rubin, J.S., Chan, A.M.-L., Bottaro, D.P., Burgess, W.H., Taylor, W.G., Cech, A.C., Hirschfield, D.W, Wong, J., Miki, T., Finch, P.W. and Aaronson, S.A. (1991) A broad spectrum human lung fibroblast-derived mitogen is a variant of hepatocyte growth factor. *PNAS USA* 88: 415–419.

Seslar, S.P., Nakamura, T. and Byers, S.W. (1993) Regulation of fibroblast hepatocyte growth factor-scatter factor expression by human breast carcinoma cell lines and peptide growth factors. *Cancer Res.* 53: 1233–1238.

Shiota, G., Rhoads, D.B., Wang, T.C., Nakamura, T. and Schmidt, E.V. (1992) Hepatocyte growth factor inhibits growth of hepatocellular carcinoma cells. *PNAS USA* 89: 373–377.

Sonnenberg, E., Meyer, D., Weidner, K.M. and Birchmeier, C. (1993) Scatter factor-hepatocyte growth factor and its receptor, the c-*met* tyrosine kinase, can mediate a signal exchange between mesenchyme and epithelia during mouse development. *J. Cell Biol.* 123: 223–235.

Stoker, M., Gherardi, E., Perryman, M. and Gray, J. (1987) Scatter factor is a fibroblast-derived modulator of epithelial cell mobility. *Nature* 327: 238–242.

Tamura, M., Arakaki, N., Tsoubouchi, H., Takada, H. and Daikuhara, Y. (1993) Enhancement of human hepatocyte growth factor production by interleukin-1 alpha and -1 beta and tumor necrosis factor-alpha by fibroblasts in culture. *J. Biol. Chem.* 268: 8140–8145.

Tominaga, O., Hamelin, R., Remvikos, Y., Salmon, R.J. and Thomas, G. (1992) P53 from basic research to clinical applications. *Crit. Rev. Oncogenesis* 3: 257–282.

Tsao, M.S., Zhu, H., Giaid, A., Viallet, J., Nakamura, T. and Park, M. (1993) Hepatocyte growth factor/scatter factor is an autocrine factor for human bronchial epithelial and lung carcinoma cells. *Cell Growth & Diff.* 4: 571–579.

Weidner, K.M., Behrens, J., Vandekerckhove, J. and Birchmeier, W. (1990) Scatter factor: Molecular characteristics and effect on invasiveness of epithelial cells. *J. Cell Biol.* 111: 2097–2108.

Yamashita, J., Ogawa, M., Yamashita, S., Nomura, K., Kuramoto, M., Saishoji, T. and Sadahito, S. (1994) Immunoreactive hepatocyte growth factor is a strong and independent predictor of recurrence and survival in human breast cancer. *Cancer Res.* 54: 1630–1633.

Yanagita, K., Nagaike, M., Ishibashi, H., Niho, Y, Matsumoto, K and Nakamura, T. (1992) Lung may have endocrine function producing hepatocyte growth factor in response to injury of distal organs. *Biochem. Biophys. Res. Commun.* 182: 802–809.

Epithelial–Mesenchymal Interactions in Cancer
ed. by I.D. Goldberg & E.M. Rosen
© 1995 Birkhäuser Verlag Basel/Switzerland

Regulation of HGF and HGFR gene expression

R. Zarnegar

*Department of Pathology, Division of Cellular and Molecular Pathology,
University of Pittsburgh, Pittsburgh, PA 15261, USA*

Summary. Hepatocyte growth factor and its receptor (the product of the c-*met* protoonco-gene) are believed to be necessary for the normal growth and development of many tissues and organs. This ligand/receptor system controls essential cellular responses such as cell prolifera-tion and motility as well as morphogenesis and differentiation. HGF mRNA is expressed primarily in mesenchymal but not in epithelial cells while its receptor is predominately expressed in epithelial cells. This pattern of HGF and HGFR gene expression in combination with the unique biological effects of HGF on its target cells has led to the postulate that HGF is one of the long-sought mediators conveying cross-talk between the epithelial and stromal compartments of a given tissue. The expression of HGF and HGFR genes are unregulated in several types of human cancer; therefore, understanding the control mechanisms governing HGF and HGFR gene expression is of great clinical interest. Toward this goal, we have analyzed the effects of various physiological agents such as cytokines and hormones on the expression of HGF and the HGFR in a multitude of cell types *in vitro*. Moreover, we have cloned and analyzed the HGF promoter and its 5'-flanking region to uncover the basis for its inducible and cell-type specific expression at the transcriptional level. Our results indicate that HGF and HGFR gene expression is inducible and their expression is orchestrated in stromal and epithelial cells, respectively, by extracellular signals derived from steroid hormones as well as cytokines such as IL-1, IL-6, and TNFα.

Introduction

Research within the last two decades has clearly demonstrated that polypeptide growth factors not only play an essential role in normal cell growth and differentiation but also in the development and progression of various human cancers. (For a review, see Aaronson, 1991.) In addition to the widely-accepted autocrine model of tumor growth induction by growth factors such as TGFα and those of the FGF family, paracrine growth mechanisms are of equal importance to this process. It is believed that interaction between tumor cells and various other cell types within the tumor microenvironment via paracrine growth factors contributes significantly to the overall growth of the tumor mass.

One such factor is suspected to be hepatocyte growth factor (HGF, also known as scatter factor). This factor was originally identified as the main mitogenic component of serum and platelets for normal adult rat hepatocytes and was purified from these sources. (For a review, see Rubin et al., 1993.) Subsequent studies revealed that HGF is a mes-enchymally derived heparin-binding glycoprotein with unique structural properties (Nakamura et al., 1989). The mature bioactive molecule is an $\alpha\beta$ heterodimer held together by a single disulfide bridge and is derived from a single-chain precursor by proteolytic cleavage (Rubin et al., 1991; Bussolino et al., 1992; Mars et al., 1993; Shimomura et al., 1993; Miyazawa et al., 1994). The α chain has an Mr of about 60 000 and contains a hairpin loop and four kringle domains [double looped structures found in proteins involved in blood coagulation (eg., thrombin) and fibrinolysis (eg., plasmin)], while the β chain has high structural homology to the serine protease domain of plasmin (Nakamura et al., 1989).

In vitro experiments have shown that HGF mainly affects epithelial cells derived from various tissues (Rubin et al., 1991), although endothelial cells (Grant et al., 1993; Bussolino et al., 1992) and melanocytes (Rubin et al., 1991) are also targets. Independent studies have revealed that scatter factor (Stoker et al., 1987), a protein secreted by fibroblasts inducing dispersion (scattering) and motility in many epithelial cells, is identical to HGF (Weidner et al., 1991; Bhargava et al., 1992). In addition to these mitogenic and motogenic properties, HGF also induces morphogenesis (branching tubules and gland-like structures) in epithelial cells *in vitro* (Soriano et al., 1995) and perhaps *in vivo* (Tsarfaty et al., 1992). Moreover, HGF has been suggested to be involved in mesenchymal–epithelial interconversion (Tsarfaty et al., 1994). HGF elicits these multiple biological responses in its target cells by binding to and activating a specific transmembrane tyrosine kinase (Bottaro et al., 1991; Naldini et al., 1991; Weidner et al., 1993), known as the c-*met* receptor (Park et al., 1987).

Recent work indicates that the HGF/HGFR expression is tightly controlled and developmentally regulated (DeFrances et al., 1992; Sonnenberg et al., 1993). Northern blot and *in situ* mRNA expression studies demonstrate that HGF expression is inducible *in vivo* (Kinoshita et al., 1989; Zarnegar et al., 1991; Kono et al., 1992) and restricted to mesenchymal cells (Noji et al., 1990), while HGFR expression is predominantly restricted to epithelial cells (Sonnenberg et al., 1993). These unique properties have made HGF one of the suspected mediators of stromal–epithelial interactions. (For review, see Rosen et al., 1994a.) Experimental (Rong et al., 1994; Giordano et al., 1993) and clinical studies (Rong et al., 1993; Di Renzo et al., 1991; Ebert et al., 1994; Wang et al., 1994; Yamashita et al., 1994) have also shown that abnormal expression of the HGF/HGFR system results in the development and/or progression of various human neoplasms. The elucidation of the mechanisms controlling HGF and HGFR expression are critical for understanding the molecular basis of the pathogenesis of this ligand/receptor system in human cancer. The mechanisms controlling HGF and HGFR availability through transcriptional regulation, alternative splicing and posttranslational alterations such as proteolytic cleavage are all potential steps that may be perturbed, contributing to the abnormal function of this ligand/receptor system. The following discussion will mainly address the regulation of HGF and HGFR expression at the transcriptional levels.

Characterization of the HGF gene and its promoter region

To understand the molecular mechanisms controlling HGF gene expression, the entire human and mouse HGF genes have been cloned and partially characterized. These studies have shown that the HGF gene exists as a single copy per genome, spans approximately 70 kb of DNA, and consists of 18 exons interrupted by 17 introns (Seki et al., 1991; Miyazawa et al., 1991; Liu et al., 1994d). The promoter regions of the human, rat and mouse HGF genes have also been cloned and sequenced (Seki et al., 1991; Miyazawa et al., 1991; Okajima et al., 1993; Liu et al., 1994b). Figure 1 shows an alignment of the nucleotide sequences of the promoter regions of the HGF gene from human, rat and mouse. Evidently, a high degree of sequence conservation exists among the species (95% between mouse and rat and about 90% between rodents and human), indicating the possibility that common control mechanisms are employed for the transcriptional regulation of this gene. This high degree of sequence homology is mainly confined to the region surrounding the TATA box to approximately − 700 bp upstream of the promoter, with the degree of homology decreasing beyond − 700 bp (Liu et al., 1994b). In particular, the sequence from the TATA box to − 130 bp which constitutes part of the minimal promoter is virtually identical among mouse, rat and human (Fig. 1). The major transcription start site is identical in the mouse and rat HGF genes, located approximately 26 bp downstream from the TATA box (Liu et al., 1994b; Okajima et al., 1993). Although the transcription start site in the human HGF gene is located within the same general region, the site is altered due to the insertion of twenty nucleotides (Okajima et al., 1993).

Sequence analysis has revealed that the 5′-flanking region of the HGF gene from all three species contains several putative regulatory elements. Among these DNA elements are AP1 and NF-IL-6 (also known as C/EBPβ) binding sites, several IL-6 response elements, a TGF-β inhibitory element (TIE) and a glucocorticoid response element (GRE). In addition to these putative binding sites, our detailed analysis of approximately 3 kb of the 5′-flanking region of the mouse HGF gene has shown the presence of many other sites including one SP1 binding site, several Egr-1/WT-1 motifs, two C/EBPα binding sites, one consensus sequence for the binding of Pu.1/ETS (a B cell- and macrophage-specific transcription factor), at least one NFκB binding site, two putative estrogen responsive elements (ERE) (one of which is located in the first

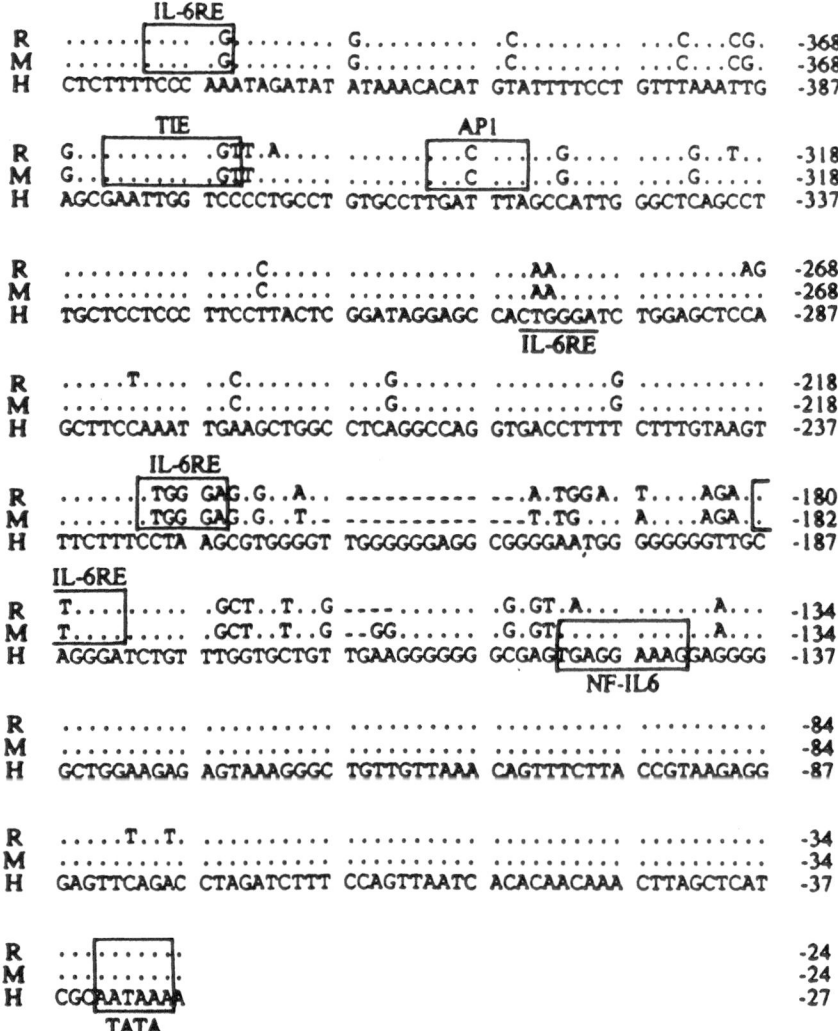

Figure 1. Alignment of the nucleotide sequences of the 5'-flanking regions of the HGF gene from rat, mouse, and human. The nucleotide sequences from rat (R) (Okajima et al., 1993), mouse (M) (Liu et al., 1994b) and human (H) (Miyazawa et al., 1991) are shown. Numbers on the right side of the sequences indicate the nucleotide positions in relation to the transcription initiation site. The putative regulatory motifs (TIE, AP1, IL-6RE and NF-IL6) are either boxed or underlined. Identical nucleotides are shown as dots, while dashes indicate nucleotide deletion.

intron), and one cAMP responsive element (CRE) (Liu et al., 1994b).

Although the functionality of these multiple putative *cis*-acting elements in the regulation of HGF gene transcription remains to be determined, these findings suggest that HGF gene expression is potentially modulated by cytokines, steroid hormones and other extracellular signals through their activation. Many pieces of evidence have been gathered in support of this possibility. For example, HGF mRNA in cultured cells is regulated by the phorbol ester, TPA (Nishino et al., 1991; Matsumoto et al., 1992a; Gohda et al., 1994; Matsunaga et al., 1994) and by serum both of which are known to regulate gene expression through AP1 and/or AP2 activation. Moreover, several cytokines such as IL-1α and IL-1β, TNFα, TGFβ, EGF, PDGF, and FGF have been shown to modulate HGF mRNA expression (Matsumoto et al., 1992b; Tamura et al., 1993; Gohda et al., 1994). Furthermore, recent studies show that HGF mRNA from cultured fibroblasts is also induced by cAMP-elevating agents such as forskolin, cholera toxin, and prostaglandin E2 (Matsunaga et al., 1994). These results indicate that the involvement of a cAMP-mediated pathway may regulate HGF gene expression, perhaps in part, through the activation of the cAMP responsive element(s) (CRE) present in the HGF promoter.

Additional data on the transcriptional regulation of the HGF gene by cytokines and hormones comes from *in vitro* analysis of chimeric constructs containing various lengths of the HGF promoter DNA sequences fused to the coding region of the bacterial chloramphenicol acetyltransferase gene. When such constructs were co-transfected with an expression vector coding for the estrogen receptor into RL95-2 cells or NIH–3T3 fibroblasts, the CAT activity was increased in the presence of 17β-estradiol, while under the same conditions, the HGF–CAT plasmids lacking the ERE sequence showed no significant response to estrogen stimulation. Further experiments utilizing electrophoretic band shift assays demonstrated that the mouse HGF ERE binds specifically to the estrogen receptor indicating that the regulation of the HGF gene by estrogen is mediated via a direct interaction of the estrogen receptor complex with *cis*-acting ERE elements (Liu et al., 1994a).

Supportive evidence that HGF gene expression is modulated *in vivo* by estrogen comes from studies in which a single injection of estradiol resulted in a transient and marked elevation of HGF mRNA levels in the ovaries of immature mice (Liu et al., 1994a). The effect of estradiol appeared to be time-dependent, since no significant change in HGF mRNA levels was noted at 1 hour after treatment, but by 6 hours post-injection, HGF mRNA levels increased by approximately 10-fold, returning to control levels by 24 hours. Furthermore, the induction of HGF gene expression by estrogen was apparently tissue-specific, since no detectable change was observed in HGF mRNA levels in the kidneys or livers of the treated animals.

The regulation of HGF gene expression by estrogen is of particular importance with regard to the role this hormone plays in epithelial–stromal interactions. Hormonally-induced epithelial growth and morphogenesis observed in many tissues such as breast, prostate, uterus and ovary may possibly be due to modulation of HGF gene expression by steroid hormones. Recently, Parrott et al. (1994) have shown that the HGF transcript is expressed in stromal cells (thecal cells) of the ovarian follicle but not in granulosa cells (which are epithelial cells). The authors demonstrate that the granulosa cells, but not the thecal cells, respond to the mitogenic effects of HGF in culture. These findings strengthen the idea that HGF mediates epithelial–mesenchymal cross-talk and perhaps participates in the development and maturation of the ovarian follicle.

The mouse HGF–CAT chimeric constructs (5′ and internal deletions of the 3 kb HGF flanking region described above) were used in other experiments to map and identify the regulatory elements which may be critical for the function (ie. cell-type specific and inducible expression) of the HGF promoter. Transient transfection of HGF–CAT chimeric plasmids containing various lengths (2.8, 2.7, 2.2, 1.7, 1.4, 1.2, 1.0, 0.7, 0.5, 0.3 and 0.1 kb) of the 5′-flanking region of the mouse HGF gene into mouse fibroblast NIH–3T3 cells which express HGF mRNA, and into human endometrial carcinoma RL95-2 cells which do not express this gene, identified multiple upstream elements which modulate the basal activity of the HGF promoter. The HGF core promoter was mapped to within 100 bp upstream of the transcription start site and may be considered 'promiscuous' in that it is active in both cell types (although to a lesser extent in epithelial cells). However, the activity of this minimal promoter in epithelial cells is totally suppressed when upstream sequences are present. These studies mapped three strong negative cell type-specific elements to positions -964 to -699; -274 to -70, and -17 to $+2$ (Liu et al., 1994b, 1994c). Additionally, one strong positive element was also located within region -538 to -274 which influenced HGF promoter activity in the fibroblast cells. These findings indicate that HGF gene expression is tightly regulated in fibroblast cells, and its expression is silent in epithelial cells due to the utilization of multiple negative elements.

One of the cell-type specific repressor elements in the HGF promoter region was further characterized by DNase I footprinting, electrophoretic mobility band shift assays, site directed mutagenesis and transient transfection assays. These studies identified a 19 bp (-17 to $+2$) sequence containing a unique palindromic structure (5′-AAC-CGA$\underline{C}$CGGTTTGCAACA-3′). Mutation of one (underlined) of the four bases which are in contact with the binding protein resulted in total abrogation of binding of the nuclear protein and a concomitant increase in the transcriptional activity of the HGF–CAT reporter gene in an epithelial cell line. This DNA element specifically binds to a nuclear protein that is present at high levels in epithelial cells but not in fibroblasts (Liu et al., 1994c). The repressor was characterized to be a nuclear protein consisting of a single polypeptide with Mr of ~ 70 kDa. Studies on the function of the repressor within the context of the HGF promoter revealed that the repressor element is required but not sufficient to completely repress HGF gene expression in epithelial cells. These findings support the conclusion that the HGF gene contains multiple cell-type-specific negative elements which are required for silencing transcription in HGF non-expressing cells. The mechanisms by which this repressor element and cognate binding factor exert effects on HGF gene expression remain to be understood. However, due to its unique location within the HGF gene (− 17 to + 2), it may function by blocking the binding of the transcription machinery (Liu et al., 1994c).

It appears that repression of HGF gene expression in epithelial cells is critical for establishing a proper directional paracrine cross-talk between the stromal and epithelial compartments of a given tissue, and hence, for subsequent normal tissue growth and development. Therefore, escape from suppression of HGF expression in epithelial cells may result in an autocrine loop and contribute to abnormal growth and aberrant differentiation. This is supported by the fact that, in experimental models, co-expression of HGF and its receptor in fibroblasts results in transformation (Giordano et al., 1993; Rong et al., 1994). Recently, clinical studies have also shown that HGF mRNA is expressed in malignant, but not normal, epithelium of human breast tissue (Wang et al., 1994).

Regulation of HGF mRNA expression in stromal cells by extracellular signals derived from epithelial cells

It has been proposed that the interaction between the epithelium and stroma is bidirectional, implying that not only do stromal cells release soluble factors such as HGF to instruct proliferation and morphogenesis in epithelial cells, but epithelial cells, in turn, convey signals to the stromal compartment to regulate (HGF) gene expression. Therefore, it is reasonable to assume that unregulated release of such factor(s) by tumor tissues amplifies signals sent to HGF-expressing stromal cells. To identify such molecule(s), we used HGF-promoter–CAT chimeric plasmids and Northern blot analysis to screen several human carcinoma cell lines for the production of such factors. When conditioned media from several human carcinoma cell lines such as DU-145, A-431, and HUH-7 were added to NIH–3T3 fibroblasts that had been transfected with an HGF–CAT construct containing 2.7 kb of the 5′-flanking region of the HGF promoter region, a significant increase in HGF promoter activity was obtained (Fig. 2). To confirm whether activation of the HGF promoter indeed resulted in the induction of endogenous HGF mRNA, Northern blot analysis was performed on total RNA isolated from MRC-5 fibroblasts that had been cultured in the presence or absence of conditioned medium from each carcinoma cell line. The highest induction (5-fold) in the 6 kb HGF mRNA was obtained from conditioned media of A-431 and DU-145 cells, which interestingly also express a high level of HGFR mRNA (Fig. 3).

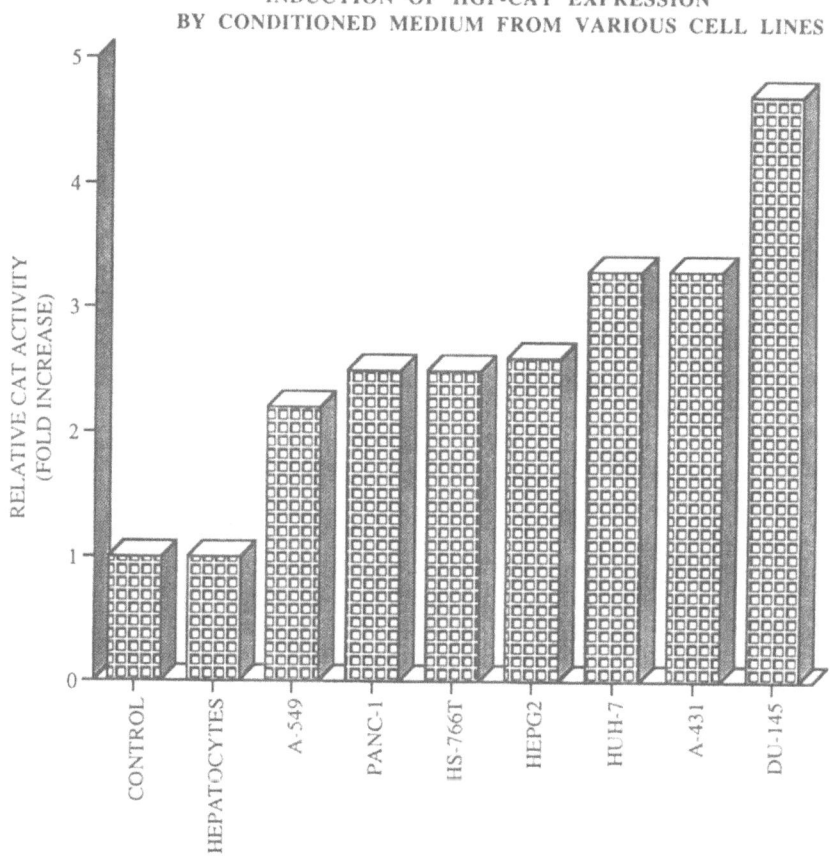

Figure 2. Induction of the HGF promoter by conditioned media from various human carcinoma cell lines. A stably transfected NIH–3T3 cell line harboring 2.8 kb of the mouse HGF 5'-flanking region fused to the CAT reporter gene was established as described previously (Liu et al., 1994b). The cells were seeded at 1×10^5/ml 24 h prior to treatment and treated with conditioned medium from the indicated cell lines. After incubation for an additional 24 h, cells were harvested and CAT assays were performed. Relative CAT activity was reported after normalization with protein concentration, and results are presented as fold-increase over control (cells treated with fresh medium only).

Further characterization of the molecule(s) responsible for the induction of HGF gene expression in these conditioned media revealed that the HGF-inducing activity is heat sensitive and has an Mr of greater than 10 000. Since we previously reported that IL-6 induces the HGF promoter activity (Liu et al., 1994b), we tested whether the HGF-inducing activity present in the conditioned media of the various carcinoma cell lines is due to IL-6 released by these cells. Neutralizing antibody against IL-6 dramatically reduced the HGF-inducing activity of the conditioned media as determined by CAT assays. Moreover, addition of pure IL-6 markedly and transiently increased the exogenous level of the

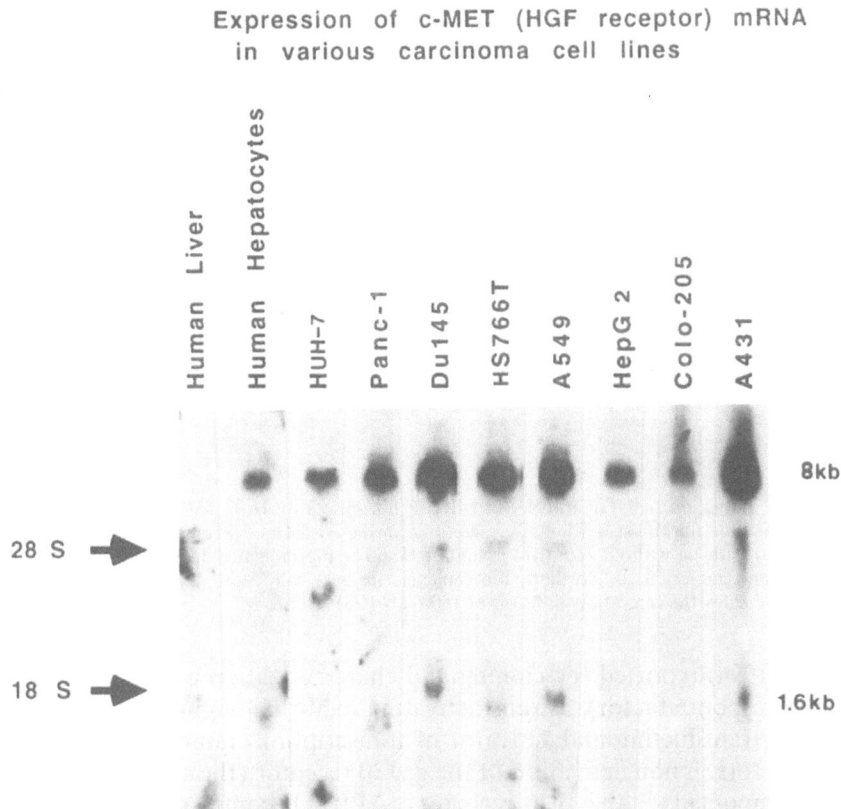

Figure 3. Expression of the HGF receptor mRNA in various human carcinoma cell lines. Total cellular RNA was isolated from different cell lines and subjected to Northern blot analysis using a ^{32}P-labeled human HGFR cDNA probe corresponding to the extracellular domain of the receptor.

6 kb HGF transcript in the MRC-5 fibroblast cell line (Fig. 4). These results indicate that IL-6 which is known to be produced by a variety of normal and malignant cells is capable of inducing HGF gene expression at the transcriptional level. Although the molecular mechanisms of the IL-6 signal transduction pathway have just begun to unfold, it is known that IL-6 modulates gene expression by regulating the synthesis of several transcription factors (at the mRNA level) such as the CCAAT/enhancer-binding-proteins (C/EBP), Octamer-binding-proteins, and the AP-1 family of transcription factors (Chen-Kiang et al., 1993). Most of these transcription factors have response elements present in the 5'-flanking region of the HGF gene (Liu et al., 1994b) that may potentially be utilized.

Another unique signaling pathway for the induction of genes responsive to IL-6 is through the activation of the IL-6 response element which is also present in the promoter region of HGF (Fig. 1). Recently, Akira

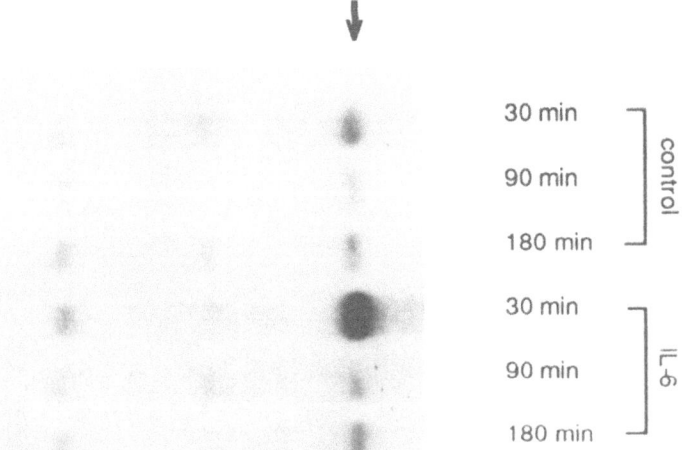

Figure 4. Induction of HGF mRNA in MRC-5 fibroblasts by IL-6. MRC-5 cells, a human embryonic lung fibroblast cell line, were cultured in the absence of serum overnight and then challenged with or without (control) 10 ng/ml IL-6 for the indicated time. Total RNA was isolated and subjected to Northern blot analysis using a full-length human HGF cDNA. Arrow indicates the migration position of the 6 kb HGF transcript.

et al. (1994) reported the cloning and characterization of APRF (acute phase response factor), a trans-activating factor belonging to the STAT (signal transducting and activator of transcription) family of transcription factors. Upon activation of the gp130 receptor (the signal transducing component of the IL-6 receptor), APRF becomes phosphorylated on a tyrosine residue, moves into the nucleus, binds to the IL-6 response element (CTGGGA) present in the promoter region of various IL-6 responsive genes (such as acute phase protein), and rapidly induces their transcription. As shown in Figure 4, IL-6 substantially and quickly upregulates HGF mRNA expression. Whether HGF gene induction by IL-6 is due to the activation of the IL-6 response elements, AP1 sites, NF-IL-6 binding sites or a combination remains to be determined.

Other studies have also described HGF-inducing factors. Matsumoto et al. (1992c) was the first to report the presence of an HGF-inducing activity in the serum of rats that had been subjected to tissue injury (hepatectomy or nephrectomy). This factor which was named 'injurin' induced HGF mRNA expression *in vivo* (when injected into normal rats) as well as *in vitro* (when added to cultured MRC-5 fibroblasts). Preliminary characterization of this molecule revealed that the HGF-inducing activity is a glycoprotein which is heat and acid stable and has an Mr of about 12 000 (Matsumoto et al., 1992c). Another factor was identified in the conditioned media from human breast carcinoma cell lines such as MCF-7 and T47D which markedly enhanced HGF mRNA expression in MRC-5 fibroblasts. This factor appears to be distinct from 'injurin' based on the fact that it is heat sensitive and has a molecular mass of greater than 30 000 (Seslar et al., 1993). Rosen et al. (1994b) also reported the presence of HGF-inducing activities in the conditioned media from a variety of normal and tumor human cell lines. One of these factors was further purified and shown to be heat stable having an apparent molecular weight 12 kDa (Rosen et al., 1994b). This factor is apparently distinct from IL-1 or TNFα which are known to induce HGF.

Based on these and other findings, it is clear that HGF mRNA expression in stromal cells is controlled by factors released from either epithelial cells (acting in paracrine mode) or from fibroblasts themselves in an autocrine fashion. It is likely that some of these partially character-ized HGF-inducing activities are due to one or more known inflamma-tory cytokines and growth factors. Novel factors may also exist and be important regulators despite a lack of information on their structure and mechanisms of action.

The HGF receptor (c-*met*)

The c-*met* protooncogene encodes a transmembrane tyrosine kinase growth factor receptor (Park et al., 1987) and was originally discovered as an activated oncogene in a chemically treated human osteosarcoma cell line (Cooper et al., 1984). Subsequent studies revealed that the ligand for this receptor is hepatocyte growth factor (Bottaro et al., 1991; Naldini et al., 1991). The multiple biological responses to HGF are all transmitted through the activation of this receptor (Weidner et al., 1993). c-*met* is encoded by a major transcript of approximately 8 kb in size, and is expressed mainly in epithelial cells. Although mesenchymal cells also express this receptor, c-*met* expression is at much lower levels (Faletto et al., 1993; Sonnenberg et al., 1993). Several investigations have demon-strated that c-*met* mRNA and protein are over-expressed in a variety of human carcinoma and sarcoma tissues (Di Renzo et al., 1991; Rong et al., 1993). Moreover, high expression of a cDNA encoding the c-*met* receptor in normal NIH–3T3 fibroblast cells resulted in the induction of invasiveness and metastatic behavior in these cells *in vitro* and *in vivo* (Giordano et al., 1993; Rong et al., 1994). Therefore, understanding the molecular mechanisms of c-*met* gene expression is also essential for delineating the role of this receptor in epithelial–mesenchymal interac-tions in normal and neoplastic growth and development.

Regulation of c-*met* gene expression by cytokines and hormones

We recently reported that the gene for the HGF receptor is inducible and that inflammatory cytokines such as IL-1α, IL-6 and TNF-α, as well as

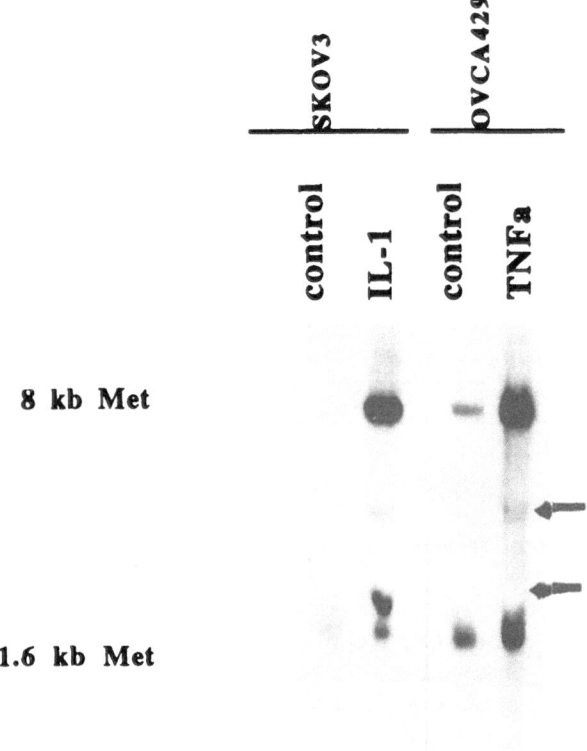

Figure 5. Induction of c-*met* mRNA in human carcinoma cell lines by IL-1 or TNFα. Two human ovarian carcinoma cell lines (SK OV3) and OVCA 429 were cultured in the absence of serum for 12 hours and then treated with the indicated cytokine for an additional 24 hours. Total RNA was isolated and subjected to Northern blot analysis using a human HGFR cDNA as described previously (Moghul et al., 1994). The upper arrow to the right represents the migration of the 28S rRNA band, while the lower arrow represents that of the 18S.

TGF-β1, EGF, HGF and the steroid hormones (estrogen, progesterone, tamoxifen and dexamethasone) markedly influence the steady-state level of the 8 kb c-*met* mRNA in several human carcinoma cell lines (Moghul et al., 1994). For instance, as shown in Figure 5, addition of IL-1 or TNFα to two human ovarian carcinoma cell lines dramatically induced c-*met* mRNA expression. As depicted in the figure, two c-*met* transcripts, 8 kb and 1.6 kb transcripts (see below), were induced by these treatments. Progesterone showed similar inductive effects on these cells (Moghul et al., 1994). In the case of the RL95-2 cells, a human endometrial carcinoma cell line, IL-6 and estradiol were the most potent cytokine and steroid hormone, respectively, inducing HGFR mRNA expression (Moghul et al., 1994). We have recently cloned and partially characterized the promoter region of the mouse c-*met* gene and have shown that several functional *cis*-acting DNA response elements inducible by cytokines exist within the 5'-flanking region of c-*met* promoter (Seol and Zarnegar, unpublished observations). Together, these results support the belief that c-*met* gene expression is regulated at the transcriptional level by extracellular stimuli. Other investigators have also shown that the c-*met* gene is inducible by serum, HGF, and TPA and that the induction is at the transcriptional level (Boccaccio et al., 1994; Gambarotta et al., 1994).

We have observed that most of the human carcinoma cell lines thus far examined express a 1.6 kb c-*met* mRNA, in addition to expressing the 8 kb c-*met* transcript (Figs 3 and 5). Interestingly, the smaller transcript was preferentially upregulated by progesterone (Moghul et al., 1994). This transcript corresponds to the amino-terminal region of the receptor which encodes the extracellular domain of the receptor. A cDNA for a 1.6 kb splice variant of the c-*met* transcript was cloned and sequenced by Rodrigues and Park (1993) from a human carcinoma cell line. This variant also encodes a C-terminally truncated c-*met* lacking the intracellular tyrosine kinase domain. The variant discovered by Rodrigues and Park may be identical to the transcript we describe above, but this has not been confirmed. Nevertheless, the variant, if translated, may result in a protein which potentially competes with full-length c-*met* for binding to HGF neutralizing HGF's biological effects.

Conclusion

HGF is the prototype of a new class of stromally-derived peptide growth factors having a broad spectrum of biological activities. It plays a pivotal role in normal growth and development by controlling cell proliferation, motility, morphogenesis, and differentiation of various epithelial tissues. Expression of HGF and HGFR genes is tightly regulated, and disregulation of these genes undoubtedly contributes to the development and progression of various forms of human cancer. Both HGF and HGFR genes are inducible, and their expression in the stromal and epithelial compartments of a given tissue is coordinated by several similar cytokines and hormones.

As depicted in Figure 6, cytokines such as IL-1, IL-6, TNFα, or TGFβ derived from neighboring epithelial cells, from stromal cells or from the migration of inflammatory cells such as macrophages into the microenvironment during tissue repair or tumor infiltration will stimulate HGF gene expression in stromal cells, while simultaneously regulating HGFR expression in epithelial cells. This may involve paracrine and/or autocrine mechanisms. Steroid hormones derived locally or from the circulation may also coordinate HGF and HGFR gene expression in stromal and epithelial cells. As shown in the diagram, another level of control may be the bioavailability of HGF protein. For example, pro-HGF, which is

Model for regulation of HGF and c-Met expression and actions via epithelial-stromal communication

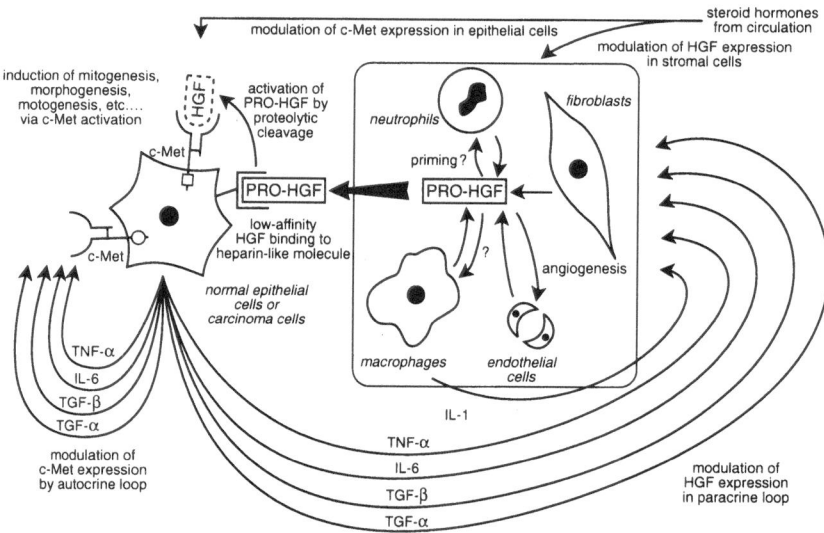

Figure 6. Proposed model for the coordinated expression of HGF and HGFR (c-*met*) mRNA in the stromal and epithelial compartments via a cytokine/hormone network. Cytokines and steroid hormones, as indicated in the figure, can modulate the levels of HGF (and HGFR) mRNA in mesenchymal cells generating an autocrine/paracrine loop leading to various biological responses induced by HGF receptor activation. Similarly, these factors can also orchestrate the expression of the HGFR receptor in epithelial cells thus establishing a paracrine loop between the stromal and epithelial compartments of a given tissue.

biologically inert, has high affinity (in the nM range) for heparin-like molecules present in the extracellular matrix and on the cell surface but has little affinity for the HGFR. After proteolytic cleavage [most likely by uPA, which is itself upregulated by IL-1β and TNFα (Marshal et al., 1992)], the heterodimeric bioactive HGF undergoes conformational changes, acquires a very high affinity (in the pM range) for the c-*met* receptor, binds and activates biochemical pathways resulting in myriad biological responses (mitogenesis/mitoinhibition, motogenesis, metastogenesis, and morphogenesis) commonly attributed to this receptor/ligand system.

Acknowledgments
Some of the experimental studies reviewed in this chapter have resulted from the collective effort of several individuals from the laboratory of the author, all of whom are gratefully acknowledged including Dr. Y. Liu, Dr. Q. Chen, A. Moghul, D.W. Seol, G. Jiang, L. Lin, A. Bell, and A. Beedle. Special thanks should also be given to M.C. DeFrances for critical reading of this review. Work in the laboratory of the author was supported by grants from the American Cancer Society (CN # 55B) and NIH (#R01ES06109-01).

References

Aaronson, S.A. (1991) Growth factors and cancer. *Science* 254: 1146–1153.

Akira, S., Nishio, Y., Inoue, M., Wang, X., Wei, S., Matsusaka, T., Sudo, T., Naruto M. and Kishimoto, T. (1994) Molecular cloning of APRF, a novel IFN-stimulated gene factor3 p91-related transcription factor involved in the gp130-mediated signaling pathway. *Cell* 77: 63–71.

Bhargava, M., Joseph, A., Kensel, J., Halaban, R., Li, Y., Pang, S., Goldberg, I.D., Setter, E., Donovan, M.A., Zarnegar, R., Michalopoulos, G.K., Nakamura, T., Faletto, D. and Rosen, E.M. (1992) Scatter factor and hepatocyte growth factor: Activities, properties, and mechanism. *Cell Growth & Diff.* 3: 11–20.

Boccaccio, C., Gaudino, G., Gambarotta, G., Galimi, F. and Comoglio, P.M. (1994) Hepatocyte growth factor receptor expression is inducible and is part of the delayed-early response to HGF. *J. Biol. Chem.* 269: 12846–12851.

Bottaro, D.P., Rubin, J.S., Faletto, D.L., Chan, A.M.-L., Kmiecik, T.E., Vande Woude, G.F. and Aaronson, S.A. (1991) Identification of the hepatocyte growth factor receptor as the c-*met* proto-oncogene product. *Science* 251: 802–804.

Bussolino, F., Di Rezo, M.F., Ziche, M., Bocchietto, E., Olivero, M., Naldini, L., Gaudino, G., Tamagnone, L., Coffer, A. and Comoglio, P.M. (1992) Hepatocyte growth factor is a potent angiogenic factor which stimulates endothelial cell motility and growth. *J. Cell Biol.* 119: 629–641.

Chen-Kiang, S., Hsu, W., Natkunman, Y. and Zhang, X. (1993) Nuclear signaling by interleukin-6. *Current Opinion in Immunology* 5: 124–128.

Cooper, C.S., Park, M., Blair, D., Tainsky, M.A., Huebner, K., Croce, C.M., Vande Woude, G.F. (1984) Molecular cloning of a new transforming gene from a chemically transformed human cell line. *Nature* 311: 29–33.

DeFrances, M. C., Wolf, H., Michalopoulos, G.K. and Zarnegar, R. (1992) The presence of hepatocyte growth factor in the developing rat. *Development* 116: 38–395.

Di Renzo, M.F., Narsimhan, R.P., Olivero, M., Bretti, S., Giordano, S., Medico, E., Gaglia, P., Zara, P. and Comoglio, P.M. (1991) Expression of the met/HGF receptor in normal and neoplastic human tissues. *Oncogene* 6: 1997–2003.

Ebert, M., Yakoyama, M., Friess, H., Buchler, M.W. and Korc, M. (1994) Coexpression of the c-*met* protooncogene and hepatocyte growth factor in human pancreatic cancer. *Cancer Res.* 54: 5775–5778.

Faletto, D.L., Kaplan, D.R., Halverson, D.O., Rosen, E.M. and Vande Woude, G.F. (1993) Signal transduction in c-*met* mediated motogenesis. *In*: I.D. Goldberg and E.M. Rosen (eds): *Hepatocyte Growth Factor-Scatter Factor (HGF-SF) and C-*Met *Receptor.* Birkhäuser Verlag, Basel, pp 107–130.

Gambarotta, G., Pistoi, S., Giordano, S., Comoglio, P.M. and Santoro, C. (1994) Structure and inducible regulation of the human *met* promoter. *J. Biol. Chem.* 269: 12852–12857.

Giordano, S., Zhen, Z., Medico, E., Gaudino, G., Galimi, F. and Comoglio, P.M. (1993) Transfer of motogenic and invasive response to scatter factor/hepatocyte growth factor by transfection of human *met* protooncogene. *Proc. Natl. Acad. Sci. USA* 90: 649–653.

Gohda, E., Matsunaga, T., Kataoka, H., Takebe, T. and Yamamoto, I. (1994) Induction of hepatocyte growth factor in human skin fibroblasts by EGF, PDGF and FGF. *Cytokine* 6: 633–640.

Grant, D.S., Kleinman, H.K., Goldberg, I.D., Bhargava, M.M., Nickoloff, B.J., Kinsella, J.L., Polverini, P. and Rosen, E.M. (1993) Scatter factor induces blood vessel formation *in vitro*. *J. Cell Biol.* 90: 1937–1941.

Kinoshita, T., Tashiro, K. and Nakamura, T. (1989) Marked increase of HGF mRNA in non-parenchymal liver cells of rats treated with hepatotoxins. *Biochem. Biophys. Res. Commun.* 165: 1229–1234.

Kono, S., Nagaike, M., Masumoto, K., Nakamura, T. (1992) Marked induction of HGF mRNA in intact kidney and spleen in response to injury of distant organs. *Biochem. Biophys. Res. Commun.* 186: 991–998.

Liu, Y., Lin, L. and Zarnegar, R. (1994a) Modulation of hepatocyte growth factor gene expression by estrogen in mouse ovary. *Mol. and Cell. Endo.* 104: 173–181.

Liu, Y., Michalopoulos, G.K. and Zarnegar, R. (1994b) Structural and functional characterization of the mouse hepatocyte growth factor gene promoter. *J. Biol. Chem.* 269: 4152–4160.

Liu, Y., Beedle A.B., Lin, L., Bell, A.W. and Zarnegar, R. (1994c) Identification of a cell-type-specific transcriptional repressor in the promoter region of the mouse hepatocyte growth factor gene. *Mol. Cell. Biol.* 14: 7046–7058.

Liu, Y., Bell, A.W., Michalopoulos, G.K. and Zarnegar, R. (1994d) The mouse hepatocyte growth factor-encoding gene: Structural organization and evolutionary conservation. *Gene* 144: 179–187.

Mars, W., Zarnegar, R. and Michalopoulos, G.K. (1993) Activation of HGF by plasminogen activators. *American J. of Pathology* 143: 949–958.

Marshal, B., C., Xu, P.Q., Rao, N.V., Brown, B.R. and Hoidal, J.R. (1992) Pulmonary epithelial cell uPA: induction by IL-β and TNFα. *J. Biol. Chem.* 267: 11462–11469.

Matsumoto, K., Okazaki, H. and Nakamura, T. (1992a) Up-regulation of HGF gene expression by IL-1 in human skin fibroblasts. *Biochem. Biophys. Res. Commun.* 188: 235–243.

Matsumoto, K., Tajima, H., Okazaki, H. and Nakamura, T. (1992b) Negative regulation of HGF gene expression in human lung fibroblasts and leukemic cells by transforming growth factor *b* and glucocorticoids. *J. Biol. Chem.* 267: 24917–24920.

Matsumoto, K., Tajima, H., Hamanoue, M., Kohno, S., Kinoshita, T. and Nakamura, T. (1992c) Identification and characterization of 'injurin' an inducer of expression of the gene for hepatocyte growth factor. *Proc. Natl. Scad. Sci. USA* 89: 3800–3804.

Matsunaga, T., Gohda, E., Takebe, T., Wu, Y.L., Iwao, M., Kataoka, H. and Yamamoto, I. (1994) Expression of hepatocyte growth factor is up-regulated through activation of a cAMP-mediated pathway. *Exp. Cell Res.* 210: 326–335.

Miyazawa, K., Kitamura, A. and Kitamura, N. (1991) Structural organization and the transcriptional initiation site of the human HGF gene. *Biochemistry* 30: 9170–9176.

Miyazawa, K., Shimomura, T., Naka, D. and Kitamura, N. (1994) Proteolytic activation of hepatocyte growth factor in response to tissue injury. *J. Biol. Chem.* 269: 8966–8970.

Moghul, A., Lin, L., Beedle, A., Kanbour-Shakir, A., DeFrances, M.C., Liu, Y. and Zarnegar, R. (1994) Modulation of c-*met* proto-oncogene (HGF receptor) mRNA abundance by cytokines and hormones: evidence for rapid decay of the 8 kb c-MET transcript. *Oncogene* 9: 2045–2052.

Nakamura, T., Nishizawa, T., Hagiya, M., Seki, T., Shimonishi, M., Sugimura, A., Tashiro, K. and Shimizu, S. (1989) Molecular cloning and expression of hepatocyte growth factor. *Nature* 342: 440–443.

Naldini, L., Vigna, E., Narsimhan, R., Guadino, G., Zarnegar, R., Michalopoulos, G. and Comoglio, P.M. (1991) Hepatocyte growth factor stimulates the tyrosine kinase activity of the receptor encoded by the proto-oncogene, c-*met*. *Oncogene* 6: 501–504.

Noji, S., Tashiro, K., Koyama, E., Nohno, T., Ohyama, K., Taniguchi, S. and Nakamura, T. (1990) Expression of hepatocyte growth factor gene in endothelial and Kupffer's cells of damaged rat livers as revealed by *in situ* hybridization. *Biochem. Biophys. Res. Commun.* 173: 42–47.

Nishino, T., Kaise, N., Sindo, Y., Nishino, N., Nishida, T., Yasuda, S. and Masui, Y. (1991) Promyelocytic leukemia cell line, HL-60, produces human hepatocyte growth factor. *Biochem. Biophys. Res. Commun.* 181: 323–330.

Okajima, A., Miyazawa, K. and Kitamura, N. (1993) Characterization of the promoter region of the rat hepatocyte growth factor/scatter factor gene. *Eur. J. Biochim.* 213: 113–119.

Park, M., Dean, M., Kaul, K., Braun, M.J., Gonda, M. and Vande Woude, G.F. (1987) Sequence of *met* protooncogene cDNA has features characteristic of the tyrosine kinase family of growth factor receptors. *Proc. Natl. Acad. Sci. USA* 84: 6379–6383.

Parrott, J.A., Vigne, J-L., Chu, B.Z. and Skinner, M.K. (1994) Mesenchymal–epithelial interactions in the ovarian follicle involve keratinocyte growth factor and hepatocyte growth factor production by thecal cells and their action on granulosa cells. *Endocrinology* 135: 569–575.

Rodrigues, G.A. and Park, M. (1993) Isoforms of the *met* receptor tyrosine kinase. *In*: I.D. Goldberg and E.M. Rosen (eds): *Hepatocyte Growth Factor-Scatter Factor (HGF-SF) and C-Met Receptor*. Birkhäuser Verlag, Basel, pp 167–179.

Rong, S., Jeffers, M., Resau, J.H., Tsarfaty, I., Okarsson, M. and Vande Woude, G.F. (1993) *met* expression and sarcoma tumorigenicity. *Cancer Res.* 53: 5355–5360.

Rong, S., Segal, S., Anver, M., Resau, J.H. and Vande Woude, G.F. (1994) Invasiveness and metastasis of NIH-3T3 cells induced by Met–HGF/SF autocrine stimulation. *Proc. Natl. Acad. Sci. USA* 91: 4731–4735.

Rosen, E., Nigam, S.K. and Goldberg, I.D. (1994a) Scatter factor and the c-*met* receptor: A paradigm for mesenchymal/epithelial interaction. *J. Cell Biol.* 127: 1783–1787.

Rosen, E.M., Joseph, S., Jin, L., Rockwell, S., Elias, J.A., Knesel, J., Wines, J., McClellan, J., Kluger, M.J., Goldberg, I.D. and Zitnik, R. (1994b) Regulation of scatter factor production via soluble inducing factor. *J. Cell Biol.* 127: 225–234.

Rubin, J.S., Chan, M.L., Bottaro, D., Burgess, W., Taylor, W.J., Cech, A.C., Hirschfield D.W., Wong, J., Miki, T., Finch, P. and Aaronson, T. (1991) A broad-spectrum human lung fibroblast-derived mitogen is a variant of hepatocyte growth factor. *Proc. Natl. Acad. Sci. USA* 88: 415–419.

Rubin, J.S., Bottaro, D.P. and Aaronson, S.A. (1993) Hepatocyte growth factor/scatter factor and its receptor, the c-*met* proto-oncogene product. *Biochimica et Biophysica Acta* 1155: 357–371.

Seki, T., Hagiya, M., Shimonishi, M., Nakamura, T. and Shimizu, S. (1991) Organization of the HGF encoding gene. *Gene* 102: 213–219.

Sesler, S.P., Nakamura, T. and Byers, W. (1993) Regulation of fibroblast hepatocyte growth factor/scatter factor expression by human breast carcinoma cell lines and peptide growth factors. *Cancer Res.* 53: 1233–1238.

Shimomura, T., Kondo, J., Ociai, M., Naka, D., Miyazawa, K., Morimoto, Y. and Kitamura, N. (1993) Activation of the zymogen of hepatocyte growth factor activity by thrombin. *J. Biol. Chem.* 268: 22927–22932.

Sonnenberg, E., Meyer, D., Weidner, K.M. and Birchmeier, C. (1993) Scatter factor/hepatocyte growth factor and its receptor, the c-*met* tyrosine kinase, can mediate a signal exchange between mesenchyme and epithelia during mouse development. *J. Cell Biol.* 123: 223–235.

Soriano, J.V., Pepper, M.S., Nakamura, T., Orci, L. and Montesano, R. (1995) Hepatocyte growth factor stimulates extensive development of branching duct-like structures by cloned mammary gland epithelial cells. *J. Cell Science*; *in press.*

Stoker, M., Gherardi, E., Perryman, M. and Gray, J. (1987) Scatter factor is a fibroblast-derived modulator of epithelial cell mobility. *Nature* 327: 239–242.

Tamura, M., Arakaki, N., Tsubouchi, H., Takada, H. and Daikuhara, Y. (1993) Enhancement of human hepatocyte growth factor production by IL-1α and IL-β and tumor necrosis factor by fibroblasts in culture. *J. Biol. Chem.* 268: 8140–8145.

Tsarfaty, I., Resau, J.H., Rulong, S., Keydar, I., Faletto, D.L. and Vande Woude, G.F. (1992) The *met* protooncogene receptor and lumen formation. *Science* 257: 1258–1261.

Tsarfaty, I., Rong, S., Resau, J.H., Rulong, S., Pinto da Silva, P. and Vande Woude, G.F. (1994) The *met* proto-oncogene mesenchymal to epithelial cell conversion. *Science* 263: 98–101.

Wang, Y., Selden, A.C., Morgan, N., Stamp, G.W.H. and Hodgson, H.J.F. (1994) Hepatocyte growth factor/scatter factor expression in human mammary epithelium. *Am. J. Path.* 144: 675–682.

Weidner, K.M., Arakaki, N., Hartmann, G., Vandekerckhove, J., Weingart, S., Rieder, H., Fonatsch, C., Tsubouchi, H., Hishida, T., Daikuhara, Y. and Birchmeier, W. (1991) Evidence for the identity of human scatter factor and human hepatocyte growth factor. *Proc. Natl. Acad. Sci. USA* 88: 7001–7005.

Weidner, K.M., Sachs, M. and Birchmeier, W. (1993) The *met* receptor tyrosine kinase transduces motility, proliferation and morphogenic signals of scatter factor/hepatocyte growth factor in epithelial cells. *J. Cell Biol.* 121: 145–154.

Yamashita, J., Ogawa, M., Yamashita, S., Nomura, K., Kuramoto, M. and Saishoji, T. (1994) Immunoreactive hepatocyte growth factor is a strong and independent predictor of recurrence and survival in human breast cancer. *Cancer Res.* 54: 1630–1633.

Zarnegar, R., DeFrances, M.C., Kost, K., Lindroos, P. and Michalopoulos, G.K. (1991) Expression of hepatocyte growth factor (HGF) in regenerating rat liver after partial hepatectomy. *Biochem. Biophys. Res. Commun.* 177: 559–565.

Epithelial–Mesenchymal Interactions in Cancer
ed. by I.D. Goldberg & E.M. Rosen
© 1995 Birkhäuser Verlag Basel/Switzerland

The role of scatter factor and the c-*met* proto-oncogene in angiogenic responses

P.J. Polverini[1] and B.J. Nickoloff[2]

[1]*Laboratory of Molecular Pathology and*
[2]*Department of Pathology, University of Michigan Schools of Dentistry and Medicine, Ann Arbor, Michigan 48109-1078, USA*

Introduction

Angiogenesis, the formation of new capillary blood vessels from pre-existing vessels, is one of the most pervasive and fundamentally important biological processes encountered in mammalian organisms. Angiogenesis is an important event in a number of physiological settings and is central to the etiology and/or pathogenesis of a number of developmental disorders, inflammatory diseases, and solid tumors. Under normal conditions, this process is driven by a complex array of stimulatory and inhibitory mediators that function coordinately to initiate, sustain, and abruptly terminate angiogenesis. However, when positive stimulatory signals persist or fail to be counteracted by inhibitory signals, disease often results. Recent work from several laboratories suggests that scatter factor (SF) also known as hepatocyte growth factor, a multifunctional mesenchymal derived cytokine, and its receptor, the c-*met* proto-oncogene, may play a pivotal role in the development of angiogenic responses in certain physiological and pathological settings. In this review we will examine the evidence that implicates SF as a mediator of angiogenesis and describe recent findings which suggest that inappropriate expression of SF or c-*met* may contribute to aberrant angiogenesis and disease pathogenesis.

Angiogenesis in physiological and pathological processes

Tissue regeneration and the repair of wounds, the cyclical proliferation of the nutrient-rich endométrium in preparation for implantation of the fertilized egg, and the development of the embryo and its supporting tissues are biological events that are strictly dependent on the rapid yet temporary ingrowth of new capillary blood vessels (Folkman and Cotran, 1976; Folkman, 1985a,b; Folkman and Klagsbrun, 1987). In

Table 1. Physiological angiogenesis

Ovulation
Development of the corpus luteum
Embryogenesis
Immune responses
Inflammation & wound repair
Lactating breast

---

adult organisms capillary endothelial cells divide relatively infrequently. Turnover times for endothelial cells are typically of the order of several months or years (Engerman et al., 1967; Tannock and Hayashi, 1972). Yet when called upon, as in response hormonal sgnals during menstruation, following the release of proangiogenic cytokines from inflammatory cells or as a consequence of the activity of proteolytic enzymes that release angiogenic mediators sequestered in the extracellular matrix, endothelial cells lining venules will systematically degrade their basement membrane and proximal extracellular matrix, migrate directionally, divide, and organize into new functioning capillaries; all within a matter of days. This dramatic amplification of the microvasculature is nevertheless temporary; for as rapidly as the new capillaries are formed they virtually disappear within a matter of days or weeks, returning the tissue microvasculature to its *status quo*. It is this feature of transient growth and regression of capillaries that primarily distinguishes physiological from pathological angiogenesis (DiPietro and Polverini, 1993).

As shown in Table 1, angiogenesis is a feature of a limited number of physiological processes. In contrast, the etiology and pathogenesis of a much larger and increasingly expanding number of pathologic conditions (Tab. 2) have been shown to be a consequence of a vasoproliferative or angiogenic response that is persistent; due either to the overproduction of normal or aberrant forms of angiogenic mediators, or to the underproduction or decreased responsiveness of endothelial cells to physiological inhibitory signals. Before describing those physiological and disease processes where there appears to be a recently recognized link between SF and c-*met* expression and angiogenesis, a few words must be said about the general characteristics of some of the more well known angiogenic mediators, that both drive and temper this response.

Properties of proangiogenic cytokines and inhibitors of neovascularization

Angiogenic mediators can be broadly divided into two groups: stimulators (Tab. 3) and inhibitors (Tab. 4) (Folkman and Klagsbrun, 1987; Klagsbrun and D'Amore, 1991; Moses and Langer, 1991). It should be

Table 2. Angiogenesis and vasoproliferative-dependent diseases

Neoplasia
  Solid tumors
Vascular malformations & cardiovascular disorders
  Angiofibroma
  Arteriovenous malformations
  Atherosclerosis
  Hemangiomatosis
  Restenosis/reperfusion injury
  Vascular adhesions
Syndromes
  Dyschondroplasia with vascular hamartomas (Maffucci's Syndrome)
  Hereditary hemorrhagic telangiectasia (Rendu-Osler-Weber Syndrome)
  Von Hippel–Lindau Syndrome
Ocular disorders
  Corneal graft neovascularization
  Diabetic retinopathy
  Neovascular glaucoma
  Retrolental fibroplasia
  Trachoma
Chronic inflammatory diseases and aberrant wound repair
  Diabetes
  Granulations, burns
  Hemophiliac joints
  Hypertrophic scars
  Non-union fractures
  Osteoradionecrosis
  Psoriasis
  Pyogenic granuloma
  Rapidly progressing adult and juvenile periodontitis
  Rheumatoid arthritis
  Systemic sclerosis

See Moses and Langer (1991).

Table 3. Proangiogenic cytokines & mediators of angiogenesis

Growth factors	Other proteins and peptides
Acidic fibroblast growth factor	Angiogenin
Basic fibroblast growth factor	Angiotensin II
Epidermal growth factor	Ceruloplasm
Interleukin 1	Fibrin
Interleukin 2	Human angiogenic factor
Scatter factor	Interleukin 8
Transforming growth factor alpha	Plasminogen activator
Transforming growth factor beta	Polyamines
Tumor necrosis factor alpha	Substance P
Vascular endothelial growth factor	Urokinase
Carbohydrates and lipids	Other agents
12(R)-hydroxyeicosatrienoic acid	Adenosine
Hyaluronan fragments	Angiotropin
Lactic acid	Copper
Monobutyrin	Heparin
Prostaglandins E1 and E2	Nicotinamide
	ESAF

See reviews by Bouck (1990) and DiPietro and Polverini (1993) for references.

Table 4. Endogenous inhibitors of angiogenesis

Angiostatic steroids and sulfated polysaccharides
Eosinophilic major basic protein
High molecular weight hyaluronan
γ Interferon
Interleukin-1 (IL-1)
Laminin peptides
Placental RNase (angiogenin) inhibitor
Platelet factor 4
Prostaglandin synthesis inhibitors
Protamine
Somatostatin
Thrombospondin 1
Tissue inhibitors of metalloproteinases
Vitamin A and retinoids
Vitreous fluid

Adapted from Moses and Langer (1991).

noted however, that this distinction is not absolute since some of these mediators have been reported to exhibit proangiogenic as well as angiostatic activity (Heinmark et al., 1986; Sato et al., 1987; Fajardo et al., 1992; Nicosia and Tuszynski, 1994). The majority of the stimulatory molecules are multifunctional glycoproteins that induce endothelial cells to divide, migrate directionally toward the inducing stimulus, and differentiate into tubular structures. They are secreted by a variety of epithelial and mesenchymal cells including endothelial cells themselves in response to exogenous or endogenous stimuli. Most are produced locally and function in a paracrine or intercrine manner. These mediators can stimulate angiogenesis directly by interacting with receptors on the endothelial cell surface, or indirectly by attracting and activating accessory cells, ie., inflammatory macrophages, and inducing them to produce angiogenic mediators (Polverini et al, 1977; Thakral et al., 1979; Polverini and Leibovich, 1984; DiPietro and Polverini, 1993). Others such as copper may function as cofactors in key interstitial enzyme systems or in the case of plasminogen activator can activate latent enzymes such as transforming growth factor-β to reveal its angiogenic or angiostatic activity (Bouck, 1990; Klagsbrun and D'Amore, 1991). Still others play a key role in stabilizing and/or enhancing the function of stimulatory molecules normally sequestered in the extracellular matrix surrounding blood vessels, as heparin does, which when bound to basic fibroblast growth factor facilitates its interaction with high affinity receptors on the endothelial cell surface.

While the mediators responsible for initiating new capillary growth have been the subject of extensive investigation for a number of years (Folkman and Klagsbrun, 1987; Klagsbrun and Folkman, 1990; Klagsbrun and D'Amore, 1991), only recently has attention focused on the mechanisms and mediators responsible for the timely down-regulation of angiogenesis (Bouck, 1990; Klagsbrun and D'Amore, 1991; Moses and Langer, 1991) (Tab. 4). A common property of many of these inhibitors is that they can influence the ability of cells to produce, interact with or degrade their extracellular matrix. Alterations in the organization and composition of the extracellular matrix has been shown to have a profound effect on the growth and function of endothelial cells and in determining whether endothelial cells will differentiate and organize into a three-dimensional capillary network (Ingber and Folkman, 1988, 1989; Ingber, 1991).

An important feature of these complementary mediator systems is that with rare exception none of them are endothelial cell specific or unique to the process of angiogenesis. Most of these mediators have a wide range of functions and target cells. This is perhaps one of the most remarkable features of the angiogenic response; the ability of populations of endothelial cells to respond in an identical fashion to a phylogenetically diverse group of mediators. It would appear that the angiogenic phenotype has evolved as a highly conserved response without the need for its own mediator system. Rather, endothelial cells are able to utilize whatever growth stimulators and inhibitors are made available to it to produce new capillaries. For example during embryonic development basic fibroblast growth factor (Risau, 1991) has been shown to be an important mediator of vasculogenesis and angiogenesis. In contrast, in adult organisms this same mediator appears to have a much more restricted role in angiogenesis where an entirely different complement of angiogenic mediators come into play as for example in wound repair (Polverini et al., 1977; Polverini, 1989; Sunderkotter et al., 1991). Whether angiogenic stimulators and inhibitors are tissue or process specific is the subject of much speculation. Clearly the great redundancy in positive and negative regulators capable of orchestrating an angiogenic response attests to its fundamental importance in pathophysiological processes subsequently found to be a potent angiogenic.

Rosen et al. (1990a,b) reported that scatter factor induces motility and directed migration of vascular endothelials, processes that are often correlated with an angiogenic response. Bussolino et al. (1992) showed that physiologic concentrations of SF stimulated endothelial cell receptor kinase activity, cell proliferation, and motility. SF was also able to induce the repair of wounds in endothelial cell monolayers, stimulate the scatter of endothelial cells grown in three dimensional collagen gels, and induce neovascularization in the rabbit cornea. Grant et al. (1993) reported that SF stimulated capillary-like tube formation in Matrigel and induced angiogenesis in rat corneas. Before discussing the mechanisms underlying SF's angiogenic activity and its potential role in physiological and pathological angiogenic responses, some of the structural and biological properties of SF and the c-*met* proto-oncogene that suggest it is ideally suited as a mediator of angiogenesis will be described.

Structure and function of scatter factor and the c-*met* proto-oncogene

SF was first identified by Stoker et al. (1985) as a fibroblast-derived
cytokine that induces dispersion and spreading, and enhanced the
motility of normally cohesive epithelial cell populations (Stoker, 1984;
Stoker and Perryman, 1985; Stoker et al., 1987; Rosen et al., 1989). SF
is a heparin-binding glycoprotein consisting of a 60 kDa heavy a chain
and a 30 kDa light b chain linked by disulfide bonds (Gherardi et al.,
1989; Rosen et al., 1990a; Weidner et al., 1990). SF is a member of the
family of kringle-containing proteins, and exhibits about 38% aino acid
sequence homology to the proenzyme plasminogen (Nakamura et al.,
1989; Naldini et al., 1992). SF is also closely related to macrophage
stimulating protein (MSP), a serum protein that renders macrophages
responsive to chemoattractants (Yoshimura et al., 1993). It has been
demonstrated by functional, biochemical and sequence analysis that SF
and hepatocyte growth factor are one and the same protein and are
indistinguishable ligands for the c-*met* proto-oncogene tyrosine kinase
receptor (Furlong et al., 1991; Weidner et al., 1991; Bottaro et al., 1991;
Naldini et al., 1991; Bhargava et al., 1992).

In addition to stimulating cell motility, SF can influence the growth
and differentiation of a variety of epithelial cell types including mammary and renal tubular epithelial cells, keratinocytes, bronchial epithelia, and biliary epithelial cells (Rubin et al., 1991; Kan et al., 1991;
Rosen and Goldberg, 1994). SF has also been shown to function as a
morphogen capable of inducing kidney and mammary epithelial cells to
organize into branching tubules and mammary duct-like structures
respectively (Montesano et al., 1991; Tsarfaty et al., 1992). Similarly,
when vascular endothelial cells are grown on the reconstituted basement
membrane Matrigel, SF induces endothelial cells to organize into capillary-like tubes (Grant et al., 1993) and is angiogenic *in vivo* (Bussolino
et al., 1992; Grant et al, 1993; Naidu et al., 1994).

The SF receptor, the c-*met* proto-oncogene, is a tyrosine kinase
growth factor receptor that is found predominantly on epithelial cells
and on some mesenchymal cells, ie. endothelial cells (Park et al., 1987;
Naidu et al., 1994; Rosen and Goldberg, 1994). Although mesenchymal
in origin, endothelial cells display certain features characteristic of
epithelial cells including the formation of gap and tight junctions, a
flattened squamous-like morphology, and the ability to organize into
tubular structures. Thus endothelial cells have the potential of serving
both as a source and target of SF. The significance of this conversion of
SF/c-*met* from a paracrine to an autocrine signaling system in tumor
development and neovascularization will be discussed later in this
review. Binding of SF to c-*met* induces phosphorylation of an intracellular tyrosine acceptor site, activation of tyrosine kinase, inititiation of
a signal and transduction of the signal to the nucleus. However, the

molecules that transduce the c-*met* signal and the target genes induced by that signal are largely unknown (Rosen and Goldberg, 1994).

Mechanisms underlying the angiogenic activity of SF and the c-*met* proto-oncogene

The production of SF by cells in close proximity to vascular endothelium originally led Rosen and his colleagues to speculate on a role for SF in angiogenesis (Rosen et al., 1989, 1990a). They found that bovine aortic and human iliac artery smooth muscle cells produced SF in concentrations comparable to several lines of fibroblasts. These workers subsequently found that SF was a potent stimulator of motility for both large and small vessel endothelial cells motility using a variety of assays that measure different aspects of motility (Rosen et al., 1990a,b,c). Using a microcarrier bead migration assay (Rosen et al., 1990a) they showed that stimulated migration was abolished in the presence of cycloheximide, but was unaffected by hydroxyurea, indicating a requirement for protein but not DNA synthesis. Also agents that activate the adenylate cyclase signaling pathway, transforming growth factor-β, a variety of protein kinase inhibitors and anti-microfilament and anti-microtubule agents block SF induced migration. Thus the detachment of endothelial cells from carrier beads and reattachment to culture surfaces appeared to require protein phosphorylation and an intact cytoskeletal system.

An important early step in the formation of new capillaries is the focal degradation by endothelial cells of the subendothelial basement membrane and their subsequent invasion across this matrix barrier. Both mouse and human SF are able to markedly enhance the proteolytic and invasive properties of endothelial cells. Not unexpectedly, treatment of endothelial cells with SF induces large increases in both secreted and cell-associated urokinase type plasminogen activator (uPA) activity (Rosen et al., 1991; Grant et al., 1993). SF is also mitogenic for endothelial cells (Morimoto et al., 1991) and can induce endothelial cells to differentiate into capillary-like structures *in vitro*. The ability of SF to induce endothelial cells to organize into vessel-like structures is consistent with its previously described morphogenetic-inducing potential where it has been shown to induce kidney epithelial cells (Montesano et al., 1991) and mammary epithelial cells to form duct-like structures in suspension culture (Tsarfaty et al., 1992). Thus it would appear that SF is able to influence several endothelial cell functions that are associated with the angiogenic phenotype.

Role of scatter factor in physiological angiogenesis and the pathogenesis of vasoproliferative/angiogenesis-dependent disorders

The above data strongly implicate SF as a potential mediator of angiogenesis. Recent observations suggest a role for SF in angiogenic responses that accompany certain physiological and pathological processes. Normal wound repair is an essential biologic process that is driven by the cooperative interaction of multiple cell types and mediator systems. It is well established that monocytes and macrophages are key regulators of wound neovascularization. Activated macrophages or culture supernatants have been shown to induce new capillary growth (Polverini et al., 1977; Koch et al., 1986), and numerous macrophage derived products have been demonstrated to influence angiogenesis (for a review, see Folkman and Klagsbrun, 1987; Sunderkotter et al., 1991). Macrophages are also able to down-regulate capillary growth, as they have been shown to produce substances which inhibit endothelial cell growth *in vitro*, and/or angiogenesis *in vivo* (Jaffe et al., 1985; Polverini, 1989; Besner and Klagsbrun, 1991). We have preliminary evidence that both primary cultures of human and murine macrophages and

Figure 1. Colloidal carbon perfused rat cornea 7 days after implantation of a Hydron pellet containing culture media (5 µg total protein) from phorbol ester activated human monocyte-derived macrophages. Note the vigorous ingrowth of capillaries.

Figure 2. An attenuated angiogenic response following incorporation of anti-SF antibody with macrophage conditioned media. Note the markedly suppressed angiogenic response when compared with Figure 1.

macrophage-like cell lines produce biologically active SF (Rosen and Polverini, unpublished). Furthermore, exposure of macrophages to activating agents that normally enhance their angiogenic activity (Polverini et al., 1977) also up-regulates SF production (Fig. 1). When the conditioned culture media of activated macrophages is incubated with antibodies to SF, angiogenic activity in the rat cornea is substantially reduced although not completely abolished (Polverini and Rosen, unpublished) (Fig. 2). Thus SF appears to be a newly recognized macrophage-derived mediator of angiogenesis.

Chronic inflammatory diseases are frequently associated with prominent components of angiogenesis and/or proliferation of microvascular endothelium. We have reported previously the presence of immunoreactive SF in the skin of patients with psoriasis, a common inflammatory skin disorder characterized by epidermal hyperplasia and prominent elongation and proliferation of capillary endothelial cell within the dermal papillae and papillary dermis (Folkman, 1972; Braverman and Sibley, 1982; Morganroth et al., 1991; Nickoloff, 1991). Antisera to human SF stained spindle-shaped and mononuclear cells surrounding microvessels in psoriatic plaques (Grant et al., 1993). The cells of the microvessel wall (endothelium and pericytes) did not stain for SF; and normal skin from psoriatic patients and from normal subjects showed little or no immunoreactivity for SF. These results suggest a possible role for SF as a paracrine mediator of pathological angiogenesis in a human inflammatory disease.

Inappropriate expression of SF and/or c-*met* contribute to tumorigenicity by altering cellular differentiation and by stimulating tumor neovascularization (Faletto et al., 1991; Rong et al., 1993a,b; Tsarfaty et al., 1994). SF and the c-*met* receptor normally function as a paracrine signaling system where mesenchymal cells produce the SF ligand that binds to its receptor on epithelial cells. This is in agreement with the proposed role of SF/c-*met* in epithelial/mesenchymal cell communication. Recent studies have suggested that conversion of SF and c-*met* from a normal paracrine signaling system to an autocrine pathway can lead to dramatic alterations in cell differentiation and neoplasia (Faletto et al., 1991; Rong et al., 1992; Rong et al., 1993a,b; Tsarfaty et al., 1994). These workers demonstrated that c-*met* proto-oncogene is overexpressed in most spontaneously transformed NIH/3T3 cells and can mediate NIH/3T3 tumorigenicity (Cooper et al., 1986; Hudziak et al., 1992). While c-*met* is normally expressed at low levels in primary fibroblasts, in the presence of endogenously expressed SF this results in an autocrine interaction and disregulated cell growth. Rong et al. (1993a) showed that over-expression of c-*met* occurs frequently in various types of human sarcomas, thus implicating c-*met* in tumorigenicity. In a more direct test of this hypothesis, Tsarfaty et al. (1994) demonstrated that when c-*met* and SF were co-expressed in NIH/3T3 fibroblasts the cells became tumorigenic in nude mice. Furthermore, they found that the resultant tumors displayed a lumen-like morphology, contained carcinoma-like focal areas with intercellular junctions resembling desmosomes and co-expressed the epithelial and mesenchymal cytoskeletal markers cytokeratin and vimentin. We have recently examined the significance of conversion of SF and c-*met* from a paracrine to an autocrine signaling system in the pathogenesis of Kaposi's sarcoma (Naidu et al., 1994; Polverini and Nickoloff, 1995).

Kaposi's sarcoma (KS) is a complex mesenchymal neoplasm of suspected vascular endothelial cell origin (Rutgers et al., 1986; Roth et al., 1988; Zhang et al., 1994). It presents in several distinct pathological settings, with AIDS-associated KS being the most severe and life threatening form of the disease. It is especially lethal in homosexual and bisexual men with AIDS where it is reported to account for up to 95% of all AIDS-associated malignancies (Friedman-Kien, 1981; Friedman-Kien et al., 1981; Buchbinder and Friedman-Kien, 1991; Tappero et al., 1993).

During the course of our studies of KS tumors, we found that when HTLV-II conditioned media was added to cultures of human umbilical vein endothelial cells (HUVEC), within 24 h the normal epitheloid cobblestone morphology of HUVEC changed to a spindle-shaped appearance with some cells demonstrating prominent dendritic processes. This 'phenotypic conversion' to a Kaposi's sarcoma tumor cell-like phenotype was also associated with a dramatic up-regulation in the expression of cell surface antigens including: factor XIIIa, ICAM-1, and several cytokines that are unique to KS cells (Naidu et al., 1994; Polverini and Nickoloff, 1995).

We examined the components in KS growth medium (KSGM) that might be responsible for this conversion phenomenon and found high titers of SF in KSGM (Naidu et al., 1994). Two different batches of HTLV-II CM were found to contain elevated levels of SF. Scatter factor production rates for T cells infected with HTLV-II was estimated to be 120 units (U)/10^6 cells/48 h as compared to 20–80 U/10^6 cells/48 h for six different fibroblastic lines. PCR analysis confirmed that the HTLV-II infected T cells expressed SF mRNA (Naidu et al., 1994). Neither resting peripheral blood T cells nor the HUT 78 T cell line produced detectable SF by bioassay or ELISA. Interestingly, when HUT 78 T cells were infected by the HIV-1 IIIb virus, PCR analysis revealed induction of SF mRNA. Furthermore, when KSGM was assayed in the rat corneal bioassay for angiogenic activity, it induced prominent neovascularization. This angiogenic activity was almost completely abrogated by pre-incubating KSGM with neutralizing antibody (Ab) to SF, indicating that SF was the major angiogenic mediator in KSGM. The phenotypic conversion of normal endothelial cells induced by KSGM could be mimicked with the addition of purified native or recombinant SF to HUVEC medium. Thus, it became clear to us that SF present in KSGM was responsible for both the angiogenic activity in KSGM and the transdifferentiation of normal HUVECs to KS-like tumor cell phenotype.

To explore the *in vivo* relevance of these findings, we analyzed biopsies of KS tumors for expression of SF and the c-*met* receptor. We observed positive immunostaining of SF in lymphoid cells, perivascular dendritic cells, and interstitial spindle cells. Samples showed strong positive c-*met* staining of pili-erector smooth muscle bundles, pericytes, dermal dendritic cells, and interstitial spindle-shaped tumor cells. Cytospin preparations of KS tumor cells and HUVEC exposed to SF were also strongly positive for c-*met*. After verifying that KS cells expressed c-*met* mRNA, we asked whether SF had any mitogenic effect on KS cells. Three KS tumor lines were examined and all were stimulated to proliferate when purified native or recombinant SF was added to the culture media. Even when compared to optimal concentrations of the two other well characterized mitogens, IL-6 and oncostatin M, SF was at least as potent if not slightly better in stimulating KS tumor cell growth. Mice that have been genetically engineered to over-express the HIV-*tat* protein spontaneously develop cutaneous KS-like lesions and exhibit a high incidence of hepatocellular carcinoma in the absence of detectable HIV-*tat* transcripts or protein in the tumors (Vogel et al., 1988; Ensoli et al., 1990; Vogel et al., 1991). It has been proposed that the liver tumors were mediated by extrahepatic growth factors, perhaps SF. Taken together, these results suggest that SF may play an important role in the pathogenesis of KS. We envisage the following scenario to explain how SF participates in the pathogenesis of AIDS-KS.

Because endothelial cells exhibit a low turnover rate, the chances of a somatic mutation occurring in this quiescent cell population are exceedingly low. In AIDS-associated KS however, T cells infected with the HIV virus would provide a constant supply of SF that could function as a potent 'promoting' agent for endothelial cells that have been 'initiated' as a result of exposure to any one of several transforming viral sequences. This would be the scenario if the EC of an HIV infected individual acquired a carcinogenic strain of an organism such as human papillomavirus (HPV E6/E7) or the HIV virus itself (HIV-*tat*). In this instance, mutation might result directly or indirectly in production of SF along with over-expression of c-*met*. EC would now be both a source of and target for SF. What is normally a tightly regulated paracrine system would now be converted to a disregulated autocrine signaling system. These events would drive the tumorigenic process by promoting the growth of 'phenotypically converted', 'initiated' endothelial cells. A similar mechanism could be envisaged for an individual whose EC already contained a transforming viral sequence and who then acquired the HIV virus. In this setting, the HIV infected T cells would provide a rich source of SF that would promote the growth of initiated, mitoticly dormant EC. As a result of co-expressing SF and c-*met*, the phenotypically converted, initiated endothelial cells would be able to stimulate their own growth (autocrine) and promote the growth and recruit adjacent host cells (paracrine) into the developing KS tumor. Not only is SF able to act directly on endothelial cells to promote their growth, but it also acts indirectly by stimulating neovascularization. We speculate that the acquisition of SF-mediated angiogenic activity by EC occurs early in this process, at the stage of phenotypic conversion. In this instance, EC would behave as 'preneoplastic cells', where they would express angiogenic activity, a trait that is necessary although not sufficient for tumor formation (Gimbrone et al., 1976a; Gimbrone et al., 1976b; Brem et al., 1977; Brem et al., 1978; Folkman, 1985a,b; 1989; Moroco et al., 1990).

Conclusion

In this review we have described the evidence implicating SF and its receptor the c-*met* proto-oncogene in angiogenic responses. Its role as

Role of Scatter Factor in Angiogenic Responses

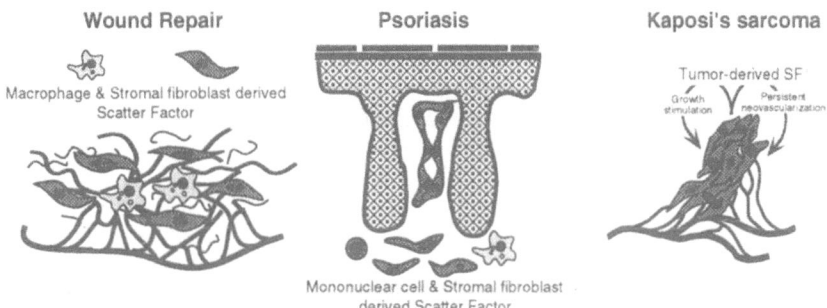

Figure 3. Diagram depicting the potential sources of SF in the induction of angiogenesis in certain physiological and pathological settings.

an important morphogen that mediates epithelial/mesenchymal interactions can now be extended to include the initiation and maintenance of neovascularization. Although there is much work yet to be done to define the mechanism of action of SF and c-*met* in angiogenesis, there is already ample evidence implicating this novel mediator system in angiogenic responses that occur in several physiological and pathological settings (Fig. 3). Perhaps the most compelling evidence implicating SF and c-*met* in angiogenesis is during the development of solid tumors, where inappropriate expression of SF and c-*met* has been linked to tumor promotion and neovascularization. Our own studies suggest that SF and c-*met* is important not only in the induction of tumor neovascularization but also in orchestrating as yet poorly understood series morphogenetic events that can lead to conversion of preneoplastic endothelial cells to malignant KS tumor cells. If our premise is correct, one could envisage several approaches whereby the modulation of SF and c-*met* expression might lead to a reduction in the incidence and severity of tumors or alter the course of angiogenesis-dependent chronic inflammatory disease. Antibody therapy directed against either SF-producing tumor cells or against the c-*met* receptor might decrease the incidence of new tumors by limiting their clonal expansion and lead to regression of established tumors by blocking SF-mediated tumor cell proliferation and neovascularization. It might also be possible to suppress production of SF or accessory cytokines involved in the induction of SF production and thus short circuit the SF/c-*met* growth promoting effects. This strategy could also be extended to chronic inflammatory diseases such as psoriasis, where modulation of SF or c-*met* might attenuate the persistent vasoproliferation that contributes to this chronic debilitating skin disease. Although considerably more work is needed in order to better understand the mechanisms of action of SF and c-*met* in physiological and pathological angiogenic responses, there is now little doubt that this novel mediator system warrants further attention if strategies aimed at controlling the pathological effects that results from aberrant angiogenesis are to be developed.

Acknowledgments
This work was supported by Grants HL39926 and CA64416 (P.J.P.) and AR0065, AR01823, and AR40488 (B.J.N.) from the National Institutes of Health.

References

Besner, G.E. and Klagsbrun, M. (1991) Macrophages secrete a heparin-binding inhibitor of endothelial cell growth. *Microvasc. Res.* 42: 187–197.
Bhargava, M., Joseph, A., Knesel, J., Halaban, R., Li, Y., Pang, S., Goldberg, I., Setter, E., Donovan, M.A., Zarnegar, R., Michalopoulos, G.A., Nakamura, T., Faletto, D. and Rosen, E.M. (1992) Scatter factor and hepatocyte growth factor: activities, properties, and mechanism. *Cell Growth & Diff.* 3: 11–20.
Bottaro, D.P., Rubin, J.S., Faletto, D.L., Chan, A.M.-L., Kmiecik, T.E., Vande Woude, G.F. and Aaronson, S.A. (1991) Identification of the hepatocyte growth factor receptor as the c-*met* proto-oncogene product. *Science* 251: 802–804.
Bouck, N. (1990) Tumor angiogenesis: The role of oncogenes and tumor suppressor genes. *Cancer Cells* 2: 179–185.
Braverman, I.M. and Sibley, J. (1982) Role of the microcirculation in the treatment and pathogenesis of psoriasis. *J. Invest. Dermatol.* 78: 12–17.
Brem, S.S., Gullino, P.M. and Medina, D. (1977) Angiogenesis: A marker for neoplastic transformation of mammary papillary hyperplasia. *Science* 195: 880–882.
Brem, S.S., Jensen, H.M. and Gullino, P.M. (1978) Angiogenesis as a marker of preneoplastic lesions of the breast. *Cancer* 41: 239–244.
Buchbinder, A. and Friedman-Kien, A.E. (1991) Clinical aspects of epidemic Kaposi's sarcoma. *Cancer Surveys* 10: 39–52.
Bussolino, F., DiRenzo, M.F., Ziche, M., Bocchietto, E., Olivero, M., Naldini, L., Gaudino, G., Tamagnone, L., Coffer, A. and Comoglio, P.M. (1992) Hepatocyte growth factor is a potent angiogenic factor which stimulates endothelial cell motility and growth. *J. Cell Biol.* 119: 629–641.
Cooper, C.S., Tempest, P.R., Beckman, P.M., Neldin, C.-H. and Breakers, P. (1984) Amplification and overexpression of met gene in spontaneously transformed NIH/3T3 mouse fibroblasts. *EMBO J.* 5: 2623–2628.
DiPietro, L.A. and Polverini, P.J. (1993) Role of the macrophage in the positive and negative regulation of wound neovascularization. *Behring Inst. Mitt.* 92: 238–247.
Engerman, R.L., Pfaffenbach, D. and Davis, M.D. (1967) Cell turnover of capillaries. *Lab. Invest.* 17: 738–743.
Ensoli, B., Barillari, G., Salahuddin, S.Z. and Gallo, R.C. (1990) Tat protein of HIV-1 stimulates growth of cells derived from Kaposi's sarcoma lesions of AIDS patients. *Nature* 345: 84–86.
Fajardo, L.F., Kwan, H.H., Kowalski, J., Prionas, S.D. and Allison, A.C. (1992) Dual role of tumor necrosis factor-α in angiogenesis. *Am. J. Pathol.* 140: 539–544.
Faletto, D.L., Tsarfaty, I., Kmiecik, T.E., Gonzatti, M., Suzuki, T. and Vande Woude, G.F. (1991). Evidence for noncovalent clusters of the c-met proto-oncogene product. *Oncogene* 7: 1149–1157.
Folkman, J. (1972) Angiogenesis in psoriasis. *J. Invest. Dermatol.* 59: 40–48.
Folkman, J. and Cotran, R.S. (1976) Relation of vascular proliferation to tumor growth. *Int. Rev. Exp. Pathol.* 16: 207–248.
Folkman, J. (1985a) Tumor angiogenesis. *Adv. Cancer Res.* 43: 175–203.
Folkman, J. (1985b) Toward an understanding of angiogenesis: Search and discovery. *Perspectives in Biol. and Med.* 9: 10–36.

Folkman, J. and Klagsbrun, M. (1987) Angiogenic factors. *Science* 235: 442–447.

Folkman, J. (1989) What is the evidence that tumours are angiogenesis dependent. *J. Natl. Cancer Inst.* 82: 4–6.

Friedman-Kien, A.E. (1981) Disseminated Kaposi's sarcoma in young homosexual men. *J. Am. Acad. Dermatol.* 5: 468–470.

Friedman-Kien, A.E., Laubenstien, L. and Marmor, M. (1981) Kaposi's sarcoma and pneumocystis pneumonia among homosexual men – New York and California. *Mortal. Morbid. Rpts.* 30: 305–308.

Furlong, R.A., Takheara, T., Taylor, W.G., Nakamura, T. and Rubin, J.S. (1991) Comparison of biologic and immunochemical properties indicate that scatter factor and hepatocyte growth factor are indistinguishable. *J. Cell Sci.* 100: 173–177.

Gherardi, E., Gray, J., Stoker, M., Perryman, M. and Furlong, R. (1989) Purification of scatter factor, a fibroblast-derived basic protein which modulates epithelial interactions and movement. *Proc. Natl. Acad. Sci. USA* 86: 5844–5848.

Gimbrone, M.A., Jr. and Gullino, P.M. (1976a) Angiogenic capacity of preneoplastic lesions of murine mammary gland as a marker of neoplastic transformation. *Cancer Res.* 36: 2611–2620.

Gimbrone, M.A., Jr. and Gullino, P.M. (1976b) Neovascularization induced by intraocular xenografts of normal, preneoplastic, and neoplastic mouse mammary tissue. *J. Natl. Cancer Inst.* 55: 305–318.

Grant, D.S., Kleinman, H.K., Goldberg, I.D., Bhargava, M.M., Nickoloff, B.J., Kinsella, J.L., Polverini, P.J. and Rosen, E.M. (1993) Scatter factor induces blood vessel formation in vivo. *Proc. Natl. Acad. Sci. USA* 90: 1937–1941.

Heinmark, R.L., Twardzik, D.R. and Schwartz, S.M. (1986) Inhibition of endothelial regeneration by type-beta transforming growth factor from platelets. *Science* 233: 1078–1080.

Hudziak, R., Lewis, G.D., Holmes, W.E., Ullrich, A. and Shepard, H.M. (1990) Selection for transformation and met proto-oncogene amplification in NIH/3T3 fibroblasts using tumor necrosis factor α. *Cell Growth & Diff.* 1: 129–134.

Ingber, D.E. and Folkman, J. (1988) Inhibition of angiogenesis through modulation of collagen metabolism. *Lab. Invest.* 59: 44–51.

Ingber, D.E. and Folkman, J. (1989) Mechanochemical switching between growth and differentiation during fibroblast growth factor-stimulated angiogenesis in vitro: Role of extracellular matrix. *J. Cell. Biol.* 109: 317–330.

Ingber, D.E. (1991) Extracellular matrix and cell shape: Potential control points for the inhibition of angiogenesis. *J. Cell Biochem.* 47: 236–241.

Jaffe, E.A., Ruggiero, J.T. and Falcone, D.J. (1985) Monocytes and macrophages synthesize and secrete thrombospondin. *Blood* 65: 79–84.

Kan, M., Zhang, G.H., Zarnegar, R., Michalopoulos, G., Myoken, Y., McKeehan, W.L. and Stevens, J.L. (1991) Hepatocyte growth factor–hematopoietin A stimulates the growth of rat proximal tubular epithelial cells (rpte), rat non-parenchymal liver cells, human melanoma cells, mouse keratinocytes, and stimulates anchorage-independent growth of SV40-transformed rpte. *Biochem. Biophys. Res. Commun.* 174: 331–337.

Klagsbrun, M. and Folkman, J. (1990) Angiogenesis. *In:* M.B. Sporn and A.B. Roberts (eds): *Handbook of Experimental Pharmacology, Vol., 95/II Peptide Growth Factors and their Receptors II.* Springer-Verlag, Berlin, Heidelberg, Germany, pp 549–586.

Klagsbrun, M. and D'Amore, P.A. (1991) Regulators of angiogenesis. *Annu. Rev. Physiol.* 53: 217–239.

Koch, A.E., Polverini, P.J. and Leibovich, S.J. (1986) Stimulation of neovascularization by human rheumatoid synovial tissue macrophages. *Arth. Rheum.* 29: 471–479.

Montesano, R., Matsumoto, K., Nakamura, T. and Orci, L. (1991) Identification of a fibroblast-derived epithelial morphogen as hepatocyte growth factor. *Cell* 67: 901–908.

Morganroth, G.S., Chan, L.S., Weinstein, G.D., Voorhees, J.J. and Cooper, K.D. (1991) Proliferating cells in psoriatic dermis are comprised primarily of T cells, endothelial cells, and factor XIIIa perivascular dendritic cells. *J. Invest. Dermatol.* 96: 333–340.

Morimoto, A., Okamura, K., Hamanaka, R., Sato, Y., Shima, N., Higashio, K. and Kuwano, M. (1991) Hepatocyte growth factor modulates migration and proliferation of human microvascular endothelial cells in culture. *Biochem. Biophys. Res. Commun.* 179: 1042–1049.

Moroco, J.R., Solt, D.B. and Polverini, P.J. (1990) Sequential loss of suppressor genes for three specific functions during in vivo carcinogenesis. *Lab. Invest.* 631: 298–306.

Moses, M.A. and Langer, R. (1991) Inhibitors of angiogenesis. *Biotechnology* 9: 630–634.

Naidu, Y.M., Rosen, E.M., Zitnick, R., Goldberg, I., Park, M., Naujokas, M., Polverini, P.J. and Nickoloff, B.J. (1994) Role of scatter factor in the pathogenesis of AIDS-related Kaposi's sarcoma. *Proc. Natl. Acad. Sci. USA* 91: 5281–5285.

Nakamura, T., Nishizawa, T., Hagiya, M., Seki, T., Shimonishi, M., Sugimura, A. and Shimizu, S. (1989) Molecular cloning and expression of human hepatocyte growth factor. *Nature* 342: 440–443.

Naldini, L., Vigna, E., Narsimhan, R., Guadino, G., Zarnegar, R., Michalopoulos, G. and Comoglio P.M. (1991) Hepatocyte growth factor (HGF) stimulates tyrosine kinase activity of the receptor encoded by the proto-oncogene c-met. *Oncogene* 6: 501–504.

Naldini, L., Tamagnone, L., Vigna, E., Sachs, M., Hartmann, L., Birchmeier, W., Daikuhara, Y., Tsubouchi, H., Blasi, F. and Comoglio, P.M. (1992) Extracellular proteolytic cleavage by urokinase is required for activation of hepatocyte growth factor/scatter factor. *EMBO J.* 11: 4825–4833.

Nickoloff, B.J. (1991) The cytokine network in psoriasis. *Arch. Dermatol.* 127: 871–884.

Nicosia, R.F. and Tuszynski, G.P. (1994) Matrix-bound thrombospondin promotes angiogenesis *in vitro*. *J. Cell Biol.* 124: 183–193.

Park, M., Dean, M., Kaul, K., Braun, M.J., Gonda, M.A. and Vande Woude, G.F. (1987) Sequence of met proto-oncogene cDNA has features characteristic of the tyrosine kinase family of growth factor receptors. *Proc. Natl. Acad. Sci. USA* 84: 6379–6384.

Polverini, P.J., Cotran, R.S., Gimbrone, M.A. and Unanue, E.R. (1977) Activated macrophages induce vascular proliferation. *Nature* 269: 804–806.

Polverini, P.J. and Leibovich, S.J. (1984) Induction of neovascularization in vivo and endothelial cell proliferation in vitro by tumor-associated macrophages. *Lab. Invest.* 51: 635–642.

Polverini, P.J. (1989) Macrophage-induced angiogenesis; a review. *In*: C. Sorg (ed.): *Cytokines*. S. Karger, Basel, pp 54–73.

Polverini, P.J. and Nickoloff, B.J. (1995) Role of scatter factor and the c-*met* protooncogene in the pathogenesis of AIDS-associated Kaposi's sarcoma. *Adv. Cancer Res.* 66: 235–253.

Risau, W. (1991) Embryonic angiogenesis factors. *Pharmacol. Ther.* 51: 371–376.

Rong, S., Jeffers, M., Resau, J.H., Tsarfaty, I., Oskarsson, M. and Vande Woude, G.F. (1993a) Met expression and sarcoma tumorigenicity. *Cancer Res.* 53: 5355–5360.

Rong, S., Oskarsson, M., Faletto, D.L., Tsarfaty, I., Resau, J., Nakamura, T., Rosen, E., Hopkins, R. and Vande Woude, G.F. (1993b) Tumorigenesis induced by coexpression of human hepatocyte growth factor and the human *met* proto-oncogene leads to high levels of expression of the ligand and receptor. *Cell Growth & Diff.* 4: 563–569.

Rosen, E.M., Goldberg, I.D., Kacinski, B.M., Buckholtz, T. and Vinter, D.W. (1989) Smooth muscle releases an epithelial scatter factor which binds to heparin. *In Vitro Cell Dev. Biol.* 25: 163–173.

Rosen, E.M., Carley, W. and Goldberg, I.D. (1990a) Scatter factor regulates vascular endothelial cells motility. *Cancer Invest.* 8: 647–650.

Rosen, E.M., Meromsky, L., Setter, E., Vinter, D.W. and Goldberg, I.D. (1990b) Smooth muscle-derived factor stimulates mobility of human tumor cells. *Invasion Metastasis* 10: 49–64.

Rosen, E.M., Meromsky, L., Setter, E., Vinter, D.W. and Goldberg, I.D. (1990c) Purification and migration-stimulating activities of scatter factor. *Pro. Soc. Exp. Biol. Med.* 195: 34–43.

Rosen, E.M., Grant, D., Kleinman, H., Jaken, S., Donovan, M.A., Setter, E., Luckett, P.M., Carley, W., Bhargava, M. and Goldberg, I.D. (1991) Scatter factor stimulates migration of vascular endothelium and capillary-like tube formation. *In*: I.D. Goldberg (ed.): *Cell Motility Factors*, Birkhäuser Verlag, Basel, pp 76–88.

Rosen, E.M. and Goldberg, I.D. (1994) Scatter factor and the c-*met* receptor: A paradigm for mesenchymal: epithelial interaction. *J. Cell Biol.* 127: 1783–1787.

Roth, W.K., Werner, S., Risau, W., Remberger, K. and Hofschneider, P.H. (1988) Cultured, AIDS-related Kaposi's sarcoma cells express endothelial cell markers and are weakly malignant *in vitro*. *Int. J. Cancer* 42: 767–773.

Rubin, J.S., Chan, A.M.-L., Bottaro, D.P., Burgess, W.H., Taylor, W.G., Cech, A.C., Hirschfield, D.W., Wong, J., Miki, T., Finch, P.W. and Aaronson, S.A. (1991) A broad spectrum human lung fibroblast-derived mitogen is a variant of hepatocyte growth factor. *Proc. Natl. Acad. Sci. USA* 88: 415–419.

Rutgers, J.L., Weiczorek, R., Bonetti, F., Kaplan, K., Posnett, D.N., Friedman-Kien, A.E. and Knowles II, D.M. (1986) The expression of endothelial cell surface antigens by AIDS-associated Kaposi's sarcoma: Evidence for a vascular endothelial cell origin. *Am. J. Pathol.* 122: 493–499.

Sato, N., Fukuda, K., Nariuchi, H. and Sagara, N. (1987) Tumor necrosis factor inhibiting angiogenesis in vitro. *J. Natl. Cancer Inst.* 79: 1383–1391.

Stoker, M. (1984) Junctional competence in clones of mammary epithelial cells, and modulation by conditioned medium. *J. Cell Physiol.* 171: 174–183.

Stoker, M. and Perryman, M. (1985) An epithelial scatter factor released by embryo fibroblasts. *J. Cell Sci.* 77: 209–233.

Stoker, M., Gherardi, E., Perryman, M. and Gray, J. (1987) Scatter factor is a fibroblast-derived modulator of epithelial cell motility. *Nature* 327: 238–242.

Sunderkötter, C., Goebeler, M. Schulze-Osthoff, K., Bhardwaj, R. and Sorg, C. (1991) Macrophage derived angiogenesis factors. *Pharmac. Ther.* 51: 195–216.

Tannock, I.F. and Hayashi, S. (1972) The proliferation of capillary endothelial cells. *Cancer Res.* 32: 77–82.

Tappero, J.W., Conant, M.P.H., Wolfe, S.F. and Berger, T.G. (1993) Kaposi's sarcoma: Epidemiology, pathogenesis, histology, clinical spectrum, staging criteria and therapy. *Am. Acad. Dermatol.* 28: 371–395.

Thakral, K.K., Goodson, W.H. and Hunt, T.K. (1979) Stimulation of wound blood vessel growth by wound macrophages. *J. Surg. Res.* 26: 430–436.

Tsarfaty, I., Resau, J.H., Rulong, S., Keydar, I., Faletto, D.L. and Vande Woude, G.F. (1992) The *met* proto-oncogene receptor and lumen formation. *Science* 257: 1258–1261.

Tsarfaty, I., Rong, S., Resau, J.H., Rulong, S., da Silva, P.P. and Vande Woude, G.F. (1994) The *met* proto-oncogene mesenchymal to epithelial conversion. *Science* 263: 98–101.

Vogel, J., Hinrichs, H., Reynolds, R.K., Luciw, P.A. and Jay, G. (1988) The HIV *tat* gene induces dermal lesions resembling Kaposi's sarcoma in transgenic mice. *Nature* 335: 606–611.

Vogel, J., Hinrichs, S.H., Napolitano, L.A., Ngo, L. and Jay, G. (1991) Liver cancer in transgenic mice carrying the human immunodeficiency virus *tat* gene. *Cancer Res.* 51: 6686–6690.

Weidner, K.M., Behrens, J., Vandekereckhove, J. and Birchmeier, W. (1990) Scatter factor: Molecular characteristics and effect on invasiveness of epithelial cells. *J. Cell Biol.* 111: 2097–2108.

Weidner, K.M., Arakaki, N., Vandekereckhove, J., Weingart, S., Hartmann, G., Reider, H., Fonatsch, C., Tsubouchi, H., Hishida, T., Diakuhara, Y. and Birchmeier, W. (1991) Evidence for the identity of human scatter factor and hepatocyte growth factor. *Proc. Natl. Acad. Sci. USA* 88: 7001–7005.

Yoshimura, T., Yuhki, N., Wang, M.H., Skeel, A. and Leonard, E.J. (1993) Cloning, sequencing, and expression of human macrophage stimulating protein (MSP, MSTI) confirms MSP as a member of the family of kringle proteins and locates the MSP gene on chromosome 3. *J. Biol. Chem.* 268: 15461–15468.

Zhang, Y.-M., Bachmann, S., Hemmer, C., van Luzen, J., von Stemm, A., Kern, P., Detrich, M., Zeigler, R., Waldherr, R. and Narwroth, P.P. (1994) Vascular origin of Kaposi's sarcoma. Expression of leukocyte adhesion molecule-1, thrombomodulin, and tissue factor. *Am. J. Pathol.* 144: 51–59.

Epithelial–Mesenchymal Interactions in Cancer
ed. by I.D. Goldberg & E.M. Rosen
© 1995 Birkhäuser Verlag Basel/Switzerland

Modulation of intercellular junctions of epithelia by scatter factor (hepatocyte growth factor)

A. Nusrat and J.L. Madara

Division of Gastrointestinal Pathology, Department of Pathology, Brigham and Women's Hospital and Harvard Medical School and the Harvard Digestive Diseases Center, Boston, MA 02115, USA

Introduction

Columnar epithelia such as those which line the renal tubules and the gastrointestinal and respiratory tracts are composed of cells joined by intercellular contacts. These contacts, or intercellular junctions are important in determining the function and three dimensional architecture of such organs. Intimately underlying columnar epithelia are mesenchymal cells including fibroblasts and smooth muscle cells. Indeed, a subepithelial sheath of fibroblasts is common to several organs with lumens lined by columnar epithelial cells. Constant exchange of information occurs between these mesenchymal and epithelial compartments and is mediated in part by factors produced by these cell types. Such communications play an important role in diverse physiologic and pathologic events that include organ development in embryogenesis, wound repair and tumor metastasis.

Hepatocyte growth factor/scatter factor represents one such mesenchymal-derived signal that influences epithelial growth, motility and morphogenesis (Gherardi and Stoker, 1990; Montesano et al., 1991; Stern et al., 1990). Hepatocyte growth factor, also known as hepatopoetin A, was first described as a hepatotropic factor present in the serum of rats after partial hepatectomy (Nakamura et al., 1984; Thaler and Michalopoulos, 1985). HGF, a potent mitogen for hepatocytes also has these effects on other cell types such as keratinocytes and endothelial cells (Bussolino et al., 1992; Kan et al., 1991). Scatter factor was characterized originally as a fibroblast and smooth muscle derived cytokine that 'scatters' (disperses) and stimulates motility of epithelial cells (Rosen et al., 1989; Stoker and Perryman, 1985; Weidner et al., 1990). Sequence analysis and cDNA cloning revealed that HGF and SF were indeed identical polypeptides (Gherardi and Stoker, 1990; Naldini et al., 1991b). HGF and SF are therefore interchangeable and effective to the same extent in promoting hepatocyte growth, epithelial cell

Model columnar epithelium

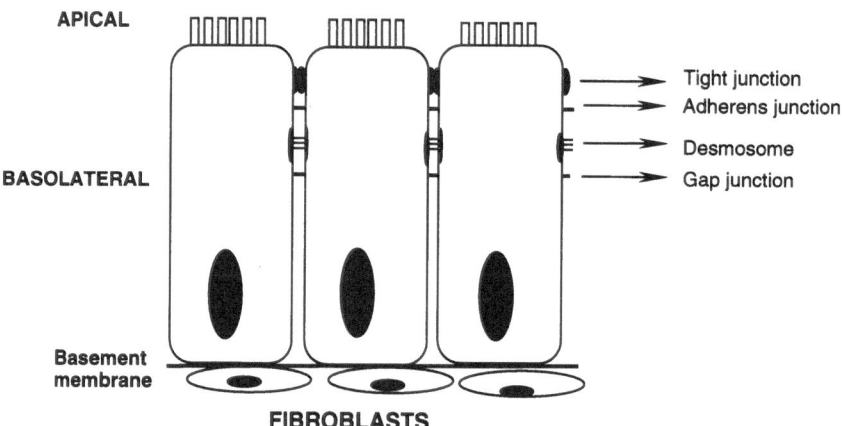

Figure 1. Polarity of epithelial cells. This figure demonstrates intercellular junctions, apical and basolateral membranes of columnar epithelial cells.

dissociation and movement. We will therefore refer to this factor as HGF/SF.

HGF/SF was originally described (Stoker et al., 1987; Stoker and Perryman, 1985) as a factor that induces Madin-Darby canine kidney (MDCK) cells in adherent colonies to flatten out, assume a fibroblast-like morphology and disperse. HGF/SF has similar effects on several other epithelial cells in culture such as carcinoma cell lines (Weidner et al., 1990) and primary mammary epithelial cells. As described below polarized epithelial cells in a colony are joined together by organized series of intercellular junctions. Loosening of these junctions is required for epithelial cells to change shape and migrate. Since HGF/SF induces these morphologic events it is likely to influence intercellular adhesion events. Below we describe intercellular junctions of epithelial cells and modulation of these structures by HGF/SF.

Organization and regulation of intercellular junctions in epithelia

Cells of columnar epithelia are polarized and joined by a series of intercellular junctions that are physiologically regulated and important for maintaining the vectoral transport of ions, water, solutes and macromolecules between the different biologic compartments. Specialized intercellular junctional complexes that include the following are illustrated in Figure 1: a) zonula occludens or tight junctions, b) zonula adherens or intermediate junctions, c) desmosomes and d) gap junctions. These junctional complexes perform important biologic functions such as cell adhesion (adherens junctions and desmosomes) and cell–cell communication (gap junctions). They also form a permeability barrier adjoining cells (tight junction) that regulates the flux of molecules through the paracellular space (Farquhar and Palade, 1963). In so doing the tight junction serves as a permeability barrier which restricts diffusion of hydrophilic solutes from the threatening lumenal space (airway, renal tubular lumen, alimentary tract lumen) into the subepithelial tissue compartment. The tight junction also separates the plasma membrane bilayer into functionally and biochemically distinct apical and basolateral membrane domains which are essential for epithelial function (Rodriguez-Boulan and Nelson, 1989). For example it is this asymmetric distribution of transporters, pumps and channels between apical and basolateral membrane domains which permits net vectoral movement of water and electrolytes.

Tight junctions are the most apical of these intercellular junctional complexes and comprise of a region where membranes of adjoining cells come in close contact (membrane fusions or 'kisses'). The tight junction runs as a circumferential apical belt, approximately 100–300 nm wide, around each epithelial cell. This junctional belt actually consists of a series of linear anastamosing fusion sites and thus the permeability barrier afforded by the junction is a series of restrictive barriers (corresponding to the number of fusion sites within the junction) (Gumbiner, 1987). F-actin microfilaments associate with the cytoplasmic face of the plasma membrane at such fusion sites and it is believed that the cytoskeleton participates in the reversible regulation of the permeability and charge selective states of the tight junction (Madara et al., 1988). It has recently been noted that these fusion sites also harbor several junction-specific and/or related proteins. These include occludin (membrane associated), ZO-1 a 220 kDa phosphoprotein (submembranous), ZO-2 (160 kDa protein associated with ZO-1) and cingulin which is also a phosphoprotein (140–108 kDa protein), and appears located furthest from the membrane, perhaps near the putative attachment of the tight junction components to microfilaments (Citi et al., 1988; Gumbiner et al., 1991; Stevenson et al., 1986). Lastly, a small cytoplasmic GTPase (rab 13) has recently been demonstrated to be selectively located at the intercellular tight junction (Zahraoui et al., 1994). Occludin, by being an integral membrane protein and exhibiting topological similarities to connexins (see below), is currently envisioned as the sealing element of the junction which allows adjacent cells to closely appose thus restricting diffusion in the paracellular pathway (Furuse et al., 1993). ZO-1 and ZO-2 belong to an emerging *discs large* family of proteins which appear to be crucial to intercellular contact sites and tissue morphogenesis. As implied above, the tight junction is not a static structure but is dynamic and regulated by various stimuli in response to physiologic demands

(Madara, 1992). Second messenger pathways and tension in the actin cytoskeleton may regulate these proteins and therefore the permeability properties of the tight junction. ZO-1 and cingulin are phosphoproteins and can potentially be regulated by kinases and phosphatases in the second messenger pathways. Although the weight of the evidence suggests that the molecular architecture of the tight junction includes a complex array of regulated (or at least regulatable) proteins, the exact nature of the interactions between these proteins and the physiological consequences of particular forms of regulation of specific proteins is poorly understood.

Adherens junctions or zonula adherens also mediate adhesion between cells and are extremely important for the establishment and maintenance of an epithelium. Given the fact that antibodies to surface proteins at this site can result in dissolution of intercellular contacts, it is likely that adherens junctions provide forceful interactions between cells that permit more delicate and regulated interactions, such as those at the tight junction, to ensue (Behrens et al., 1989; Gumbiner et al., 1988). Cellular events that require cell dispersion (wound resealing, embryogenesis, tumor metastasis) or cell impaction (embryogenesis) in an epithelium are believed to be regulated by the assembly and disassembly of proteins in this junctional complex (Takeichi, 1987). The adherens junctions, a multiprotein complex is composed of cadherins (calcium regulated adhesion molecules) and other proteins that putatively link cadherins to actin microfilaments. E-cadherin is a 120 kDa transmembrane protein with an extracellular domain that mediates homotypic adhesion with cadherins on neighboring epithelial cells, and contains an intercellular domain that associates with cytoplasmic proteins likely to mediate association with microfilaments. These latter linker proteins include the alpha, beta, gamma catenins as well as plakoglobin (Inke et al., 1994; Ozawa and Kemler, 1992). The central role played by these proteins in maintaining not only the adherens junctions but also the other intercellular junctions and therefore the polarized architecture of epithelial cells has been shown by several studies. Expression of E-cadherin in non-epithelial cells such as fibroblasts by cDNA transfection induces intercellular junction formation (Nagafuchi et al., 1987). Moreover the use of specific antibodies to E-cadherin in Madin-Darby canine kidney cells (MDCK) results in the loss of intercellular junctions and induction of a fibroblast like morphology (Behrens et al., 1985; Takeichi, 1990). Tyrosine phosphorylation of E-cadherin and β-catenin complex in MDCK cells transformed with a temperature-sensitive v-SRC gene induces a loss of epithelial differentiation and gain of invasive phenotype (Behrens et al., 1993). It thus appears that the effects of antibodies to E-cadherin or enhanced tyrosine phosphorylation of E-cadherin/catenin complex is associated with morphologic changes in the polarized architecture of epithelial cells.

Desmosomes (spot desmosome) are another junctional complex also important in providing force interactions between adjacent epithelial cells. The component proteins include desmogleins and desmoplakins and it shares proteins such as plakoglobin with the adherens junctions. The desmosomes associate with the intermediate filaments inside the cells. Thus intermediate filament cables of one cell associate with the plasma membrane in register with those of adjacent cells. It is thought that such an arrangement co-ordinates responses to and distributes sheer stress with epithelial monolayers.

Thus cells within an epithelium are polarized with specific distributions of intercellular (lateral) and matrix (basal) interactions and biochemically distinctive apical and basolateral domains. This segregation of the plasma membrane and its associated proteins into apical and basolateral domains is important for the regulated uptake and secretion of ions and macromolecules, signal reception and transduction, interaction of cells with each other and the underlying matrix. For example, in the intestine the apical plasma membrane that faces the lumen is able to transport ions and nutrients via specific channels and transporters. Transport proteins, receptors and enzymes in the basolateral membrane play a role in multiple events that include generation of ion gradients across the apical membrane and interaction with factors produced by underlying mesenchymal derived cells such as fibroblasts, circulating hormones and neurotransmitters (Rodriguez-Boulan and Nelson, 1989). Not only are these surface membrane proteins polarized, but cytoplasmic organelles and components of the cytoskeleton are also specifically distributed. For example actin microfilaments are associated with different cell–cell and cell matrix contacts where they play an important role in maintaining the three-dimensional architecture of polarized cells.

Perturbation of intestinal epithelial barrier function by HGF/SF

The polarized architecture of epithelial cells described above is also present in the intestine. Columnar (or cylindrical) shaped epithelial cells in this location rest on a basement membrane that overlies subepithelial fibroblasts (Fig. 1). The basolateral surfaces of these intestinal epithelial cells is therefore exposed to potential paracrine mediated mesenchymal–epithelial interactions. HGF/SF, a known modulator of epithelial function is secreted by mesenchymal cells such as fibroblasts therefore represents a candidate paracrine factor that influences epithelial function (Rosen et al., 1989a; Rosen et al., 1990; Rosen, 1989).

The receptor for HGF/SF is the protein encoded by the c-*met* protooncogene (Naldini et al., 1991a). This receptor protein is a heterodimer composed of an extracellular alpha (45–50 kDa) and a transmembrane beta (145 kDa) chain that has tyrosine kinase activity.

It would therefore be logical that HGF/SF produced by fibroblasts interacts with its receptor located in the basolateral domain of epithelial cells. This has been shown to be indeed true in the native human intestine, in an intestinal epithelial cell line, T84 and in the kidney epithelial cell line MDCK (Crepaldi et al., 1994; Nusrat et al., 1994).

Morphologically HGF/SF appears to modulate junctional contacts in polarized epithelial cell colonies as if to facilitate its motogenic effect (Stoker and Gherardi, 1989; Stoker et al., 1987; Stoker and Perryman, 1985b). To study this observation we examined the effects of HGF/SF on barrier function of the polarized epithelial cell lines, T84 (human intestinal) and MDCK (dog kidney). Barrier function refers to the ability of an epithelium to resist passive movement of solutes and other particles across it and is largely contributed by the tight junction, although as outlined above, the integrity of this junction depends not only on local regulation of components within it, but on the state of other cell surface junctions as well. There exist two major pathways by which solutes passively penetrate the epithelium: the transcellular pathway and paracellular pathway. Intercellular junctions determine the characteristics of passive paracellular flow of hydrophilic solutes. An electrical assay of transepithelial resistance is commonly used by cell biologists as an assay of tight junction permeability (Madara and Dharmasthaphorn, 1985). T84 cells (a model human intestinal cell line which grows as a monolayer and exhibits many of the phenotypic, biochemical, and physiological attributes naturally displayed by intestinal crypt epithelium) and MDCK cells (a canine renal cell line which, like T84, also grows as a polarized physiologically confluent monolayer) used in these studies develop intercellular junctions and high transepithelial resistance to passive ion flow. 24–48 h exposure of these monolayers to HGF/SF results in a dose-dependent decrease in transepithelial resistance (Fig. 2). In addition to the state of the intercellular junctions, transepithelial resistance is influenced by diverse events such as cell lysis (ie., opening of the transcellular pathway) and even physiologically regulated opening of channels on apical and basolateral membranes of epithelial cells (Parkos et al., 1992). However, once supplementary flux data have ruled out other such possibilities, this resistance assay is generally a reasonable one for detecting alterations in passive ion flow through the paracellular pathway for which the tight junction is rate limiting. Such use of this assay takes a view of the epithelium in which paracellular pathways essentially represent individual resistors in parallel circuit. The reciprocal of the total resistance across this circuit will equal the sum of the reciprocals of the individual resistors. The resistance assay is thus highly skewed toward detection of resistances within the circuit which are low relative to others. At high resistance values, such as those seen in T84 monolayers, such skewing provides for extremely sensitive detection of even minor alterations in passive para-

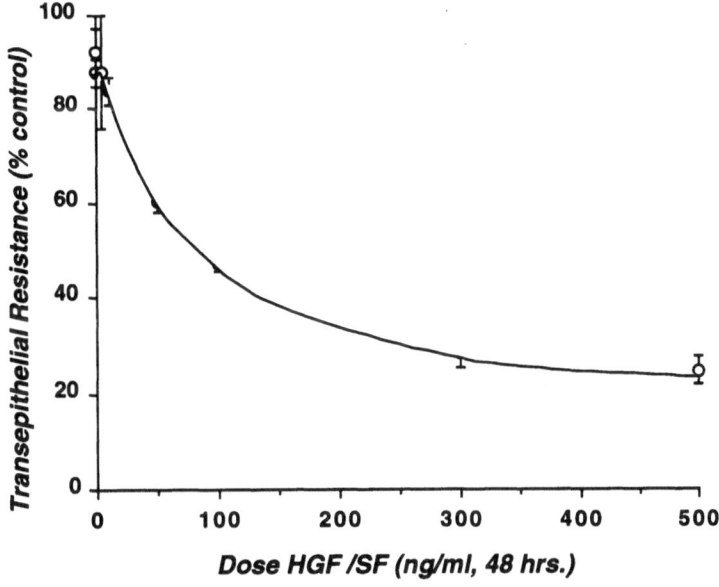

Figure 2. The effects of HGF/SF on tranepithelial resistance of T84 monolayers. T84 intestinal epithelial monolayers were exposed to varying concentrations of recombinant HGF/SF and transepithelial resistance to passive ion flow was measured 48 h later. Negative controls included monolayers that were exposed to vehicle (media) only. The transepithelial resistance of the experimental group expressed as a % of control (unexposed) monolayers is plotted on the vertical axis and dose of HGF/SF in ng/ml on the horizontal. As demonstrated here HGF/SF attenuates transepithelial resistance to passive ion flow across T84 monolayers in a dose dependent fashion with maximal attenuation at 300 ng/ml. This figure is obtained from: Hepatocyte growth factor/scatter factor effects on epithelia. *J. Clin. Invest.* (1994) 93: 2056–2065. (Prepared with permission from the Rockefeller University Press).

cellular ion flow. Thus perturbations in paracellular pathways small enough to yield barely detectable changes in flux of inert solutes and no alterations in distribution of junctional proteins such as the tight junction specific protein ZO-1, in high resistance monolayers, yield easily detectable resistance signals. This fact makes this assay useful in screening for general aspects of cell–cell associations (Madara and Dharmasthaphorn, 1985). While HGF/SF diminished resistance in these cell lines, it did not do so by affecting transcellular pathways or by enhancing proliferation. Thus, attenuation of transepithelial resistance induced by HGF/SF probably represents effects on intercellular junctions. These findings suggest that the decrease in transepithelial resistance observed were specifically due to altered cell–cell interactions with resulting changes in the rate-limiting resistance barrier in the paracellular pathway, the intercellular tight junction. Given the relationship between alterations in junctional resistance and junctional sieving of inert hydrophilic molecules, transepithelial flux of such solutes would only be expected to change marginally for the end resistances observed with

HGF/SF. As expected, fluxes of ^{3}H mannitol and ^{14}C inulin were affected only marginally by HGF/SF. Moreover, the effect of HGF/SF on resistance occurred in both transformed (T84) and non-transformed (MDCK) lines which originated from different species (human, dog) and tissues (intestine, kidney) suggesting that effects seen are likely general features which follow HGF/SF interactions with columnar epithelia.

In keeping with the basolateral morphologic localization of HGF/SF receptor (Fig. 3-II), the functional effect of HGF/SF, ie., attenuation of transepithelial resistance is also polarized to the basolateral membranes of T84 cells (Fig. 3-I).

Molecular effects of HGF/SF on intercellular junctional proteins

The receptor for HGF/SF, the c-*met* proto-oncogene, is a tyrosine kinase. In addition, as outlined above, several proteins of the intercellu-

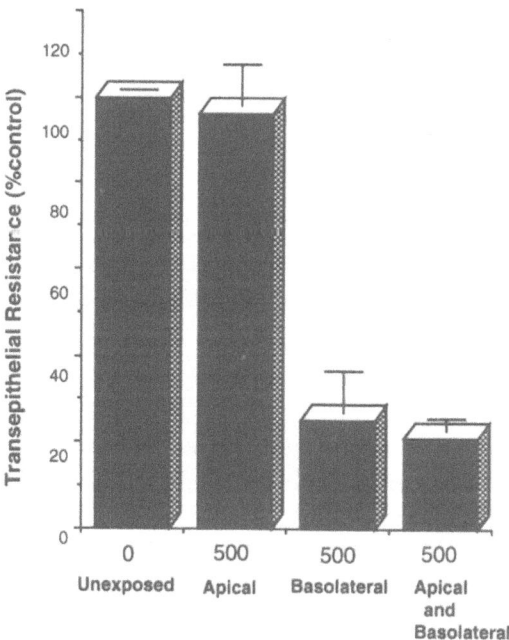

Figure 3-I. Effects of apical *versus* basolateral exposure of T84 monolayers to HGF/SF. The apical, basolateral or both surfaces of T84 epithelial monolayers were exposed to HGF/SF. Transepithelial resistance to passive ion flow across these monolayers was measured 48 h later and is expressed as a % of resistance in control monolayers exposed to media only. Like unexposed controls, monolayers exposed to HGF/SF in the apical compartment maintain high resistance. In contrast monolayers exposed to basolateral HGF/SF exhibit a major fall in resistance that is not further attenuated with simultaneous apical exposure. This figure is obtained from: Hepatocyte growth factor/scatter factor effects on epithelia. *J. Clin. Invest.* (1994) 93: 2056–2065. (Prepared with permission from the Rockefeller University Press).

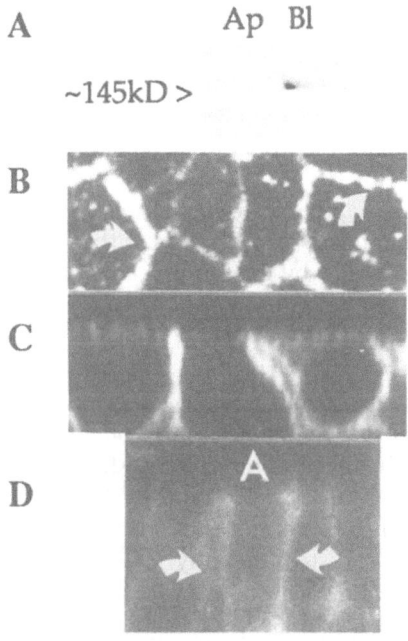

Figure 3-II. Polarity of c-*met* in T84 cells. (A) demonstrates results of immunoprecipitation of T84 monolayers with antibodies to c-*met*. These monolayers were selectively biotinylated on the apical (Ap) or basolateral membranes (Bl). Shown here is a 145 kDa protein corresponding to the molecular mass of the beta chain of c-*met* in the immunoprecipitates of the basolateral but not the apical membranes. (B) and (C) reveal immunofluorescent confocal images of c-*met*. (B) demonstrates an en face image of a confluent T84 monolayer demonstrating perijunctional/basolateral staining of c-*met* (arrows). (C) reveals basolateral staining of c-*met* in a reconstructed confocal image taken in the xz plane. This photomicrograph demonstrates strong labeling of basolateral membranes of T84 cells (arrows) and not apical membranes. (D) demonstrates basolateral distribution of c-*met* in native intestinal crypt epithelial cells (arrows). Frozen sections of the native intestine were immunostained with antibodies to c-*met*. The apical membrane of the crypt epithelium is designated as A. This figure is obtained from: Hepatocyte growth factor/scatter factor effects on epithelia. *J. Clin. Invest.* (1994) 93: 2056–2065. (Prepared with permission from the Rockefeller University Press).

lar junctional complex may be phosphorylated on tyrosine groups (including E-cadherin, ZO-1, and β-catenin). Evidence also points to the fact that these proteins may, by direct or indirect means, regulate the state of intercellular junctional contacts. We have observed that transepithelial resistance, that is sensitive to the state of intercellular junctional contacts is diminished by HGF/SF. In aggregate this information raises the strong possibility that HGF/SF affects resistance by modulating the phosphorylation state of junctional proteins. Certainly evidence exists that other tyrosine kinases, such as v-*src*, may affect interactions between epithelial cells via this mechanism (Warren and Nelson, 1987) (although the v-*src* transfection systems may create unusually strong and non-selective tyrosine phosphorylation pressure).

As alluded to above, E-cadherin at the adherens junction plays an important role in assembly and maintenance of epithelial junctions. Since E-cadherin plays a central role in cell–cell associations and in the putative initiation of the basolateral cytoskeletal cortex assembly (Mc-Neill et al., 1990) a number of studies have looked HGF/SF effects on cell surface distribution, dynamics and phosphorylation status of E-cadherin. Weidner et al. incubated MDCK cells with and without HGF/SF and assayed for changes in *de novo* synthesis and total amount of E-cadherin in these cells by metabolic labeling, immunoprecipitation and Western blotting (Weidner et al., 1990). However, they were unable to demonstrate any influence of HGF/SF on dynamics of E-cadherin or its associated 102 and 98 kDa proteins (undoubtedly α- and β-catenins) in these cells. They also could not detect any effects of HGF/SF on the phosphorylation status of E-cadherin. They therefore concluded that HGF/SF did not influence the rate of E-cadherin synthesis and complex formation with its associated 102 and 98 kDa proteins. By immunofluorescence and confocal microscopy we were unable to demonstrate changes in E-cadherin distribution by HGF/SF (Nusrat, unpublished). Despite these compelling negative results, other possibilities exist that might lead to altered E-cadherin interactions induced by HGF/SF. For example other E-cadherin associated proteins might be modulated by HGF/SF and subsequently influence E-cadherin homeotype interactions and therefore cell–cell adhesion.

As described above E-cadherin is noncovalently associated with cytoplasmic proteins that include α-, β-catenin and plakoglobin (Ozawa and Kemler, 1992). Effective E-cadherin/catenin complex formation is required for adhesive functions of E-cadherin. α-Catenin has sequence similarity to vinculin and is a candidate linker protein between the cadherin/catenin complex and cytoskeleton (Herrenknecht et al., 1991; Nagafuchi et al., 1991). β-Catenin shares 70% amino acid identity with both plakoglobin and the product of the *Drosophila* segment polarity gene *armadillo* (Butz et al., 1992; McCrea et al., 1991). Plakoglobin was originally described as a constituent of desmosomes (Cowin et al., 1986) and localizes on the cytoplasmic face of both adherens junctions and desmosomes. Cadherin/catenins are tyrosine phosphorylated, events that may influence their associations and therefore cell–cell adhesion. HGF/SF thus may modulate this interaction between the cadherin and one or more catenins, thereby influencing the pathways that regulate intercellular interactions. Recent studies have demonstrated that cells of a scattered-type carcinoma line, express E-cadherin and β-catenin but not α-catenin and are unable to aggregate. These cells acquired cell–cell adhesiveness following transfection of these cells with the α-catenin cDNA (Shimoyama et al., 1992). Tyrosine phosphorylation of β-catenin in v-*src* and RSV transformed 3Y1 and fibroblasts respectively has been correlated with perturbation of cadherin mediated adhesion

(Behrens et al., 1993; Hamaguchi et al., 1993). Three cell lines that include HT29 (intestine), MKN7 and MKN74 (human gastric cell line) were used in a study by Shibamoto et al. to investigate the influence of HGF/SF on these proteins (Shibamoto et al., 1994). In their studies morphologic loosening of intercellular junctions was observed in the epithelial cell colonies. The three cell types responded to a different degree with the strongest response in MKN74 cells (human gastric cell line) that revealed a complete epithelial to fibroblast like change in morphology. The amount of E-cadherin, α-, β-catenin and plakoglobin was not altered in these cell types following HGF/SF exposure as judged by SDS-PAGE and Western blotting of total cell lysates. However when E-cadherin immunoprecipitates were immunoblotted with phosphototyrosine antibodies three bands at 115 kDa, 98 kDa and 85 kDa were observed. HGF/SF exposure enhanced the intensity of these bands in all cell types though to a different degree (Fig. 4). The 98 kDa and 85 kDa bands comigrate with β-catenin and plakoglobin respectively. The 115 kDa band co-migrated with none of the known components of E-cadherin–catenin complex. The authors state that the enhancement in tyrosine phosphorylation roughly correlated with morphologic changes in cell–cell adhesion. The course of events that follow binding of HGF/SF to its receptor followed by activation of its tyrosine kinase domain and phosphorylation of these proteins is unknown. Whether the receptor kinase domain directly phosphorylates β-catenin and plakoglobin or these events are mediated by other second messenger molecules and kinases is also not known.

Although the adherens junctions play a pivotal role in maintaining other junctional structures it is possible HGF/SF could influence the surface expression or phosphorylation status of proteins in other junctional complexes such as ZO-1, cingulin of tight junctions or desmoplakin of desmosomes. A study on the effect of HGF/SF on cadherin mediated cell–cell adhesion reported complete disruption of cell–cell boundaries in a keratinocyte cell line following exposure of these cells to HGF/SF in combination with anti-cadherin antibodies. In these cells the cadherins and desmoplakin were lost at cell–cell contact sites. HGF/SF alone caused loosening of intercellular junctions but not their disruption and the authors postulate that HGF/SF may modulate activities of desmosomal cadherins.

Biologic implications of HGF/SF modulation of interepithelial junctions/associations

Loosening of intercellular junctions eventually resulting in cell scattering underlies important phenomena *in vivo*. These events include migration of cells during wound healing and embryogenesis. Under

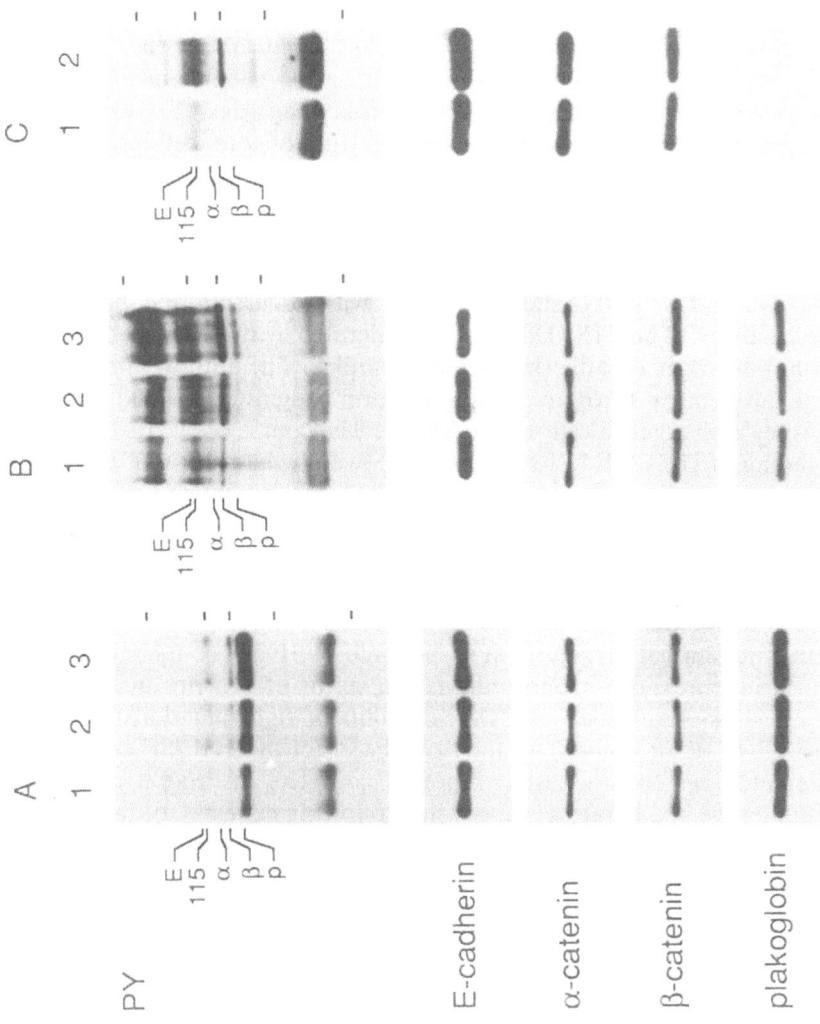

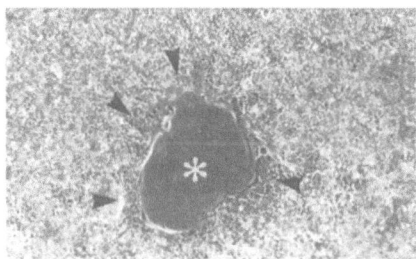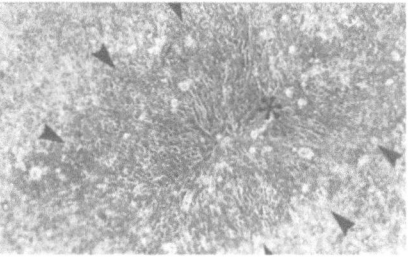

Figure 5. Influence of HGF/SF on resealing of wounds in T84 monolayers. Equivalent wounds were made 20 h prior to these photographs. The wound (asterisk) in the left panel recovered in media and is not resealed as yet. In contrast, equivalent HGF/SF wounds have resealed (right panel). Note the enlarged perimeter of surrounding the HGF/SF exposed wounds (arrowheads) compared with control wounds. Cells in this perimeter exhibit flattened phenotype (with wider intercellular spaces) and are participating in the migration events. This figure is obtained from: Hepatocyte growth factor/scatter factor effects on epithelia. *J. Clin. Invest.* (1994) 93: 2056–2065. (Prepared with permission from the Rockefeller University Press).

pathologic states such events logically might promote metastasis and/or local spread of tumor cells.

Rapid reepithelialization of wounds in columnar epithelia occurs by migration of cells into the denuded area (Feil et al., 1987; Moore et al., 1989). This process of rapid resealing, termed epithelial restitution, represents flattening of columnar cells bordering the wound by the extension of lamellipodia and filopodia into the denuded site (Feil et al., 1987; Moore et al., 1989; Nusrat et al., 1992; Rutten and Ito, 1983). While the phenotype of migrating epithelial cells at the leading edge is similar to that of other migrating cell types such as macrophages and neutrophils (Omann et al., 1987; Singer and Kupfer, 1986), migration of these epithelial cells exhibits one major difference. In contrast to neutrophils, macrophages and other cells that may migrate independently, the borders of the migrating epithelial cells at the leading edge retain junctional contacts with adjoining cells. It is likely that such rearward contacts constrain the ability of these cells to migrate into wounds, effects that could be related either directly to cell–cell adhesive interactions or indirectly to the cortical organization signals likely conferred by

Figure 4. Immunoblot analysis of E-cadherin immunoprecipitates. (A) HT29, (B) MKN7 and (C) MKN74 cells. Cells were untreated (lane 1) or treated with 10 ng/ml of HGF/SF (lane 2) or 10 ng/ml EGF (lane 3). E-cadherin immunoprecipitates obtained from these cells were resolved by SDS-PAGE, and subjected to immunoblot analysis for detection of Phosphotyrosine (PY), E-cadherin, α- and β-catenin, and plakoglobin. Molecular weight markers of 200, 116, 97.4, 66.2 and 45×10^3 are indicated by bars. This figure was obtained from: Tyrosine phosphorylation of β-catenin and plakoglobin enhanced by hepatocyte growth factor and epidermal growth factor in human carcinoma cells. *Cell Adhes. Commun.* (1994) 1: 295–305. (Prepared with permission of S. Shibimoto and Harwood Academic Publishers).

such adhesive interactions. It is therefore probable that HGF/SF could loosen up intercellular junctions and facilitate wound resealing. We examined this possibility in the model intestinal epithelial monolayer comprised of T84 cells (described above) (Nusrat et al., 1994). Wounds in T84 monolayers exposed to HGF/SF recovered 5 times faster than control unexposed wounds (Fig. 5). Not only did these wounds reseal significantly faster than control wounds, but the perimeter of cells surrounding the wound that had looser intercellular junctions also markedly increased (Fig. 5, arrows). It therefore appears that modulations of intercellular contacts afforded by HGF/SF aid in releasing constraints between cells, thus facilitating wound repair by permitting multiple waves of cells encircling the wound to participate in the spreading event. By exerting similar effects on endothelial cells, HGF/SF may enhance wound healing by promoting migration of these cells and therefore angiogenesis (Bussolino et al., 1992).

Development of organs during embryogenesis is associated with migration and restructuring of epithelial cells and the surrounding mesenchyme. Such morphogenetic events require the coordinate interaction of epithelial cells with the associated mesenchyme. These interactions are mediated in some cases by soluble factors while direct contact between the epithelium and mesenchyme is required in others. Madin-Darby canine kidney (MDCK) epithelial cell line that forms monolayers of polarized epithelial cells on flat surfaces when cultured in three dimensional collagen gels, forms spherical fliud filled epithelial cysts. In the presence of fibroblast conditioned medium these cells form long branching structures reminiscent of kidney tubules. This effect has been shown to be mediated by HGF/SF (Montesano et al., 1991). HGF/SF is postulated to be an important factor produced by mesenchymal cells that mediates at least some of the restructuring of epithelial cells during organogenesis. To understand the dynamics of HGF/SF production during development, Sonnenberg et al. investigated the expression of genes encoding HGF/SF and its receptor during mouse embryogenesis (Sonnenberg et al., 1993). They used *in situ* hybridization and RNase protection assays to show a distinct pattern of HGF/SF and its receptor expression in developing mice. The receptor c-*met* was expressed in epithelial cells of many developing organs such as the urinary tract, lung, pancreas, salivary glands, while its ligand was expressed in the surrounding mesenchyme. These authors made an interesting observation on the gastrointestinal tract. Formation of villi from the multilayered epithelium of the intestinal anlagen is associated with dissociation of tight epithelial complexes, formation of secondary lumina and invasion of mesenchymal cells into loosened epithelium. Patches of HGF/SF expression were observed in the invading mesenchyme. The authors postulate that this specific pattern of HGF/SF expression by mesenchymal cells is important in loosening of junctions between epithelial cells required for reorganization of these cellular structures.

Loosening of intercellular junctions that promotes movement of cells in a cluster plays an important role in tumor cell metastasis. The c-*met* receptor, originally described as an oncogene, revealed rearrangement in human gastric carcinoma and precursor lesions (Soman et al., 1991). It is amplified and over-expressed in the gastric carcinoma cell line GTL-16 and various human carcinomas (Di Renzo et al., 1991; Ponzetto et al., 1991). Enhanced activation of c-*met* either as a result of its aberrant expression or enhanced susceptibility to HGF/SF that is produced in the vicinity of tumor cells may play a role in enhanced tumor progression or metastasis. It is becoming clear that genes which encode junctional proteins may themselves represent proto-oncogenes. For example, the protein product of the APC gene that is mutated in both familial and sporadic colorectal carcinomas is associated with β-catenin that binds to the intercellular junctional protein, E-cadherin (Rubinfeld et al., 1993; Su et al., 1993). E-cadherin plays an important role in intercellular adhesion and the interaction of APC with catenins may therefore indirectly influence such events. Another possibility is that this complex is important in mediating contact inhibition signal into the cell. Such observations emphasize the importance of junctional proteins in tumor formation and spread and, in so doing, highlight the potential importance of molecules, such as HGF/SF, which putatively regulate these junctional proteins to events in the development of epithelial-derived cancers.

Conclusion

Exchange of information between the epithelial and mesenchymal compartments is important in diverse physiologic and pathologic events such as organ development, wound repair and tumor metastasis. Hepatocyte growth factor/scatter factor (HGF/SF) represents one such mesenchymal derived signal important in mediating mesenchymal epithelial interactions. Epithelial cells such as those of the intestine and renal tubules are polarized and joined by well developed intercellular junctions. When exposed to HGF/SF, loosening of intercellular junctions between epithelial cells is observed both morphologically and by measurement of transepithelial resistance across these cells.

The receptor for HGF/SF is a protein product of the c-*met* proto-oncogene. It is polarized to the basolateral membrane of epithelial cells, a surface that is exposed to the underlying mesenchyme derived tissue. This receptor is a tyrosine kinase and may directly or indirectly modulate the phosphorylation of junctional proteins that may in turn influence the functional state of intercellular junctions. Candidate junctional proteins that are modulated by HGF/SF include the E-cadherin/catenin complex. Enhanced tyrosine phosphorylation of β-catenin and plakoglobin by

HGF/SF have been demonstrated in intestinal and gastric epithelial cell lines and this phosphorylation correlates with morphologic changes in cell–cell adhesion. Although these junctional proteins play a central role in cell–cell adhesion, other likely candidate junctional proteins that may be modified by HGF/SF include those of the tight junction and desmosomes. Further studies are required to clearly sort out the biologic effects of HGF/SF.

Acknowledgments

We thank our colleagues P.J. Godowski, E.M. Rosen, C.A. Parkos for collaboration on our work with HGF/SF.

References

Behrens, J., Birchmeier, W., Goodman, S.L. and Imhof, A. (1985) Dissociation of Madin-Darby canine kidney epithelial cells by the monoclonal antibody anti-Arc-1: mechanistic aspects and identification of the antigen as a component related to uvomorulin. *J. Cell Biol.* 101: 1307–1315.

Behrens, J., Mareel, M.M., Van Roy, F.M. and Birchmeier, W. (1989) Dissecting tumor cell invasion: Epithelial cell acquire invasive properties after the loss of uvomorulin-mediated cell–cell adhesion. *J. Cell Biol.* 108: 2435–2447.

Behrens, J., Vakaet, L., Friis, R., Winterhager, E., Van Roy, F., Mareel, M.M. and Birchmeier, W. (1993) Loss of epithelial differentiation and gain of invasiveness correlates with tyrosine phosphorylation of the E-cadherine/β-catenin complex in cell transformed with a temperature-sensitive v-*src* gene. *J. Cell Biol.* 120: 757–766.

Bussolino, F., Di Renzo, M.F., Ziche, M., Bocchietto, E., Olivero, M., Naldini, L., Gaudino, G., Tamagnone, L., Coffer, A. and Comoglio, P.M. (1992) Hepatocyte growth factor is a potent angiogenic factor which stimulates endothelial cell motility and growth. *J. Cell Biol.* 119: 629–641.

Butz, S., Stappert, J., Weissig, H. and Kemler, R. (1992) Plakoglobin and beta-catenin: distinct but closely related. *Science* 257: 1142–1143.

Citi, S., Sabanay, H., Jakes, R., Geiger, B. and Kendrick-Jones, J. (1988) Cingulin, a new peripheral component of tight junctions. *Nature* 333: 272–276.

Cowin, P., Kapprell, H.P., Franke, W.W., Tamkun, J. and Hynes, R.O. (1986) Plakoglobin: a protein common to different kinds of intercellular adhering junctions. *Cell* 46: 1063–1073.

Crepaldi, T., Pollack, A.L., Prat, M., Zborek, A., Mostov, K. and Comoglio, P.M. (1994) Targeting of the HGF/SF receptor to the basolateral domain of polarized epithelial cells. *J. Cell Biol.* 125: 313–320.

Di Renzo, M.F., Narsimhan, R.P., Olivero, M., Bretti, S., Giordano, S., Medico, S., Gaglia, P., Zara, P. and Comoglio, P.M. (1991) Expression of the Met/HGF receptor in normal and neoplastic human tissues. *Oncogene* 6: 1997–2003.

Farquhar, M.G. and Palade, G.E. (1963) Junctional complexes in various epithelia. *J. Cell Biol.* 17: 375–412.

Feil, W., Wenzl, E., Vattay, P., Starlinger, M., Sogukoglu, T. and Schiessel, R. (1987) Repair of rabbit duodenal mucosa after acid injury *in vivo* and *in vitro*. *Gastroenterology* 92: 1973–1986.

Furuse, M., Hirase, T., Itoh, M., Nagafuchi, A., Yonemura, S., Tsukita, S. and Tsukita, S. (1993) Occludin: A novel integral membrane protein localizing at tight junctions. *J. Cell Biol.* 123: 1777–1788.

Gherardi, E. and Stoker, M. (1990) Hepatocytes and Scatter factor. *Nature* 346: 228.

Gumbiner, B. (1987) Structure, biochemistry, and assembly of epithelial tight junctions. *Am. J. Physiol.* 87: C749–758.

Gumbiner, B., Stevenson, B. and Grimaldi, A. (1988) The role of the cell adhesion molecule, uvomorulin, in the formation and maintenance of the epithelial junctional complex. *J. Cell Biol.* 107: 1575–1587.

Gumbiner, B., Lowenkopf, T. and Apatira, D. (1991) Identification of a 160-kDa polypeptide that binds to the tight junction protein ZO-1. *Proc. Natl. Acad. Sci. USA* 88: 3460–3464.

Hamaguchi, M., Matsuyoshi, N., Ohnishi, Y., Gotoh, B., Takeichi, M. and Nagai, Y. (1993) p60v-*src* causes tyrosine phosphorylation and inactivation of the N-cadherin-catenin cell adhesion system. *EMBO J.* 12: 307–314.

Herrenknecht, K., Ozawa, M., Eckerskorn, C., Lottspeich, F., Lenter, M. and Kemler, R. (1991) The uvomorulin-anchorage protein alpha-catenin is a vinculin homologue. *Proc. Natl. Acad. Sci. USA* 88: 9156–9160.

Inke, S.N., Hinck, L., Swedlow, J.R., Papkoff, J. and Nelson, W.J. (1994) Defining interactions and distributions of cadherin and catenin complexes in polarized epithelial cells. *J. Cell Biol.* 125: 1341–1352.

Kan, M., Zhang, G.H., Zarnegar, R., Machalopoulos, G., Myoken, Y., McKeehan, W.L. and Stevens, J.L. (1991) Hepatocyte growth factor/hepatopoietin A stimulates the growth of rat kidney proximal tubule epithelial cells (RPTE), rat nonparenchymal liver cells, human melanoma cells, mouse keratinocytes and stimulates anchorage-independent growth of SV40-transformed RPTE. *Biochem. Biophys. Res. Commun.* 174: 331–337.

Madara, J.L. and Dharmasthaphorn, K. (1985) Occluding junction structure–function relationships in a cultured epithelial monolayer. *J. Cell Biol.* 101: 2124–2133.

Madara, J.L., Stafford, J., Barenberg, D. and Carlson, S. (1988) Functional coupling of tight junctions and microfilaments in T84 monolayers. *Amer. J. Physiol.* 254: G416–G423.

Madara, J.L. (1992) Tight junction dynamics: Is paracellular transport regulated. *Cell* 53: 497–498.

McCrea, P.D., Turck, C.W. and Gumbiner, B. (1991) A homolog of the armadillo protein in *Drosophila* (plakoglobin) associated with E-cadherin. *Science* 254: 1359–1361.

McNeill, H., Ozawa, M., Kemler, R. and Nelson, W.J. (1990) Novel function of the cell adhesion molecule uvomorulin as an inducer of cell surface polarity. *Cell* 62: 309–316.

Montesano, R., Matsumoto, K., Nakamura, T. and Orci, L. (1991) Identification of a fibroblast-derived epithelial morphogen as hepatocyte growth factor. *Cell* 67: 901–908.

Moore, R., Carlson, S. and Madara, J.L. (1989) Rapid barrier restitution in an *in vitro* model of intestinal epithelial injury. *Lab. Invest.* 60: 237–244.

Nagafuchi, A., Shirayoshi, Y., Okazaki, K., Yasuda, K. and Takeichi, M. (1987) Transformation of cell adhesion properties by exogenously introduced E-cadherin cDNA. *Nature* 329: 341–343.

Nagafuchi, A., Takeichi, M. and Tsukita, S. (1991) The 102 kDa cadherin associated protein: similarity to vinculin and posttranscriptional regulation of expression. *Cell* 65: 849–857.

Nakamura, T., Nawa, K. and Ichihara, A. (1984) Partial purification and characterization of hepatocyte growth factor from serum of hepatectomized rats. *Biochem. Biophys. Res. Comm.* 122: 1450–1459.

Naldini, L., Vigna, E., Narsimhan, R.P., Gaudino, G., Zarnegar, R., Michalopoulos, G.K. and Comoglio, P.M. (1991a) Hepatocyte growth factor (HGF) stimulates the tyrosine kinase activity of the receptor encoded by the proto-oncogene c-MET. *Oncogene* 6: 501–504.

Naldini, L., Weidner, K.M., Vigna, E., Gaudino, G., Bardelli, A., Ponzetto, C., Narsimhan, R.P., Hartman, G., Zarnegar, R., Michalopoulos, G.K., Birchmeier, W. and Comoglio, P.M. (1991b) Scatter factor and hepatocyte growth factor are indistinguishable ligands for the *met* receptor. *EMBO J.* 10: 2867–2878.

Nusrat, A., Delp, C. and Madara, J.L. (1992) Intestinal epithelial restitution. *J. Clin. Invest.* 89: 1501–1511.

Nusrat, A., Parkos, C.A., Bacarra, A.E., Godowski, P.J., Delp-Archer, C., Rosen, E.M. and Madara, J.L. (1994) Hepatocyte Growth Factor/Scatter Factor effects on epithelia: Regulation of intercellular junctions in transformed and nontransformed cell lines, basolateral polarization of c-*met* receptor in transformed and natural intestinal epithelia and induction of rapid wound repair in a transformed model epithelium. *J. Clin. Invest.* 93: 2056–2065.

Omann, G.M., Allen, R.A., Bokoch, G.M., Painter, R.G., Traynor, A.E. and Sklar, L.A. (1987) Signal transduction and cytoskeletal activation in the neutrophil. *Physiol. Rev.* 67: 285–322.

Ozawa, M. and Kemler, R. (1992) Molecular organization of the uvomorulin-catenin complex. *J. Cell Biol.* 116: 989–996.

Parkos, C.A., Colgan, S.P., Delp, C., Arnaout, M.A. and Madara, J.L. (1992) Neutrophil migration across a cultured epithelial monolayer elicits a biphasic resistance response representing sequential effects on transcellular and paracellular pathways. *J. Cell Biol.* 117: 757–764.

Ponzetto, C., Giordano, S., Peverali, F., Della Valle, G., Abate, M.L., Vaula, G. and Comoglio, P.M. (1991) C-*met* is amplified but not mutated in a cell line with an activated *met* tyrosine kinase. *Oncogene* 6: 553–559.

Rodriguez-Boulan, E. and Nelson, W.J. (1989) Morphogenesis of the polarized epithelial cell phenotype. *Science* 245: 718–725.

Rosen, E.M., Goldberg, I.D., Kacinski, B.M., Buckholz, T. and Vinter, D.W. (1989) Smooth muscle releases an epithelial cell scatter factor which binds to heparin. *In Vitro Cell Dev. Biol.* 25: 163–173.

Rosen, E.M., Meromsky, L., Setter, E., Vinter, D.W. and Goldberg, I.D. (1990) Purified scatter factor stimulates epithelial and vascular endothelial cell migration. *P. Soc. Exp. Biol. & Med.* 195: 34–43.

Rosen, E.M. and Goldberg, I.D. (1989) Protein factors which regulate cell motility. *In Vitro Cell Dev. Biol.* 25: 1079–1087.

Rubinfeld, B., Souza, B., Albert, I., Muller, O., Chamberlain, S.H., Masiarz, F.R., Munemitsu, S. and Polakis, P. (1993) Association of the APC gene product with beta-catenin. *Science* 262: 1731–1733.

Rutten, M.J. and Ito, S. (1983) Morphology and electrophysiology of guinea pig gastric mucosal repair *in vitro*. *Am J. Physiol.* 244G: 171–182.

Shibamoto, S., Hayakawa, M., Takeuchi, K., Hori, T., Oku, N., Miyazawa, K., Kitamura, N., Takeichi, M. and Ito, F. (1994) Tyrosine phosphorylation of beta catenin and plakoglobin enhanced by hepatocyte growth factor and epidermal growth factor in human carcinoma cells. *Cell Adhes. Commun.* 1: 295–305.

Shimoyama, Y., Nagafuchi, A., Fujita, S., Gotoh, M., Takeichi, M., Tsukita, S. and Hirohashi, S. (1992) Cadherin dysfunction in a human cancer cell line: Possible involvement of loss of alpha catenin expression in reduced cell–cell adhesiveness. *Cancer Res.* 52: 5770–5774.

Singer, S.J. and Kupfer, A. (1986) The directed migration of eukaryotic cells. *Ann. Rev. Cell Biol.* 2: 337–365.

Soman, N.R., Correa, P., Ruiz, A. and Wogan, G.N. (1991) The *tpr – met* oncogenic rearrangement is present and expressed in human gastric carcinoma and precursor lesions. *Proc. Natl. Acad. Sci. USA* 88: 4892–4896.

Sonnenberg, E., Meyer, D., Weidner, K.M. and Birchmeier, C. (1993) Scatter factor/hepatocyte growth factor and its receptor, the c-*met* tyrosine kinase, can mediate a signal exchange between mesenchyme and epithelia during mouse development. *J. Cell Biol.* 123: 223–235.

Stern, C.D., Ireland, G.W., Herrick, S.E., Gherardi, E., Gray, J., Peryman, M. and Stoker, M. (1990) Epithelial scatter factor and development of the chick embryonic axis. *Development* 110: 1271–1284.

Stevenson, B.R., Siliciano, J.D., Mooseker, M.S. and Goodenough, D.A. (1986) Identification of ZO-1: a high molecular weight polypeptide associated with the tight junction (Zonula Occludens) in a variety of epithelia. *J. Cell Biol.* 103: 755–766.

Stoker, M. and Perryman, M. (1985) An epithelial scatter factor released by embryo fibroblasts. *J. Cell Sci.* 77: 209–213.

Stoker, M., Gherardi, E., Perryman, M. and Gray, J. (1987) Scatter factor is a fibroblast derived modulator of epithelial cell mobility. *Nature* 327: 239–241.

Stoker, M. and Gherardi, E. (1989) Scatter factor and other regulators of cell mobility. *British Med. Bulletin* 45: 481–491.

Su, L.-K., Vogelstein, B. and Kinzler, K.W. (1993) Association of the APC tumor suppressor protein with catenins. *Science* 262: 1734–1736.

Takeichi, M. (1987) Cadherins: a molecular family essential for selective cell–cell adhesion and animal morphogenesis. *Trends Genet.* 3: 213–217.

Takeichi, M. (1990) Cadherins: a molecular family important in selective cell–cell adhesion. *Annu. Rev. Biochem.* 59: 237–252.

Thaler, J. and Michalopoulos, G. (1985) Hepatopoietin A: partial characterization and trypsin activation of a hepatocyte growth factor. *Cancer Res.* 45: 2545–2549.

Warren, S.L. and Nelson, W.J. (1987) Nonmitogenic morphoregulatory action of pp60[v-src] on multicellular epithelial structures. *Mol. Cell Biol.* 7: 1326–1337.

Weidner, K.M., Behrens, J., Vanderkerckhove, J. and Birchmeier, W. (1990) Scatter factor: Molecular characteristics and effect on the invasiveness of epithelial cells. *J. Cell Biol.* 111: 2097–2108.

Zahraoui, A., Joberty, G., Arpin, M., Fontaine, J.J., Hellio, R., Tavitian, A. and Louvard, D. (1994) A small rab GTPase is distributed in cytoplasmic vesicles in non polarized cells but colocalizes with the tight junction marker ZO-1 in polarized epithelial cells. *J. Cell Biol.* 124: 101–115.

Epithelial–Mesenchymal Interactions in Cancer
ed. by I.D. Goldberg & E.M. Rosen
© 1995 Birkhäuser Verlag Basel/Switzerland

The Met-HGF/SF autocrine signaling mechanism is involved in sarcomagenesis

J. Cortner, G.F. Vande Woude and S. Rong

ABL-Basic Research Program, NCI-Frederick Cancer Research and Development Center, P.O. Box B, Frederick, MD 21702, USA

Summary. Hepatocyte growth factor/scatter factor (HGF/SF) can elicit a wide variety of effects upon cells expressing its receptor, the tyrosine kinase proto-oncogene product Met, including mitogenicity, motility, and morphogenesis. Normally, *met* expression is restricted to epithelial cells and is activated in a paracrine fashion by HGF/SF secreted from cells of mesenchymal origin. In this chapter, we review data showing that: (i) *met* over-expression in HGF/SF-expressing NIH/3T3 fibroblasts leads to sarcomagenesis and metastasis via an autocrine mechanism; (ii) Met-HGF/SF autocrine signalling occurs to a low level in normal fibroblasts and to a much greater extent in human sarcomas and sarcoma cell lines; (iii) *met* expression is enhanced as p53-deficient fibroblasts are passaged *in vitro* and (iv) *met* and HGF/SF over-expression are selected for during tumorigenesis of p53-deficient late-passage fibroblasts. Thus, loss of p53 predisposes a mesenchymal cell to over-express *met* and high level Met–HGF/SF autocrine signaling in mesenchymal cells promotes both sarcomagenesis and metastasis through inappropriate induction of the pleiotropic responses to Met-HGF/SF stimulation.

Introduction

Hepatocyte Growth Factor (HGF), originally identified as a potent mitogen for hepatocytes (Nakamura et al., 1984), is a disulfide-linked heterodimer consisting of a 69-kD α-subunit and a 34-kD β-subunit, both of which are derived from cleavage of a single precursor protein. Molecular cloning of HGF cDNA revealed that HGF shares very high sequence homology with the N-terminal kringle domains and C-terminal serine protease domain of plasminogen, although HGF has no known protease activity (Nakamura et al., 1992). Independently, Scatter Factor (SF) was identified by Stoker and colleagues (Stoker et al., 1985) as a motility factor secreted by fibroblasts and smooth muscle cells capable of inducing scattering of epithelial cells such as Madin-Darby canine kidney (MDCK) cells (Stoker et al., 1987; Rosen et al., 1989; Stoker et al., 1989). SF was shown to promote the invasiveness of a number of human carcinoma cell lines *in vitro* (Weidner et al., 1990) and induce scattering in many human primary mammary epithelial cells (Stoker et al., 1985; Stoker et al., 1987). Subsequent studies showed that HGF and SF are identical (Furlong et al., 1991; Gherardi et al., 1990; Konishi et al., 1991; Naldini et al., 1991a,b; Weidner et al., 1991). Interestingly, although HGF/SF stimulates the proliferation of a variety of cell types (Igawa et al., 1991; Rubin et al., 1991). including rat and human hepatocytes (Nakamura et al., 1987; Gohda et al., 1988; Zarnegar et al., 1989; Strain et al., 1991) and some human mammary epithelial cells, it can also inhibit proliferation in other systems (Shima et al., 1991a; Shima et al., 1991b; Matsumoto et al., 1992). In addition to effects on cell proliferation and movement, HGF/SF has also been shown to mediate differentiation in some systems. For example, MDCK cells cultured in collagen gels form branching structures resembling kidney tubules when treated with HGF/SF (Montesano et al., 1991) and some breast cell lines can respond *in vitro* to HGF/SF by formation of duct-like structures (Tsarfaty et al., 1992).

The receptor for HGF/SF was shown to be the tyrosine kinase proto-oncogene product Met (Bottaro et al., 1991; Naldini et al., 1991a,b). Under physiological conditions, the kinase activity of Met is dependent on the binding of the mature heterodimeric receptor to its ligand (Park et al., 1987). The primary Met product is synthesized as a glycosylated 170-kD precursor. Cleavage of p170^{Met} yields the mature Met receptor, a disulfide-linked heterodimer of a 140-kD β subunit (p140^{Met}) and a 45-kD α subunit (p45^{Met}; Russo et al., 1989; Kleinberg et al., 1990; Coleman et al., 1988; Faletto et al., 1993). The α subunit is totally extracellular, while the β subunit includes the major portion of the extracellular domain, the transmembrane domain, and the intracellular tyrosine kinase domain (Chan et al., 1988; Giordano et al., 1988; Gonzatti-Haces et al., 1988). The α-β heterodimeric Met proteins are clustered on the cell surface (Faletto et al., 1992) and are widely distributed in adult and embryonic tissues, as well as in established cell lines (Gonzatti-Haces et al., 1988; Park et al., 1986, Di Renzo et al., 1991; Higuchi et al., 1991; Iyer et al., 1990; Tajima et al., 1992); the highest levels are found in epithelial cells (Bacus et al., 1992; Iyer et al., 1990; Tsarfaty et al., 1992). Recently, two groups have used chimeric receptors composed of the extracellular domain of the nerve growth factor receptor or the epidermal growth factor receptor and the transmembrane and cytoplasmic domains of Met to demonstrate that the Met receptor transduces the motogenic, mitogenic and morphogenic signals of HGF/SF (Weidner et al., 1993; Komada et al., 1993).

The pleiotropic biological responses induced by HGF/SF suggest that multiple complex signal transduction pathways may be activated following ligand stimulation: for example, a specific pathway may be uniquely activated in cells that have a mitogenic, as opposed to a motogenic or morphogenic, response to HGF/SF. It is known that HGF/SF induces the tyrosine phosphorylation of the 85 kD subunit of PI-3 kinase, the MAP2/ERK2 kinase, RAS–GAP and its associated proteins, PLCγ, and Src-related tyrosine kinase (Graziani et al., 1991; Faletto et al., 1993; Bardelli et al., 1992). MAP2/ERK tyrosine phosphorylation has not been detected in MDCK cells, which scatter rather than proliferate, following treatment with HGF/SF; ligand stimulation of MDCK cells leads to rapid increase of PI-3 kinase activity that is complexed with Met (Chatani et al., 1992). HGF/SF activates Ras by increasing the uptake of GTP, through the stimulation of a guanine nucleotide exchange factor (Graziani et al., 1993). However, the specific substrates that mediates distinct biological activities induced by HGF/SF are not known.

The *met* proto-oncogene was initially identified as the proto-oncogene homologue of the activated oncogene *tpr–met* in a human osteosarcoma cell line (HOS), generated by treatment with the chemical carcinogen *N*-methyl-*N*-nitro*N*-nitrosoguanidine (Park et al., 1987; Gonzatti et al., 1988; Cooper et al., 1984). This discovery was the first indication that the Met–HGF/SF signaling apparatus might be involved in tumorigenesis. *tpr–met* activation resulted from a chromosomal rearrangement that fused *tpr* sequences encoding a dimerization motif from chromosome 1 with the tyrosine kinase domain of *met* on chromosome 7 (Park et al., 1986). It has been shown that *tpr–met* is activated and hence transforming by auto-phosphorylation resulting from the *tpr*-mediated juxtapositon of two *met* kinase domains (Rodrigues et al., 1993). TPR is a common fusion partner for various oncogenes and TPR has recently been shown to be associated with the nuclear pore complex, localized to the cytoplasmic surface (Byrd et al., 1994).

Normally, HGF/SF is expressed in mesenchymal-derived fibroblasts and other stromal cells, while Met expression is restricted to epithelial cells. Over-expression of Met in epithelial cells is thought to lead to transformation in several human gastric carcinoma cell lines (Giordano et al., 1989) as well as carcinomas of the lung, pancreas, thyroid, colon and stomach (Giordano et al., 1988; Ponzetto et al., 1991; Kuniyasu et al., 1992; Prat et al., 1991; Liu et al., 1992; Di Renzo et al., 1992; Yoshinaga et al., 1992 and Tsao et al., 1993). Although the mechanism is unknown, Met activation in these tumors may occur by receptor aggregation due to over-expression or by an activating point mutation.

The *met* proto-oncogene is frequently amplified and over-expressed in spontaneous NIH/3T3 fibroblast tumors (Cooper et al., 1986; Hudziak et al., 1990), and over-expression of *met* in NIH/3T3 cells directly leads to cell transformation and tumorigenicity (Iyer et al., 1990). We have shown that NIH/3T3 cells express murine HGF/SF (HGF/SF^{mu}) endogenously and that these cells become tumorigenic by an autocrine mechanism when they over-express murine Met^{mu} (Rong et al., 1992). Additionally, co-expression of Met^{hu} with the human ligand (HGF/SF^{hu}) also efficiently transforms NIH3T3 cells, rendering the resultant HMH cells tumorigenic in nude mice (Rong et al., 1992). Furthermore, cells from both the *met*^{mu} and HMH autocrine-driven tumors are invasive *in vitro* and metastatic in nude mice (Rong et al., 1994). Additionally, we found evidence indicating that Met–HGF/SF autocrine signaling occurs at a low level in primary fibroblasts, suggesting a role for the autocrine loop in untransformed mesenchymal cells (Rong et al., 1994). Most importantly, Met is often over-expressed in human mesenchymal tumor cell lines and in primary sarcomas, suggesting that the Met-HGF/SF autocrine mechanism may play a role in mesenchymal tumor progression (Rong et al., 1993b). Sarcomas often arise both in mice and in humans (Malkin et al., 1990) as a result of p53 deficiency and we have shown that Met is frequently over-expressed in these tumors. Our studies suggest that the loss of p53 enhances the opportunity for inappropriate Met expression in mesenchymal tumors but that the loss of p53 is not a prerequisite for Met over-expression.

Materials and methods

Cell culture

NIH/3T3 cells transformed by met^{mu} (Met^{mu} cells) or by the co-expression of met^{hu} and HGF/SF^{hu} (HMH cells) have been described (Rong et al., 1992). Met^{mu} and HMH cells are either parental cells or primary (1°) or secondary (2°) tumor explants. All cells were maintained as previously described (Rong et al., 1992, 1993a). Most of the human cell lines used in this study were obtained from the American Type Culture Collection (ATCC) and grown as recommended (Rong et al., 1993b). Early (less than 10 passages) and late passage (more than 50 passages) cultures of mouse embryonic fibroblasts were established and maintained as described previously (Harvey et al., 1993b). Cell lines with different numbers are derived from different embryos (eg., 68, 70, 74). Independent fibroblast clones established from the same embryo are denoted with the same first number separated by a dash from the specific cell line number (eg., 70, 70−1, and 70−2; Tab. 1). Li-Fraumeni fibroblasts were maintained as described by Bischoff et al. (1990). Human osteosarcoma cell lines (kindly provided by Steve Friend, Massachusetts General Hospital) were grown as previously reported (Diller et al., 1990).

Immunoprecipitation analysis

Immunoprecipitation analysis for both Met and HGF/SF was carried out as previously described (Rong et al., 1992, 1993a,b). The 19S Met^{hu} monoclonal antibody was generated against a bacterially-expressed p50 form of Met^{hu} (Faletto et al., 1991). The Met^{hu}-specific C28 anti-peptide antibody was raised by immunization of rabbits with the 28-amino acid C-terminal peptide of Met^{hu} (Gonzatti-Haces et al., 1988). A3.1.2 is a monoclonal antibody directed against human recombinant HGF/SF (Rong et al., 1992; a gift from Dr. T. Nakamura, Osaka, Japan). The SP260 peptide antibody is a rabbit antiserum direct against the C-terminal 21 amino acids of Met^{mu} (Iyer et al., 1990). Anti-rhHGF is a rabbit polyclonal antibody against human recombinant HGF/SF (Montesano et al., 1991; a gift from T. Nakamura, Osaka, Japan). 4G10 is a monoclonal phosphotyrosine antibody (anti-P-Tyr) (Morrison et al., 1989). pAb 122 recognizes a pan-specific epitope on p53 (Boehringer Mannheim).

Western blot

Western immunoblot analysis was performed as described (Rong et al., 1993b).

Pulse-chase analysis

Near-confluent cells were labeled for 45 min with 0.3 mCi of Translabel (ICN, 1 ml/35 mm diameter dish) in DMEM (Dulbecco's modified Eagle medium) lacking methionine and cysteine (GIBCO). The cells were washed twice, chased with complete medium for 0.5 and 2 h, lysed in RIPA buffer (1% Triton X-100, 1% sodium deoxycholate, 0.1% SDS, 0.15 M NaCl, 0.02 M NaPO$_4$ (pH 7.2)), and their proteins analyzed by immunoprecipitation and PAGE (polyacrylamide gel electrophoresis).

HGF/SF stimulation

Near-confluent cells in 100 mm-diameter dishes were serum-starved overnight, then stimulated with 40 or 200 units of mouse HGF/SF (kindly provided by Eliot Rosen, Long Island Jewish Hospital; Rosen et al., 1990) for 20 min at 37 °C. Cells were lysed and analyzed by Western blotting as previously described (Rong et al., 1993a).

Nude mouse tumor assay

The assay was performed as previously described (Blair et al., 1982; Rong et al., 1992) using weanling athymic nude mice (Harlan Sprague Dawley, Inc.).

Experimental and spontaneous in vivo metastasis assays

These assays were carried out as previously described (Rong et al., 1994).

Immunofluorescence analysis and confocal laser scan microscopy

Spontaneous tumors generated from p53-deficient mice (Donehower et al., 1992; Harvey et al., 1993a) were fixed in formalin and embedded in paraffin. Five μm sections were mounted on superfrost slides, and stained with either SP260 or C28 peptide antibody (Tsarfaty et al., 1994) which had been pre-absorbed with an acetone powder of NIH/3T3 cells. A donkey anti-rabbit immunoglobulin complexed in fluorescein isothiocyanate (Jackson Immune Research) was used as a secondary antibody. As a negative control, sections were also stained with normal goat serum alone and analyzed as described previously (Tsarfaty et al., 1994).

Reverse Transcriptase – Polymerase Chain Reaction assays

They were performed as described by Niehrs et al. (1994) and Rupp et al. (1991) using specific primers Rong et al. (1995).

Results

Met over-expression in HGF/SF-expressing NIH/3T3 fibroblasts leads to sarcomagenesis and metastasis

NIH/3T3 cells stably transfected with a plasmid expressing *met*^{mu} from the MSV LTR (met^{mu} cells) are anchorage independent *in vitro* and are tumorigenic to athymic nude mice (Iyer et al., 1990; Rong et al., 1992; Tab. 1). Both NIH/3T3 cells and *met*^{mu} cells have been shown by Northern hybridization to express endogenous HGF/SF mRNA (data not shown), which suggests that transformation by *met*^{mu} is mediated by autocrine interaction of the endogenous ligand with the over-expressed receptor (Rong et al., 1992). *met*^{hu} was also shown to be highly tumorigenic in NIH/3T3 cells but only when co-expressed with HGF/SF^{hu} (HMH cells; Tab. 2B), suggesting that HGF/SF^{mu} binds with low affinity to the human receptor. We were able to show that transformation specifically involved ligand/receptor interaction by demonstrating that a chimeric receptor bearing the mouse ligand-binding domain linked to the human transmembrane and tyrosine kinase domains, but

Table 1. Tumorigenicity of NIH/3T3 cells transfected with met^{hu} or met^{mu} cDNA[a]

Transfected genes	Mice with tumors/No. tested	Latency (weeks)
A		
neo^r	0/19	
neo^r, met^{hu}	2^b/41	5
neo^r, met^{mu}	17/17	3-5
B		
neo^r	0/6	
neo^r, met^{hu}, $HGF/SF^{hu/c}$	17/19	4-6
neo^r, HGF/SF^{hu}	3/7	7

[a]Data from Rong et al. (1992). Cells (10^6) were washed twice with serum-free medium and injected subcutaneously on the back of weanling athymic nude mice. Tumor formation was monitored each week for up to 10 weeks.
[b]These two tumors displayed elevated levels of endogenous Met^{mu} and reduced levels of Met^{hu} compared to those of the parental line.
[c]Cells derived from this transfection are referred to as HMH cells.

not the converse, was tumorigenic in NIH/3T3 cells (data not shown; Rong et al., 1992).

Interestingly, explants of HMH primary tumors had increased levels of both Met^{hu} and HGF/SF^{hu} as compared to the parental lines (Fig. 1A and C, respectively; lanes 2, 4, and 5; Rong et al., 1992) but showed no change in the levels of endogenous Met^{mu} (Fig. 1B). This observation indicates that the trait conferring a selectable growth advantage to the HMH tumors is indeed due to the Met^{hu}-HGF/SF^{hu} autocrine signaling apparatus. After two additional tumor passages, the Met^{hu} and HGF/SF^{hu} levels did not increase further, suggesting that there is an optimal level of ligand and receptor for maximal tumor growth advantage.

Explants of HMH primary tumors also showed increased tumorogenicity in nude mice as compared with primary HMH transfectants: tumors arose from primary HMH cells in 4–6 weeks while the same number of injected cells from tumor explants produced tumors in less than 4 weeks (Rong et al., 1994; data not shown). As before, we observed a significant increase in HGF/SF^{hu} after tumor passage (Fig. 2B, lanes 3–6; Rong et al., 1993b). The Met^{hu} receptor was also highly expressed in the same tumor explants (Fig. 2A, lanes 2 and 3). Furthermore, the level of Met^{hu} detected with anti-P-Tyr (Fig. 2C, upper panel, lanes 1–3) increased with the increase in the receptor (Fig. 2C, lower panel, lanes 1–3) after tumor passage, indicating that Met^{hu} is activated in these cells. As before, no amplification of the endogenous Met^{mu} was found in these tumor explants (Fig. 2A, lanes 5 and 6).

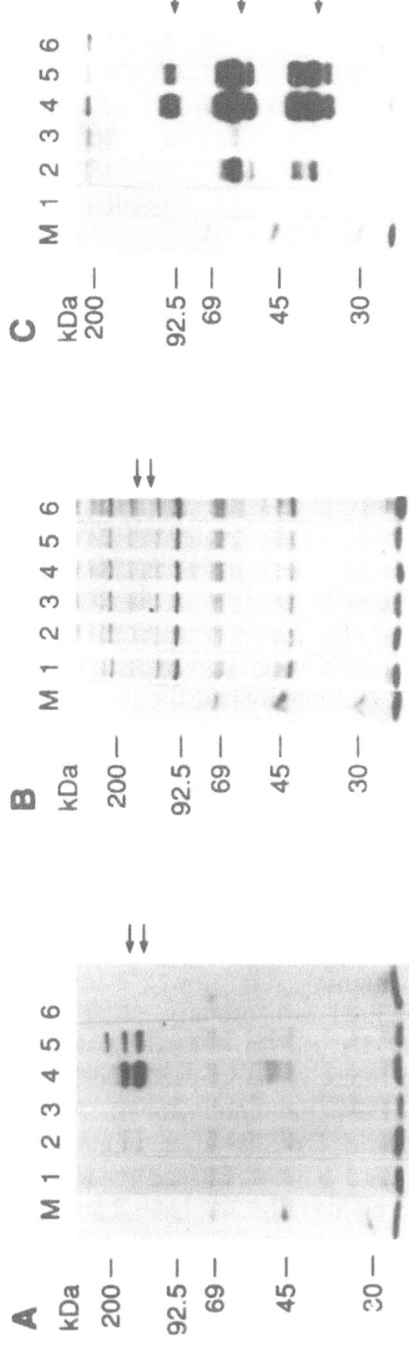

Figure 1. Characterization of Met and HGF/SF in NIH/3T3 tumor explants. Tumor cells were explanted and metabolically labeled with [³⁵S] methionine and [³⁵S] cysteine for 6 h. (A) Cell lysates were immunoprecipitated with either 19S monoclonal antibody or (B) SP260 peptide antibody. A 0.25-ml aliquot of 6-h supernatants was concentrated threefold in the Centricon apparatus (Amicon; 10 K cut-off); the volumes were adjusted to 0.3 ml with RIPA buffer and (C) the samples were immunoprecipitated with HGF monoclonal antibody A3.1.2 and autoradiographed for 16 h. Lanes 1 and 3 are samples from two different lines of cotransfected cells before injection. Lane 2 is a tumor explant derived from the cells analyzed in lane 1; lanes 4 and 5 are tumor explants derived from the cells analyzed in lane 3. Lane 6 is a sample prepared from control NIH/3T3 cells. (A + B) Arrows indicate the positions of p170^Met and p140^Met and (C) the positions of the 87-kDa (precursor), 69-kDa, and 34-kDa HGF polypeptides. (Reprinted with permission from Rong et al., 1992).

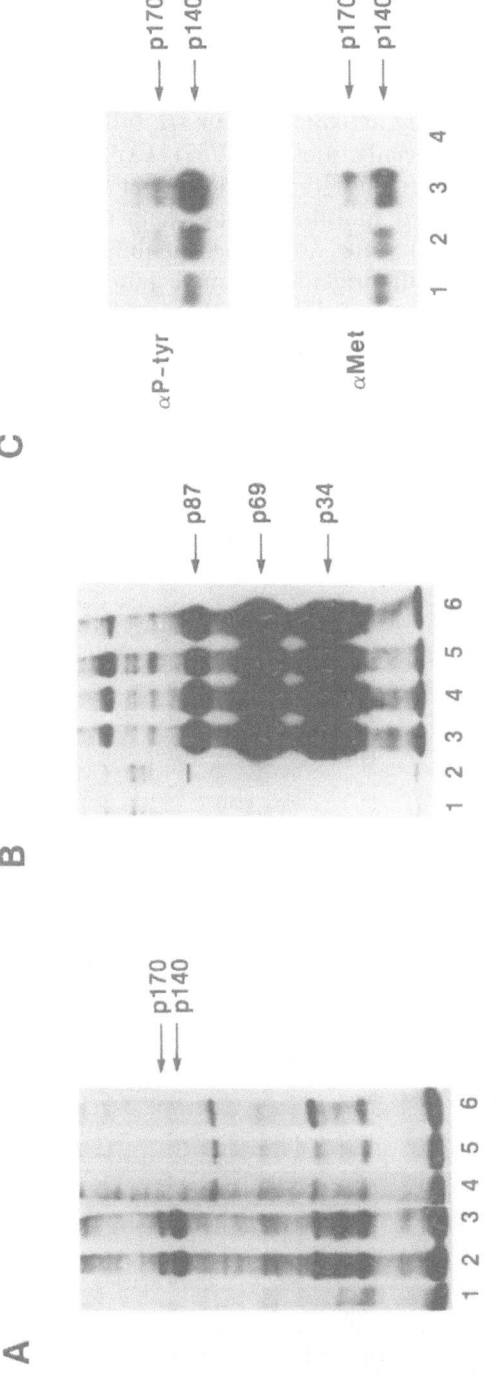

Figure 2. Characterization of Met and HGF/SF in cells of the second tumor passage. (A) Immunoprecipitation analysis for Met. Cells were metabolically labeled with [^{35}S] methionine and [^{35}S] cysteine for 6 h. Cell lysates were immunoprecipitated with C28 peptide antibody for Met^{hu} (lanes 1–3) or SP260 peptide antibody for Met^{mu} (lanes 4–6). Lanes 1 and 4, NIH/3T3 490 cells; lanes 2–3 and 5–6, cells of the second tumor passage. Arrows indicate the positions of p170^{Met} and p140^{Met}. (B) Immunoprecipitation analysis for HGF/SF. A 0.5-ml aliquot of medium was concentrated threefold in the Centricon apparatus (Amicon; 10 K cut-off); the volumes were adjusted to 0.3 ml with RIPA buffer (lacking reducing agent) to allow efficient immunoprecipitation with HGF monoclonal antibody A3.1.2 Lane 1, NIH/3T3 490; lane 2 cells cotransfected with met^{hu}: HGF/SF^{hu}; lanes 3–6, cells of the second tumor passage. Arrows indicate the positions of the 87-kDa (precursor), 69-kDa, and 34-kDa HGF polypeptides. (C) Western analysis for Met^{hu}. Near-confluent cells were lysed in lysis buffer and immunoprecipitated with C28 peptide antibody. After being dissolved in SDS-PAGE sample buffer, proteins were separated by SDS-PAGE on a 7.5% acrylamide gel, transferred to immobilon-P (Millipore) and probed with anti-P-Tyr (upper panel), or 19S monoclonal antibody (lower panel). (Reprinted with permission from Rong et al., 1994).

Histopathological examination of Met^mu and HMH tumors and the adjacent tissue revealed that the tumors were invasive (data not shown). We assayed the metastatic potential of the various *met*-transformed NIH/3T3 cells in nude mice following either subcutaneous injection or mammary fat pad injection, using activated *ras* or *src* NIH/3T3 transfectants as positive controls (Greenberg et al., 1989). Explants of both the Met^mu and HMH tumors, but not the primary Met^mu and HMH transfectants, efficiently produced multifocal lung metastases in both assays (Tab. 2; Rong et al., 1994). Thus, autocrine signaling through the over-abundant receptor is both tumorigenic and metastatic in NIH/3T3 cells. It is likely that inappropriate induction of the motility and other cellular responses to Met-HGF/SF stimulation is responsible for the metastatic phenotype.

Table 2. Met-mediate spontaneous metastasis in athymic mice*

Cells	Metastasis/ Total	Organs	Cells injected (10^5)	Latency (Weeks)
A. Subcutaneous injection				
Neo^r	0/4		10	
Met^mu	1/9	lung	10	12
	1/3	lung	3	10
	0.3		1	
Met^mu 1° explant	2/6	lung	10	5
	3/3	lung	3	6-7
	2/3	lung/salivary gland/ retroperitoneum	1	6
HMH	0/8		10	
	0/3		3	
	0/2		1	
HMH 1° explant	2/7	lung	10	6-9
	1/2	lung	3	9
	2/3	lung/diaphragm/ heart	1	5-8
ras	2/3	lung	10	4-5
src	1/3	lung	10	2.5
B. Mammary fat pad injection				
Neo^r	0.5		4	
Met^mu	1/5	lung	4	6
Met^mu 1° explant	1/5	lung	4	3-4
HMH	0/5		4	
HMH 1° explant	2/4	lung	4	1-3
HMH 3° explant	3/4	lung	4	3

*Cells were washed twice with serum-free medium and injected either subcutaneously (A) or injected into the mammary fat pad (B) of weanling athymic nude mice. When the mice seemed distressed, major organs were analyzed by histopathology for the presence of metastasis.

Met is over-expressed in many primary human sarcomas and sarcoma cell lines

To determine whether autocrine Met-HGF/SF interaction might contribute to human sarcomagenesis as it does in the mouse model described above, we examined paraffin-embedded human sarcoma sections stained for Met and HGF/SF with immunofluorescence analysis by confocal laser scan microscopy. Several were positive for both Met and HGF/SF staining: one leiomyosarcoma and one of two chondrosarcoma examined expressed both Met and HGF/SF, while three osteosarcomas showed significant Met and HGF/SF staining, consistent with the idea that the Met-HGF/SF autocrine mechanism is functioning in these tumors (data not shown; Rong et al., 1993a).

We examined human cell lines established from various human sarcomas for Met and HGF/SF expression. Low levels of Met protein, p140^{Met} and its precursor p170^{Met}, were detected in several of the primary fibroblast cultures while these same cultures secreted high levels of active HGF/SF (Tab. 3, Rong et al., 1993a). Interestingly, higher levels of Met, similar to the high levels of Met observed in the Met-transformed NIH3T3 cells (Met^{mu} cells; Rong et al., 1992), were present in most of the human sarcoma cell lines tested, but, as with the Met^{mu} cells, no HGF/SF was detected. In Met^{mu} cells, reduced steady-state levels of HGF/SF was attributed to depletion of ligand as a direct result of binding to Met and also possibly to Met-mediated repression of transcription (Rong et al., 1993b). Immunoblot analyses of Met in the sarcoma cell lines showed that Met was highly reactive with anti-P-Tyr antibody (Tab. 4), indicating that the receptor was activated, as in the Met^{mu} cells (Rong et al., 1992). Longer autoradiography showed that Met was also reactive, albeit weakly, with the anti-P-Tyr antibody in the autocrine fashion by endogenous HGF/SF (data not shown; Rong et al., 1993a).

Met-HGF/SF autocrine signaling occurs in primary fibroblasts

Similar to studies of expression levels in other primary fibroblasts (Gherardi and Stoker, 1991; Birchmeier et al., 1993), the Hems and HEL cells expressed low levels of Met and high levels of HGF/SF (Tab. 3). Since the converse situation exists in epithelial cells, the normal Met-HGF/SF interaction has been assumed to be paracrine. However, it is possible that the low level of Met expression in these cells may still be sufficient to produce some Met-HGF/SF autocrine signaling (Rong et al., 1993a). If this were so, removal of the endogenously-produced HGF/SF from the medium should slow the ligand-mediated turnover of Met. To test whether Met was being down-modulated by HGF in these

Table 3. Met and HGF/SF expression in human fibroblast and sarcoma cell lines

	Met[a]	met[b]	HGF/SF[c]	Scatter Activity[d]	P-Tyr-Met[e]
Human diploid fibroblast					
HEL299 (fetal lung)	+	+	+ + +	+ + + +	
Hems (fetal muscle)	+ +	+ + +	+ + +	+ + +	+
Hs68 (newborn fibroblast)	+ +	+ +	+ + + +	+	
Malme-3 (skin fibroblast)	+	ND[f]	+ + + +	+	ND
Fibrosarcoma					
8387	+ + + +	+ + + +	+ +	+	+ + +
HT1080	+ + + +	+ + +	+	+	+ + +
Hs913T	+ + + +	+ +	+ + + +	+	+ + + +
SW684	+ + + +	ND	+ +		+ + +
Osteogenic sarcoma					
HOS	+ + +	+			+
SAOS-2	+ + +	+ + + +	+	+	+ +
U-205	+ + +	+			+ +
Chondrosarcoma					
SW1353	+ +	ND		ND	ND
Rhabdomyosarcoma					
RD	+ + +	+ +		+	+ +
RD-1	+ +	+	+	+	
A204		ND		ND	
A673	+	ND			+
Hs729	+ +	ND	+	+	+ + +
Leiomyosarcoma					
SK-LMS-1	+ + + +	ND	+	+	+ + + +
SK-UT-1B	+	ND		+	+
Liposarcoma					
SW872	+ + +	+ + +		ND	
Mesodermal tumor					
SK-UT-1	+ + + +	ND			+ + +
Synovial sarcoma					
SW982	+	ND	+ + +		+
Melanoma					
Malme-3M	+ +	ND			+ + +
WM115	+	ND			+
WM266 − 4	+	ND			

[a]Met protein level was assessed by immunoprecipitation and Western analysis.
[b]met gene expression was detected by Northern analysis.
[c]HGF/SF protein level was assessed by immunoprecipitation analysis.
[d]Scatter activity was assayed with MDCK cells.
[e]P-Tyr-Met was analyzed by Western analysis with anti-P-Tyr.
[f]Not determined.

cultures, we titrated the HGF/SF being secreted into the medium by adding an HGF/SF-neutralizing antibody (anti-rhHGF, Montesano et al., 1991), to the growth medium. After 48 h, cell lysates were analyzed for Met expression by immunoblot analysis (Fig. 3; Rong et al., 1993a). This analysis showed that there was a significant increase in the amount of p140^{Met} in HEL299 cells (Fig. 3, lanes 1–4) and Hems cells (Fig. 3, lanes 5–6) in the presence of the antibody, indicating that Met is down-modulated in these cells by autocrine stimulation. These data further define the nature of the Met-HGF/SF autocrine interaction by demonstrating that the autocrine activation of Met by HGF/SF is

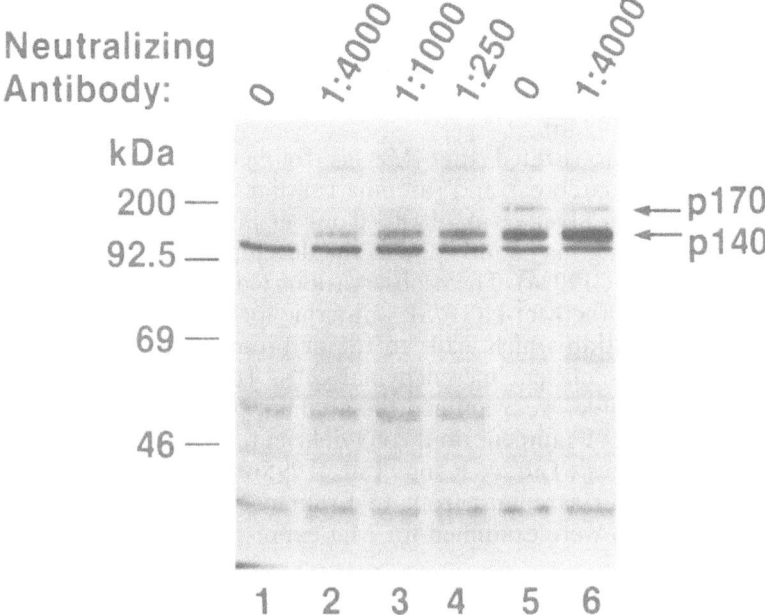

Figure 3. HGF-neutralizing antibody increases Met protein abundance. Primary fibroblast HEL299 (lanes 1–4) and Hems (lanes 5–6) cells were incubated for 48 hrs with or without HGF-neutralizing antibody (anti-rhHGF), added at 0 h and 24 h at the following dilutions: 1:4000 (lanes 2 and 6); 1:1000 (lane 3); 1:250 (lane 4); no antibody (lanes 1 and 5). After cells were lysed in RIPA buffer, 100 μg of protein was resolved by SDS-PAGE on a 7.5% acrylamide gel and immunoblotted with 19S Met^{hu} monoclonal antibody. (Reprinted with permission from Rong et al., 1993a).

mediated by external, rather than internal, binding of ligand to receptor. This conclusion is consistent with the fact that the secreted form of HGF/SF, proHGF/SF, must be cleaved by an extracellular protease before it can activate Met. (Naka et al., 1992; Miyazawa et al., 1993).

Met is over-expressed in p53-deficient human and murine sarcomas

P53 is a potent regulator of cell growth and division (Vogelstein et al., 1992; Lane et al., 1990) and loss of its wild-type function is a major step in the development and progression of many human cancers (Levine et al., 1991). For example, individuals with Li-Fraumeni syndrome inherit germline p53 mutations and develop sarcomas, as well as a variety of other neoplasms (Malkin et al., 1990; Malkin et al., 1993, Srivasta et al., 1993). Additionally, mice heterozygous and homozygous for loss-of-function mutations in p53 (p53 +/− and p53 −/− mice) appear normal at birth, but, as with the Li-Fraumeni patients, have a high propensity for the development of a variety of tumors, including sarcomas (Donehower et al., 1992; Harvey et al., 1993a). In both Li-Fraumeni patients and p53-deficient mice, the loss of p53 predisposes the individual for tumor development but does not directly result in cellular transformation; additional mutation(s) are required for tumorigenesis (Harvey et al., 1993b).

We have demonstrated that Met is frequently over-expressed in primary human sarcomas and sarcoma cell lines, suggesting that, as in the mouse model (Iyer et al., 1990; Rong et al., 1992), Met-HGF/SF autocrine signaling may contribute to the development of human sarcomas (Rong et al., 1993a). These observations led us to hypothesize that establishment of the Met-HGF/SF autocrine loop might function as an additional alteration which leads to tumor progression in both human and murine p53-deficient mesenchymal cells. To determine if this hypothesis is plausible, we examined the abundance of Met and HGF/SF expression in Li-Fraumeni and p53-deficient mouse sarcomas and fibroblast cell lines (Tab. 4; Rong et al., 1995).

Sections of four sarcomas from LI-Fraumeni patients with germline mutations in p53 were examined for Met expression by immunofluores-

Table 4. Cell lines used

1. Mouse fibroblasts cell lines used (20).		
Early Passages	p53 Genotype	
p53-1-50	$-/-$	
12	$-/-$	
320	$+/-$	
321	$+/-$	
322	$+/-$	
323	$+/+$	
Late passages		
67	$-/-$	
67-1	$-/-$	
67-2	$-/-$	
70	$-/-$	
70-1	$-/-$	
70-2	$-/-$	
74	$-/-$	
68	$(+/-)^*$	
73	$(+/-)^*$	
2. Li-Fraumeni fibroblasts (21).		
MDAH 041 (early and late passages)		
MDAH 087 (early and late passages)		
MDAH 172 (early and late passages)		
3. Human osteosarcoma cell lines (22).		
HOS	OS-IP	7922
HTLA 161	Saos-2	8842TX-20
MG63	U-2 OS	
OSH50T	143B	

*Denotes these lines were derived from $+/-$ animals, but at late passage they are $-/-$ (20).

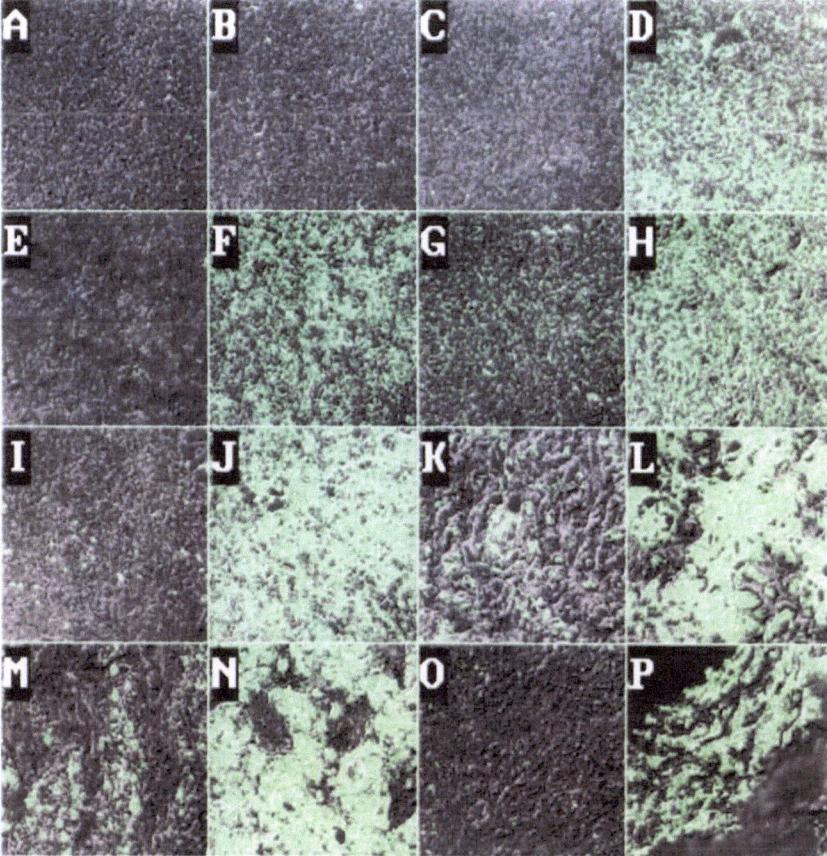

Figure 4. Immunofluorescence and confocal analysis of spontaneous tumors of p53-deficient mice. Twenty-three spontaneous tumors generated in p53-deficient mice were paraffin-embedded, sectioned and stained with either normal goat serum as negative control (A, C, E, G, I, K, M, and O) or with Met^mu-specific SP260 anti-peptide antibody (B,D,F,H,J,L,N, and P). FITC-conjugated goat anti-rabbit antibody (Boehringer Mannheim) was used as 2° antibody. The immunofluorescence was pseudo-colored (Green) and overlayed with Nomarski view of the tumor. (A) and (B) Lymphoma 1168; (C) and (D) Teratoma 1733; (E) and (F) Rhabdomyosarcoma 1577; (G) and (H) Rhabdomyosarcoma 1655; (I) and (J) Hemangiosarcoma 1696; (K-P) three Osteosarcomas (1241, 1255 and 1574). (Reprinted with permission from Rong et al., 1995).

cense and confocal laser scan microscopy using an anti-Met antibody (data not shown). Sections from three of the four tumors - a pleomorphic sarcoma, a leiomyosarcoma, and a spindle cell sarcoma - stained intensely for Met. Additionally, three independent fibroblast cell lines derived from skin biopsies of Li-Fraumeni patients (Bischoff et al., 1990) were analyzed for Met expression by immunoprecipitation from metabolically labeled cells. Met was observed in Li-Fraumeni fibroblasts

Table 5. Results of tumor staining with Met and HGF/SF

Case	Genotype	Histological type	Met staining
1226	+ / −	Lymphoma	
1102	− / −	Lymphoma	±
1168	− / −	Lymphoma	
1086	− / −	Teratoma	±
1310	− / −	Teratoma	
1733	− / −	Teratoma	+
1734	− / −	Teratoma	+ + +
1260	− / −	Rhabdomyosarcoma	+ +
1577	− / −	Rhabdomyosarcoma	+ + +
1655	− / −	Rhabdomyosarcoma	+ +
1563	− / −	Schwannoma	+ +
1698	− / −	Leiomyosarcoma	
1712	− / −	Myxosarcoma	+ +
1696	− / −	Hemangiosarcoma	+ + + +
1071	− / −	Hemangiosarcoma	+
1087	− / −	Hemangiosarcoma	±
1208	− / −	Hemangiosarcoma	+
1241	− / −	Osteosarcoma	+ + +
1255	− / −	Osteosarcoma	+ + +
1578	− / −	Osteosarcoma	±
1242	− / −	Osteosarcoma	+ +
1576	− / −	Osteosarcoma	+

at passage 14, and the level increased with cell passage (data not shown; Rong et al., 1995).

Interestingly, we observed significant Met staining in 9 of 15 sarcomas spontaneously arising in p53-deficient mice (either + / − or − / −, Tab. 5; Fig. 4, E-P; Rong et al., 1995) but little or no Met staining (Tab. 5; Fig. 4, A-D) in three lymphomas and three of four teratomas examined. These observations further suggest that Met plays a role in sarcomagenesis and reveals a link between loss of p53 and predisposition to increased Met expression.

Met is over-expressed in p53-deficient human osteosarcoma cell lines

We have previously observed that Met is overexpressed in various human sarcoma cell lines and primary sarcomas (Rong et al., 1993a), but we did not assay the p53 status of these cells and tumors. Loss-of-function p53 mutations frequently occur in these types of human tumors (Levine et al., 1991) and Diller et al. (1990) have charactrized the p53 mutations in a series of human osteosarcoma cell lines. We examined the level of Met expression in these osteosarcoma cell lines and found that Met was over-expressed in eight p53-deficient cell lines (Fig. 5, lanes 1-4, 6 and 8–10; Rong et al., 1995) as well as in two cell lines with wild-type p53 (Fig. 5, lanes 5 and 7). However, one of the two

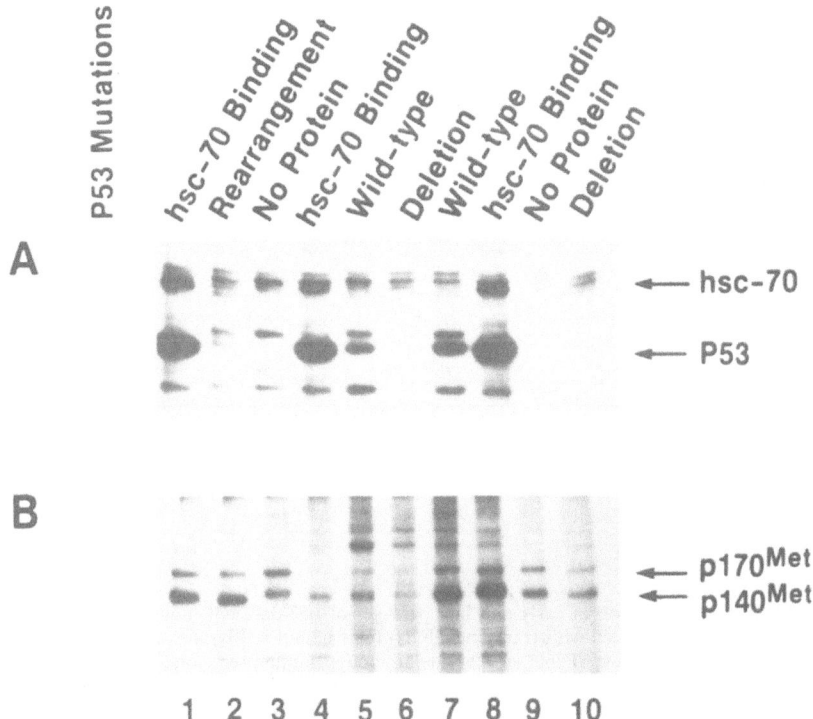

Figure 5. Met and p53 expression in human osteosarcoma cell lines. (A) Cells were metabolically labeled with Translabel (ICN) for 5 h. Half of the cell lysates were immunoprecipitated with p53 monoclonal antibody (PAb122) (Boehringer Mannheim). (B) Another half was immunoprecipitated with Met-specific 19S monoclonal antibody. Similar p53 levels were observed as previously reported (22). Lane 1, HOS; lane 2, HTLA161; lane 3, MG63; lane 4, OSH50T; lane 5, OS-IP; lane 6, Saos-2; lane 7, U-2 OS; lane 8, 143B; lane 9, 7922; lane 10, 8842TX-20. (Reprinted with permission from Rong et al., 1995).

wild-type p53 cell lines which did over-express Met had an amplification in *mdm2*, an inhibitor of p53, suggesting p53 activity in this cell line may be repressed. These data indicate that the loss of p53 may not be a prerequisite for inappropriate Met expression, but instead that loss of p53 is a powerful factor increasing the probability for enhanced Met expression.

Met is over-expressed in late passage p53-deficient mouse fibroblasts

To examine whether *met* expression occurs in fibroblast cells of p53-deficient animals, embryo fibroblast cultures established from 12–14 day wild-type, heterozygous, or p53-deficient embryos (prepared as previously described; Harvey et al., 1993) were analyzed by immunoprecipiation and quantitative RT-PCR. p140^Met was readily detected in all

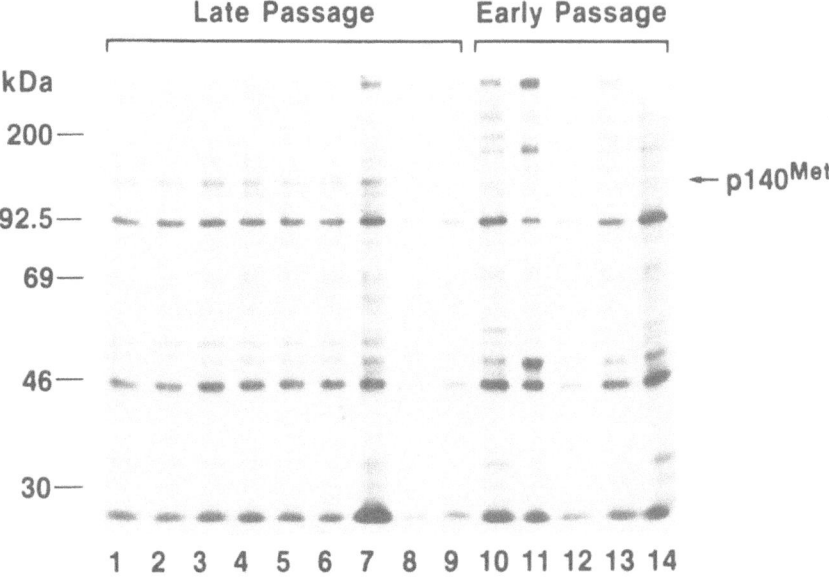

Figure 6. Met expression in early and late passages of p53-deficient mouse fibroblasts. Cells were metabolically labeled with translabel (ICN) for 5 h, and cell lysates were immunoprecipitated with Met^{mu}-specific SP260 anti-peptide antibody. Immunoprecipitates were solubilized in SDS sample buffer, subjected to 7.5% SDS-PAGE analysis and fluorography, and exposed on film. Lanes 1–9, late passage fibroblasts: lane 1, cell line 67 (− / −); lane 2, cell line 67-1 (− / −); lane 3, cell line 67-2 (− / −); lane 4, cell line 70 (− / −); lane 5, cell line 70-1 (− / −); lane 6, cell line 70-2 (− / −); lane 7, cell line 74 (− / −); lane 8, cell line 68 (+ / −); lane 9, cell line 73 (+ / −). Lanes 10–14, early passage fibroblasts: lane 10, cell line 12 (− / −) passage 3 (p3); lane 11, cell line 320(+ / −) p3. Lane 12, cell line 322 (+ / −) p3; lane 13, cell line 323 (+ / +) p3; lane 14, cell line 321 (+ / −) p3. (Reprinted with permission from Rong et al., 1995).

late passage cells (Fig. 6; Rong et al., 1995), but not in the early passage cells. However, low levels of p140^{Met} could be detected in early passage cells after two weeks of autoradiography (data not shown), consistent with our observation that primary human fibroblasts express low levels of Met (Rong et al., 1993a). Consistent with the pattern of Met protein expression, *met* mRNA, detected by RT-PCR, was significantly higher in late passage cells than the early passage cells (Rong et al., 1995; data not shown). Interestingly, only a two-fold amplification of Met gene was seen in most late passage p53-deficient cells when compared with the genomic DNA of early passage cells (Rong et al., 1994; data not shown); this level of amplification is not sufficient to account for the large increase in *met* RNA and protein expression.

Earlier we described how HGF/SF decreases in cells that over-express Met^{mu} and showed that HGF/SF is low in human sarcoma cell lines which have elevated levels of Met (Rong et al., 1993b). Similarly, the level of *HGF/SF* mRNA in all of the early passage fibroblast cells was

Table 6. Tumorigenicity of p53-deficient fibroblasts

	Cell line	p53 Genotype	# Mice tested/ # w/Tumors	Latency (wk)
early passage	12p3	$-/-$	0/3	
	321p3	$+/-$	0/3	
	322p3	$+/-$	0/3	
late passage	67	$-/-$	3/3	6
	67-1	$-/-$	3/3	4-5
	67-2	$-/-$	3/3	6-7
	70	$-/-$	0/2	
	70-1	$-/-$	2/3	6
	70-2	$-/-$	3/3	5-6
	74	$-/-$	0/3	
	68	$+/-$ *	0/3	
	73	$+/-$ *	0/3	

Cells (10^6) were washed twice with serum-free medium and injected subcutaneously on the back of weanling athymic nude mice. Tumor formation was monitored each week for up to 12 weeks.
*Late passages are ($-/-$).

very low. Our results suggest that either a Met-mediated feedback control mechanism may directly down regulate endogenous *HGF/SF* expression or that reduced *HGF/SF* expression is selected for during *in vitro* passage. As would be expected given the levels of Met and HGF/SF in the early and late passages of the p53-deficient fibroblasts, only the late passage cells respond to exogenous HGF/SF, indicting that Met is in excess of the endogenous HGF/SF: we observed that late-passage but not early-passage p53-deficient fibroblast cells display enhanced reactivity with anti-P-Tyr antibody and responded mitogenically to HGF/SF after treatment with exogenous HGF/SF (data not shown; Rong et al., 1995).

Both Met and HGF/SF are over-expressed in explants of tumors generated from late passage p53-deficient mouse fibroblasts

Five of nine late passage p53-deficient cell lines, but none of the early passage lines, were tumorigenic in athymic nude mice (Tab. 6; Rong et al., 1995). We examined cells derived from tumor explants of three of the five tumorigenic cell lines for Met expression by immunoprecipitation and found that the steady-state level of p140^{Met} was lower in the tumor explant cells than in the parental cells (Fig. 7A; Rong et al., 1995), suggesting that either Met expression is reduced or Met protein turns over more rapidly after tumorigenesis. By performing pulse-chase experiments we determined that the rate of Met turnover was elevated in the tumor explant cells: similar levels of the Met precursor, p170^{Met},

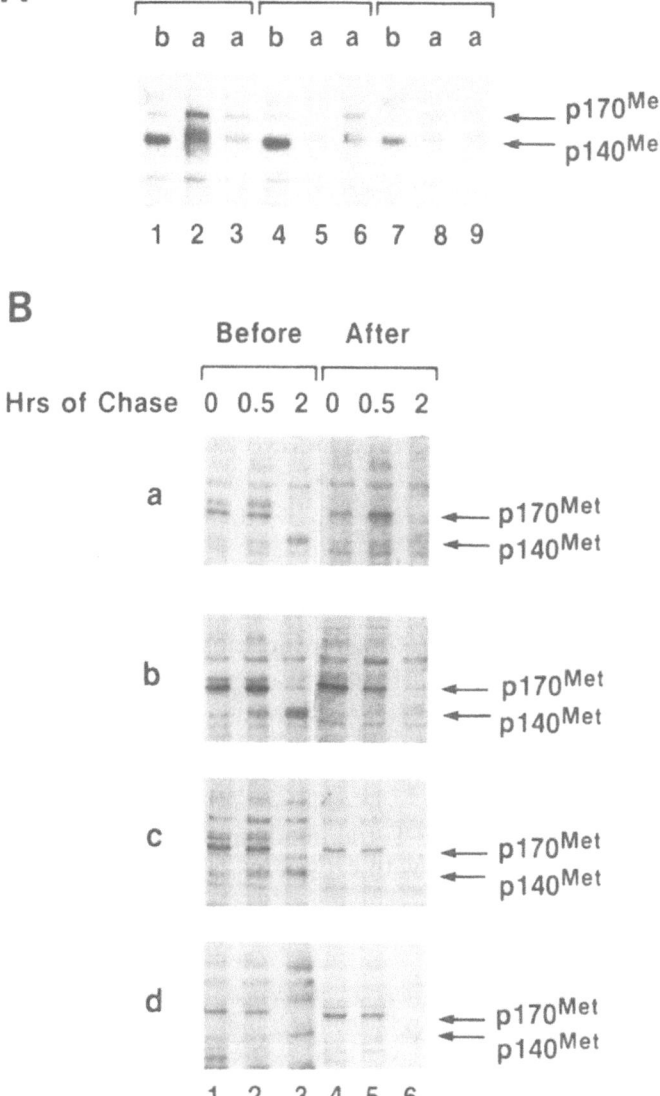

Figure 7. Met expression in late passage p53 $-/-$ fibroblasts before (b) and after (a) tumor passage. (A) Immunoprecipitation analysis. Cells were metabolically labeled with Translabel (ICN) for 6 h and cell lysates were immunoprecipitated with met^{mu}-specific SP260 anti-peptide antibody and subjected to electrophoresis as described above; b, before; a, after tumor passage. Lanes 1–3, cell line 67; lanes 4–6, cell line 67-1; lanes 7–9, cell line 67-2. The samples in lanes 2 and 3, 5 and 6, 8 and 9, respectively, were derived from two independent tumor explants. (B) Pulse chase analysis. Cells were metabolically labeled with Translabel (ICN) for 45 min (lines 1 and 4) and then chased with complete medium for 0.5 h (lanes 2 and 5), and 2 h (lanes 3 and 6). Met was immunoprecipitated with SP260 anti-peptide antibody in cell lines a, line 67; b, line 67-1; c, line 67-2; d, line 70-1. (Reprinted with permission from Rong et al., 1995).

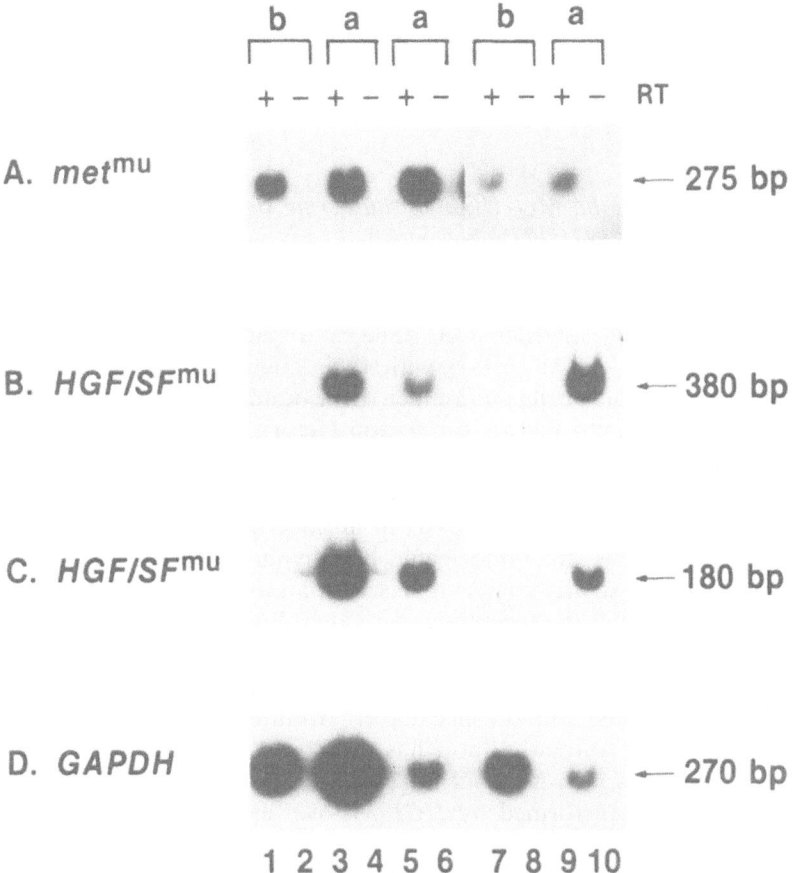

Figure 8. RT-PCR analysis of *HGF/SF* and *met* expression in late passage p53 −/ − fibroblasts before (b) and after (a) tumor passage. One μg of total RNA was reverse transcribed and PCR amplified in the presence of a ^{32}P-dCTP. (A) *met^{mu}* specific primers. (B) *HGF/SF^{mu}* specific primers (*HGF/SF^{mu}*-3). (C) *HGF/SF^{mu}* specific primers (*HGF/SF^{mu}*-2). (D) Control GAPDH specific primers. Lanes 1–6, cell line 67-1; lanes 7–10, cell line 67-2. Lanes 3–4 and 5–6 are samples from two independent tumor explants derived from cell line 67-1. (Reprinted with permission from Rong et al., 1995).

were detected in both parental and tumor explant cells, however, the mature form of Met, p140^{Met}, was more rapidly turned over in the tumor explant cells, providing an explanation for Met levels in the tumor explants (Fig. 7B). We also measured *met* and *HGF/SF* mRNA levels in two independent late passage cell lines before and after tumorigenesis and found that while abundant *met* mRNA was detected in both tumor explants and the parental late passage cells by RT-PCR, *HGF/SF* mRNA was significantly increased in the tumor explant cells as was HGF/SF protein (Fig. 8; Rong et al., 1995). These data suggest that a positive selection for Met-HGF/SF autocrine signaling mechanism occurred during tumorigenesis and resulted in increased HGF/SF expression and a concomitant increase in Met turnover.

Discussion

Establishment of the Met-HGF/SF autocrine signaling loop induces tumorigenicity of NIH/3T3 fibroblasts

Co-expression of growth factors such as epidermal growth factor (EGF; DiFore et al., 1987; Riedel et al., 1988) and colony stimulatory factor-1 (CSF-1; Roussel et al., 1987) with their cognate receptors has been shown to transform cells and render them tumorigenic as a direct result of autocrine growth factor stimulation (Sporn and Robert, 1985; Ullrich and Schlessinger, 1990). Similarly, we have shown that over-expression of met^{mu} in NIH/3T3 cells, which endogenously express mouse HGF/SF^{mu}, transforms these cells in an autocrine fashion (Iyer et al., 1990): met^{mu} cells are tumorigenic in athymic nude mice and Met^{mu} from these cells strongly cross reacts with an anti-P-Tyr antibody, thus demonstrating that the receptor is activated. We have been able to show that this autocrine mechanism is not restricted to NIH/3T3 cells, but is transforming and/or tumorigenic in other mouse and human cell lines (Jeffers and Rong, unpublished data). Additionally, the transforming potential of the autocrine loop has also been suggested by studies in hepatocytes, in which Met but not HGF/SF is normally expressed and which can be transformed by HGF/SF over-expression (Kanda et al., 1993).

Expression of met^{hu} alone in NIH/3T3 cells (Met^{hu} cells) does not lead to activation of Met^{hu} nor does it confer tumorigenicity in nude mice. However, Met^{hu} is activated when it is co-expressed with its species-specific ligand, HGF/SF^{hu} and the resultant cells, HMH cells, are tumorigenic in athymic nude mice. Thus, transformation by Met^{hu} requires the human homologue of the ligand and suggests that Met^{hu} and HGF/SF^{mu} do not interact with sufficient affinity to activate the met-HGF/SF autocrine signaling pathway (Bhargava et al., 1992; Faletto et al., 1992; Rong et al., 1992). This explanation is supported by the results of Bhargava et al. (1992), who found that purified HGF/SF^{mu} induces scattering response in the mouse mammary carcinoma cell line EMT6 but does not induce scattering in several human carcinoma cell lines. Species specificity of HGF/SF has also been reported for the activation of Met and its potential substrates in signal transduction studies (Faletto et al., 1992). Such species-specificity of ligand and receptor is not without precedent: mouse CSF-1 does not bind to the human CSF-1 receptor with high affinity (Roussel et al., 1988) and even though NIH/3T3 cells synthesize CSF-1, only co-expression of both human CSF-1 and human CSF-1 receptor can lead to transformation of NIH/3T3 cells (Roussel et al., 1987). Interestingly, NIH/3T3 cells transfected with *HGF/SF*hu are tumorigenic in this assay (Rong et al., 1992), indicating that HGF/SFhu can efficiently signal through MetMu, even though the converse is not true.

We have shown that Methu and HGF/SFhu are both over-expressed in HMH cells after tumorigenic selection in nude mice. Surprisingly, independent explants of the second tumor passage had approximately the same high levels of HGF/SFhu and Methu and these levels were not increased after a third passage, suggesting that there is an optimal level of receptor and ligand for tumorigenicity. We speculate that such high levels of Methu and HGF/SFhu are required to transform NIH/3T3 cells because Methu may interface inefficiently with the downstream signaling apparatus of mouse cells even when it is highly activated by its own ligand. Thus, high levels of Met would be required for tumorigenesis and correspondingly high levels of HGF/SFhu would be required to mediate the autocrine signal.

Met-HGF/SF autocrine signaling induces invasiveness and metastasis in addition to tumorigenicity

We have demonstrated that autocrine Met-HGF/SF signaling induces an *in vitro* invasive and *in vivo* metastatic phenotype in NIH/3T3 cells in addition to tumorigenicity (Tab. 3; Rong et al., 1994). These observations are not surprising, if one considers the broad repertoire of effects which are mediated by HGF/SF. HGF/SF was originally discovered as a paracrine mediator of epithelial cell mitogenicity and autocrine induction of the mitogenic function in the Metmu and HMH cells is sufficient to explain the tumorigenicity of these cells: the perpetual high level of growth signal transmitted in the Metmu and HMH cells overrides normal density-dependent and factor-dependent growth controls, both *in vitro* and *in vivo*. However, the mitogenic function alone cannot account for the invasive and metastatic phenotypes of the metmu and HMH cells.

Early in the study of HGF/SF, Stoker proposed that HGF/SF is involved in tumor invasion and metastasis, since the ligand elicits scattering and motility of epithelial cells *in vitro* (Stoker et al., 1989). Thus, autocrine induction of this normally paracrine function could enhance the motility of the Metmu and HMH cells, conferring a key characteristic essential to the invasive and metastatic phenotype. Subsequent to the discovery of the mitogenic and motogenic functions, HGF/SF was shown to stimulate endothelial cell migration and capillary-like tube formations *in vitro* (Rosen et al., 1991) and to induce blood vessel formation *in vivo* (Bussolino et al., 1992; Grant et al.,

1993). Since met[mu] and HMH cells secrete HGF/SF, it is possible that tumors composed of these cells can promote neovascularization via the paracrine mechanism and this effect could enhance both tumorigenesis and metastisis. Consistent with this hypothesis, we observed abundant neovascularization in Met[mu] and HMH tumors (Rong et al., 1994). HGF/SF is also postulated to be involved in inflammatory reactions, tissue repair, interactions in the immune system (Gherardi and Stoker et al., 1991) and induction of these functions could also enhance metastatic ability. Thus, the unique combination of effects mediated by HGF/SF through the Met receptor make over-activation of Met/HGF/SF autocrine signaling particularly likely to confer both the transformed and the metastatic phenotypes.

Met-HGF/SF autocrine signaling functions at a low level in untransformed fibroblasts and to a much higher level in sarcoma cell lines

We know that spontaneous transformants of NIH/3T3 fibroblasts frequently over-express the *met* proto-oncogene (Cooper et al., 1986; Hudziak et al., 1990) and we have shown that normal NIH/3T3 cells can be transformed *in vitro* and rendered tumorgenic *in vivo* by establishment of Met-HGF/SF autocrine signaling (Rong et al., 1992; Rong et al., 1993a). We have also observed high levels of Met expression in human sarcoma cell lines and tumors as compared with their non-transformed counterparts (Rong et al., 1993a), raising the possibility that the Met-HGF/SF autocrine signaling mechanism contributes to the transformation of these human malignancies.

The level of endogenous Met in many human sarcoma lines and the degree of its activation, as assessed by cross-reactivity with an anti-P-tyr antibody, is much greater than that in non-immortalized primary human fibroblasts (Rong et al., 1995) and approximates the levels of Met expression and activation seen in the Met[mu] NIH/3T3 transformants (Rong et al., 1993b). As with Met[mu] cells, the level of endogenous HGF/SF is reduced. Based on these findings, the exprssion of endogenous *HGF/SF* appears to be down-modulated when Met is over-expressed in human sarcomas. This down-modulation could be due simply to consumption of HGF/SF by receptor binding or to transcriptional down-regulation mediated by autocrine Met signaling (Rong et al., 1993a; Rong et al., 1993b). The latter hypothesis would create a negative feedback mechanism which could function to keep stimulation by the autocrine loop in check by limiting the production of ligand. It is conceivable that such a mechanism may also function in epithelial cells and is currently under investigation. However, we cannot exclude the possibility that other factors secreted by the sarcomas are responsible for their reduced expression of HGF/SF. For example, HGF/SF expression in MRC-5 cells can be modulated by transforming growth factor α, transforming growth factor β, epidermal growth factor, and conditioned medium from several carcinoma cell lines (Seslar et al., 1993; Kamalati et al., 1992).

Interestingly, we have observed that the Met-HGF/SF autocrine loop functions in the primary human fibroblast cultures which we examined. The primary human fibroblast cultures assayed had low levels of Met and high levels of HGF/SF expression, which is the expected pattern of expression. However, Met is activated in at least some of these lines, suggesting that the Met-HGF/SF autocrine loop can function in normal primary fibroblasts. Thus, similar to platelet derived growth factor (PDGF), fibroblast growth factor (FGF), and insulin-like growth factor I (IGF1) (Baserga et al., 1993), Met-HGF/SF autocrine interaction could contribute to fibroblast mitogenesis. Furthermore, in both of the cultures tested, HEL299 and Hems, the level of Met protein displayed by these cells increased when cultured in the presence of an anti-HGF/SF neutralizing antibody, indicating that the autocrine loop acts to down-modulate Met expression.

Met over-expression is linked to sarcomagenesis in individuals with LI-Fraumeni syndrome and in p53-deficient mice

Given that Met expression is very high in sarcoma cells, it appears that loss of a cellular control mechanism which normally regulates the level of Met occurs during sarcomagenesis. Since the p53 tumor suppressor gene negatively regulates cell proliferation and since p53-deficient organisms frequently develop sarcomas, as well as other neoplasias (Donehower et al., 1992; Malkin et al., 1990), we assayed the level of Met expression in p53-deficient systems to determine if loss of p53 correlates with inappropriate Met expression in mesenchymal tumors. Consistent with previous studies showing Met is involved in sarcomagenesis (Rong et al., 1993a,b; Rong et al., 1994), we found that elevated Met expression is associated with spontaneous sarcomas examined from both the animal model system and from patients with Li-Fraumeni syndrome. Interestingly, Met over-expression was not observed in other tumor types examined from the p53-deficient mice. Thus, Met over-expression in the p53-deficient organisms appears to be associated with the majority of sarcomas but is not a general phenomenon in tumors of non-mesenchymal origin.

Loss of p53 is not absolutely required for increased Met levels but imposes
a potent predisposition for Met over-expression

We have previously observed that Met is over-expressed in various
human sarcoma cell lines and primary sarcomas (Rong et al., 1993a),
but we did not assess the p53 status of these cells and tumors. Loss-of-
function p53 mutations frequently occur in these types of human
tumors (Levine et al., 1991) and so to determine if loss of p53 was
required for increased Met expression in human sarcomas, we examined
the level of Met expression in a series of human osteosarcoma cell lines
in which the p53 mutations have been characterized (Diller et al., 1990).
We found that Met was over-expressed in all of the osteosarcoma cell
lines, including eight p53-deficient lines and two cell lines with wild-type
p53. We conclude that loss of p53 is not a prerequisit for increased Met
expression but appears to enhance the frequency with which it occurs.
   Consistent with this hypothesis, we observed that Met expression
increases during *in vitro* passaging of p53-deficient fibroblasts from mice
and Li-Fraumeni patients. Although Met expression is already high in
early passages of Li-Fraumeni fibroblasts, Met expression increases in
both types of fibroblasts during *in vitro* passage. This observation
further demonstrates that the increase in Met expression is not a direct
result of loss of p53 function but instead that the loss of p53 predisposes
the cells to Met over-expression.
   The mechanism that leads to *met* over-expression in the p53-deficient
cells is unknown. We detect low level *met* amplification ($\sim$ two fold) in
both the mouse late passage p53-deficient fibroblasts, as well as in
various human sarcoma cell lines as compared with p53-deficient early
passage cells and human primary fibroblasts, respectively. This amplification could result from changes in cell ploidy (Donehower et al., 1992;
Harvey et al., 1993b) but the increase in Met expression is much greater
than can be accounted for solely by changes in ploidy. Since p53 has
been shown to transactivate the expression of a variety of cellular genes
(Levine et al., 1991) it is formally possible that wild-type p53 might
downregulate *met* gene expression; however, this possibility is not
consistent with our observation that early passage p53-deficient murine
fibroblast cells have low *met* expression or that human osteosarcomas
which have wild-type p53 can also over-express Met. Thus, there does
not appear to be a direct cause-and-effect relationship between loss of
p53 and increased *met* expression.

Increased Met expression in p53-deficient fibroblasts may promote
progression toward the tumorigenic phenotype

Although p53 $-/-$ fibroblasts passaged *in vitro* do not appear to
undergo senescence, p53 $+/-$ fibroblasts do not become immortal

upon loss of the remaining wild-type p53 allele and enter senescence in spite of conversion to the p53 $-/-$ genotype; this observation suggests that loss of p53 is not sufficient to rescue the senescent phenotype and that other genetic changes are necessary for immortalization as well as transformation (Harvey et al., 1993b). The fact that Met is over-expressed in the late passage p53-deficient cell lines and that approximately half of these lines are tumorigenic in athymic nude mice suggests that Met may be one of the growth control genes that allows tumor cell progression.

Increased HGF/SF expression in explants of tumors generated by late passage p53-deficient fibroblasts is selected for during tumorigenesis

While *met* expression increases during *in vitro* passaging of p53-deficient fibroblasts, endogenous *HGF/SF* expression decreases and is presumed to be down-regulated by a negative feed-back mechanism, as in the Met^{mu} NIH3T3 model system (Rong et al., 1993a,b). As a result of lower HGF/SF expression in late passage cells, Met is not fully activated by endogenously-produced HGF/SF, as it is in normal fibroblasts, and can signal in response to exogenous HGF/SF stimulation, as determined by increases in both mitogenicity and receptor tyrosine phosphorylation (Rong et al., 1995). Thus, it is not surprising that increased HGF/SF levels are selected during tumorigenesis of these cells. Approximately half of the late-passage cells are tumorigenic in athymic nude mice and explants of these tumors display increased *HGF/SF* expression and increased Met turnover, indicating that a high level of autocrine Met-HGF/SF signalling is necessary for tumorigenesis.

Conclusions

The data reviewed here show that establishment of a high level of Met-HGF/SF autocrine signaling renders NIH/3T3 cells tumorigenic and metastatic. We have also demonstrated that Met-HGF/SF autocrine signaling occurs at a low level in some untransformed, primary human fibroblasts and to a much greater extent in many human sarcomas and sarcoma cell lines, and in p53-deficient sarcomas and sarcoma cell lines. These observations suggests a model in which Met-HGF/SF autocrine signaling normally functions at a low level to regulate aspects of mesenchymal biology and is kept in check by maintaining low levels of receptor, either by negative feedback from the autocrine loop or by other cellular mechanisms. In this model, progression toward tumorigenesis is initiated by the loss of regulatory control mechanisms which keep Met expression low. Here we have shown that p53 appears to define one cellular mechanism for controlling Met expression since most p53-deficient sarcomas and sarcoma cell lines over-express *met*, although there must certainly be other control mechanisms since *met* over-expression also occurs in non-p53-deficient sarcomas. Once a mesenchymal cell expresses a high level of Met, endogenous expression of HGF/SF is decreased, presumably by a negative feedback mechanism. Thus, with respect to normal cell-type specific Met-HGF/SF expression, the mesenchymal cell has aquired properties of an epithelial cell. We have observed other phenotypic changes which suggest that Met-overexpressing NIH/3T3 transfectants can convert from a mesenchymal to epithelial-like cell (Tsarfaty et al., 1994).

Once a mesenchymal cell expresses a high level of Met, its ability to respond to HGF/SF is heightened and it may be targeted for tumorigenic selection via increased HGF/SF expression. This is precisely what is observed in the late-passage p53 tumor explants, in which increased HGF/SF expression appears to confer a tumorigenic advantage. We conclude that Met-HGF/SF autocrine stimulation has a role in normal fibroblast growth, that activity of the autocrine loop is controlled by negative feedback mechanisms including p53, and that inappropriate expression of Met in mesenchymal cells leads to sarcomagenesis. Further, we propose that the metastatic phenotype displayed by Met[mu] and HMH cells indicates that Met-HGF/SF autocrine stimulation is likely to enhance the metastatic potential of the human sarcomas.

References

Bacus, S.S., Huberman, E., Chin, D., Kiguchi, K., Simpson, S., Lippman, M. and Lupu, R. (1992) A ligand for the *erB*-2 oncogene product (gp30) induces differentiation of human breast cancer cells. *Cell Growth & Diff.* 3: 401–411.

Bardelli, A., Maina, F., Gout, I., Fry, M.J., Waterfield, M.D., Comoglio, P.M. and Ponzetto, C. (1992) Autophosphorylation promotes complex formation of recombinant hepatocyte growth factor receptor with cytoplasmic effectors containing SH2 domains. *Oncogene* 7: 1973–1978.

Baserga, R., Porcu, P. and Sell, C. (1993) Oncogenes, growth factors and control of the cell cycle. *Cancer Surveys* 16: 201–213.

Bhargava, M., Joseph, A., Knesel, J., Halaban, R., Li, Y., Pang, S., Goldberg, I., Setter, E., Donovan, M.A., Zarnegar, R., Michalpoulos, G.A., Nakamura, T., Faletto, D. and Rosen, E.M. (1992) Scatter factor and hepatocyte growth factor: activities, properties, and mechanism. *Cell Growth & Diff.* 3: 11–20.

Birchmeier , C., Sonnenberg, E., Weidner, K.M. and Walter, B. (1993) Tyrosine kinase receptors in the control of epithelial growth and morphogenesis during development. *Bioessays* 15: 185–190.

Bischoff, F.Z., Yim, S.O., Pathak, S., Grant, G., Siciliano, M.J., Giovanella, B.C., Strong, L.C. and Tainsky, M.A. (1990) Spontaneous abnormalities in normal fibroblasts from patients with Li-Fraumeni cancer Syndrome: Aneuploidy and immortalization. *Cancer Res.* 50: 7979–7984.

Blair, D.G., Cooper, C.S., Oskarsson, M.K., Eader, L.A. and Vande Woude, G.F. (1982) New method for detecting cellular transforming genes. *Science* 218: 1122.

Bottaro, D.P., Rubin, J.S., Faletto, D.L., Chan, A.M., Kmiecik, T.E., Vande, W.G. and Aaronson, S.A. (1991) Identfication of the hepatocyte growth factor receptor as the c-*met* proto-oncogene product. *Science* 251: 802–804.

Bussolino, F., Di, R.M., Ziche, M., Bocchietto, E., Olivero, M., Naldini, L., Gaudino, G., Tamagnone, L., Coffer, A. and Comoglio, P.M. (1992) Hepatocyte growth factor is a potent angiogenic factor which stimulates endothelial cell motility and growth. *J. Cell. Biol.* 119: 629–641.

Byrd, D.A., Sweet, D.J., Pante, N., Konstantinov, K.N., Guan, T., Saphire, A.C., Mitchell, P.J., Cooper, C.S., Aebi, U. and Gerace, L. (1994) Tpr, a large coiled coil protein whose amino teminus is involved in activation of oncogenic kinases, is localized to the cytoplasmic surface of the nuclear pore complex. *J. Cell. Biol.* 127: 1515–1526.

Chan, A.M.-L., King, H.W.S., Peakin, E.A., Tempest, P.R., Hilkens, J., Kroozen, V., Edwards, D.R., Wills, A.J., Cooper, C.S. and Brooke, P. (1988) Characterization of the mouse *met* protooncogene. *Oncogene* 2: 593–599.

Chatani, Y., Itoh, A., Tanaka, E., Hattori, A., Nakamura, T. and Kohno, M. (1992) Hepatocyte growth factor rapidly induces the tyrosine phosphorylation of 41-kDa and 43-kDa proteins in mouse keratinocytes. *Biochem. Biophys. Res. Commun.* 185: 860–866.

Coleman, S., Silberstein, G.B. and Daniel, C.W. (1988) Ductal morphogenesis in the mouse mammary gland: evidence supporting a role for epidermal growth factor. *Dev. Biol.* 127: 304–315.

Cooper, C.S., Park, M., Blair, D.G., Tainsky, M.A., Huebner, K., Croce, C.M. and Vande, W.G. (1984) Molecular cloning of a new transforming gene from a chemically transformed human cell line. *Nature* 311: 29–33.

Cooper, C.S., Tempest, P.R., Beckman, M.P., Heldin, C.H. and Brookes, P. (1986) Amplification and overexpression of the *met* gene in spontaneously transformed NIH/3T3 mouse fibroblasts. *EMBO J.* 5: 2623–2628.

DiFore, P.P., Pierce, J.H., Fleming, T.P., Hazan, R., Ullrich, A., King, C.R., Schlessinger, J. and Aaronson, S.A. (1987) Overexpression of the human EGF receptor confers an EGF-dependent transformed phenotype to NIH/3T3 cells. *Cell* 51: 1063–1070.

Diller, L., Kassel, J., Nelson, C.E., Gryka, M.A., Litwak, G., Gebhardt, M., Bressac, B., Ozturk, M., Baker, S.J., Vogelstein, B. and Friend, S.H. (1990) p53 functions as a cell cycle control protein in osteosarcomas. *Mol. Cell Biol.* 10: 5772–5781.

Di Renzo, M., Narsimhan, R.P., Olivero, M., Bretti, S., Giordano, S., Medico, E., Gaglia, P., Zara, P. and Comoglio, P.M. (1991) Expression of the Met/HGF receptor in normal and neoplastic human tissues. *Oncogene* 6: 1997–2003.

Di Renzo, M., Olivero, M., Ferro, S., Prat, M., Bongarzone, I., Pilotti, S., Belfiore, A., Costantino, A., Vigneri, R., Pierotti, M.A. and Comoglio, P.M. (1992) Overexpression of the c-MET/HGF receptor gene in human thyroid carcinomas. *Oncogene* 7: 2549–2553.

Donehower, L.A., Harvey, M., Slagle, B.L., McArthur, M., Montgomery, C.A., Butel, J.S. and Bradley, A. (1992) Mice deficient for p53 are developmentally normal but susceptible to spontaneous tumors. *Nature* 356: 215–221.

Faletto, D.L., Tsarfaty, I., Kmiecik, T.E., Gonzatti, M., Suzuki, T. and Vande Woude, G.F. (1992) Evidence for non-covalent clusters of the c-*met* proto-oncogene product. *Oncogene* 7. 1149–1157.

Faletto, D.L., Kaplan, D.R., Halverson, D.O., Rosen, E.M. and Vande Woude, D.F. (1993) Signal transduction in c-*met* mediated motogenesis. *In:* I.D. Goldberg and E.M. Rosen (eds): *Hepatocyte Growth Factor-Scatter Factor (HGF-SF) and the C-Met Receptor.* Birkhäuser Verlag, Basel, pp 107–130.

Furlong, R.A., Takehara, T., Taylor, W.G., Nakamura, T. and Rubin, J.S. (1991) Comparison of biological and immunochemical properties indicates that scatter factor and hepatocyte growth factor are indistinguishable. *J. Cell Sci.* 100: 173.

Gherardi, E. and Stoker, M. (1990) Hepatocytes and scatter factor [letter]. *Nature* 346: 228.

Gherardi, E. and Stoker, M. (1991) Hepatocyte growth factor–scatter factor: mitogen, motogen, and *met. Cancer Cells* 3: 227–232.

Giordano, S., Di, R.M., Narsimhan R.P., Tamagnone, L., Gerbaudo, E.V., Chiado, P.L. and Comoglio, P.M. (1988) Evidence for autocrine activation of a tyrosine kinase in a human gastric carcinoma cell line. *J. Cell Biochem.* 38: 229–236.

Giordano, S., Di, R.M., Narsimhan, R.P., Cooper, C.S., Rosa, C. and Comoglio, P.M. (1989) Biosynthesis of the protein encoded by the c-*met* proto-oncogene. *Oncogene* 4: 1383–1388.

Gohda, E., Tsubouchi, H., Nakayama, H., Hirono, S., Sakiyama, O., Takahashi, K., Miyazaki, H., Hashimoto, S. and Daikuhara, Y. (1988) Purification and partial characterization of hepatocyte growth factor from plasma of a patient with fulminant hepatic failure. *J. Clin. Invest.* 81: 414–419.

Gonzatti-Haces, M., Seth, A., Park, M., Copeland, T., Oroszlan, S. and Vande Woude, G.F. (1988) Characterization of the TPR–MET oncogene p65 and the MET protooncogene p140 protein-tyrosine kinases. *Proc. Natl. Acad. Sci. USA* 85: 21–25.

Grant, D.S., Kleinman, H.K., Goldberg, I.D., Bhargava, M.M., Nickloff, B.J., Kinsella, J.L., Polverini, P. and Rosen, E.M. (1993) Scatter factor induces blood vessel formation *in vivo. Proc. Natl. Acad. Sci. USA* 90: 1937–1941.

Graziani, A., Gramaglia, D., Cantley, L.C. and Comoglio, P.M. (1991) The tyrosine-phosphorylated hepatocyte growth factor/scatter factor receptor associates with phosphatidylinositol 3-kinase. *J. Biol. Chem.* 266: 22087–22090.

Graziani, A., Gramaglia, D., Dalla, Z.P. and Comoglio, P.M. (1993) Hepatocyte growth factor/scatter factor stimulates the Ras-guanine nucleotide exchanger. *J. Biol. Chem.* 268: 9165–9168.

Greenberg, A.H., Egan, S.E. and Wright, J.A. (1989) Oncogenes and metastatic progression. *Invasion Metastasis* 9: 360–378.

Harvey, M., McArthur, M.J., Montgomery, C.A., Jr., Butel, J.S., Bradley, A. and Donehower, L. (1993a) Spontaneous and carcinogen-induced tumorigenesis inp53-deficient mice. *Nature Genetics* 5: 225–229.

Harvey, M., Sands, A.T., Weiss, R.S., Hegi, M.E., Weiseman, R.W., Pantazis, P., Biovanella, B.C., Tainsky, M.A., Bradley, A. and Donehower, L.A. (1993b) *In vitro* growth characteristics of embryo fibroblasts isolated from p53-deficient mice. *Oncogene* 8: 2457–2467.

Higuchi, O. and Nakamura, T. (1991) Identification and change in the receptor for hepatocyte growth factor in rat liver after partial hepatectomy or induced hepatitis. *Biochem. Biophys. Res. Commun.* 176: 599–607.

Hudziak, R.M., Lewis, G.D., Holmes, W.E., Ullrich, A. and Shepard, H.M. (1990) Selection for transformation and *met* protooncogene amplification in NIH3T3 fibroblasts using tumor necrosis factor alpha. *Cell Growth & Diff.* 1: 129–134.

Igawa, T., Kanda, S., Kanetake, H., Saitoh, Y., Ichihara, A., Tomita, Y. and Nakamura, T. (1991) Hepatocyte growth factor is a potent mitogen for cultured rabbit renal tubular epithelial cells. *Biochem. Biophys. Res. Commun.* 174: 831–838.

Iyer, A., Kmiecik, T.E., Park, M., Daar, I., Blair, D., Dunn, K.J., Sutrave, P., Ihle, J.N., Bodescot, M. and Vande Woude, G.F. (1990) Structure, tissue-specific expression, and transforming activity of the mouse *met* protooncogene. *Cell Growth & Diff.* 1: 87–95.

Kamalati, T., Thirunavukarasu, B., Wallace, A., Holder, N., Brooks, R., Nakamura, T., Stoker, M., Gherardi, E. and Bulumela, L. (1992) Down-regulation of scatter factor in MRC-5 fibroblasts by epithelial-derived cells. A model for scatter factor modulation. *J. Cell. Sci.* 101: 323–332.

Kanda, H., Tajima, H., Lee, G.-H., Nomura, K., Ohtake, K., Matsumoto, K., Nakamura, T. and Kitagawa, T. (1993) Hepatocyte growth factor transforms immortalized mouse liver epithelial cells. *Oncogene* 8: 3047–3053.

Kleinberg, D.L., Ruan, W., Catanese, V., Newman, C.B. and Feldman, M. (1990) Non-lactogenic effects of growth hormone on growth and insulin-like growth factor-I messenger ribonucleic acid of a rat mammary gland [published erratum appears in Endocrinology 1990 Oct.; 127(4): 1977]. *Endocrinology* 126: 3274–3276.

Komada, M. and Kitamura N. (1993) The cell dissociation and motility triggered by scatter factor/hepatocyte growth factor are mediated through the cytoplasmic domain of the c-*met* receptor. *Oncogene* 8: 2381–2390.

Konishi, T., Takehara, T., Tsuji, T., Ohsato, K., Matsumoto, K. and Nakamura, T. (1991) Scatter factor from human embryonic lung fibroblasts is probably identical to hepatocyte growth factor. *Biochem. Biophys. Res. Commun.* 180: 765–773.

Kuniyasu, H., Yasui, W., Kitadai, Y., Yokozaki, H., Ito, H. and Tahara, E. (1992) Frequent amplification of the c-*met* gene in scirrhous type stomach cancer. *Biochem. Biophys. Res. Commun.* 189: 227–32.

Lane, D.P. and Benchimol, S. (1990) P53: Oncogene or anti-oncogene. *Genes and Dev.* 4: 1–8.

Levine, A.J., Momand, J. and Finley, C.A. (1991) The p53 tumor suppressor gene. *Nature* 351: 453–456.

Liu, C., Park, M. and Tsao, M.S. (1992) Overexpression of c-*met* proto-oncogene but not epidermal growth factor receptor or c-*erB-2* in primary human colorectal carcinomas. *Oncogene* 7: 181–185.

Malkin, D., Li, F.P., Strong, L.C., Fraumeni, J.F., Nelson, C.E., Kim, D.H., Kassel, J., Gryka, M.A., Bischoff, F.Z., Tainsky, M.A. and Friend, S.H. (1990) Germ line p53 mutations in a familial syndrome of breast cancer, sarcomas, and other neoplasms. *Science* 250: 1233–1238.

Malkin, D. (1993) p53 and the Li-Fraumeni syndrome. *Cancer Genet. Cytogenet.* 66: 83–92.

Matsumoto, K. and Nakamura, T. (1992) Hepatocyte growth factor: molecular structure, roles in liver regeneration, and other biological functions. *Crit. Rev. Oncog.* 3: 27–54.

Miyazawa, K., Shimomura, T., Kitamura, A. and Kondo, J. (1993) Molecular cloning and sequence analysis of the cDNA for a human serine protease responsible for activation of hepatocyte growth factor. Structural similarity of the protease precursor to blood coagulation factor XII. *J. Biol. Chem.* 10024–10028.

Montesano, R., Schaller, G. and Orci, L. (1991) Induction of epithelial tubular morphogenesis *in vitro* by fibrobast-derived soluble factors. *Cell* 66: 697–711.

Morrison, D.K., Kaplan, D.R., Escobedo, J.A., Rapp, J.R., Roberts, T.M. and Williams, L.T. (1989) Direct activation of the serine/threonine kinase activity of *raf*-1 through tyrosine phosphorylation by the PDGF-β receptor. *Cell* 58: 649–657.

Naka, D., Ishii, T., Yoshiyama, Y. and Miyazawa, K. (1992) Activation of hepatocyte growth factor by proteolytic conversion of a single chain form to a heterodimer. *J. Biol. Chem.* 267: 20114–20119.

Nakamura, T., Nawaka, K. and Ichihara, A. (1984) Partial purification and characterization of hepatocyte growth factor from serum of hepatectomized rats. *Biochem. Biophys. Res. Comm.* 122: 1450–1459.

Nakamura, T., Nawa, K., Ichihara, A., Kaise, A. and Nishino T. (1987) Subunit structure of hepatocyte growth factor from rat platelets. *FEBS Lett.* 224: 311.

Nakamura, T. (1992) Structure and function of hepatocyte growth factor. *Prog. Growth Factor Res.* 3: 67–85.

Naldini, L., Vigna, E., Narsimhan, R.P., Gaudino, G., Zarnegar, R., Michalopoulos, G.K. and Comoglio, P.M. (1991a) Hepatocyte growth factor (HGF) stimulates the tyrosine kinase activity of the receptor encoded by the proto-oncogene c-*met*. *Oncogene* 6: 501–504.

Naldini, L., Weidner, K.M., Vigna, E., Guadino, G., Bardelli, A., Ponzetto, C., Narsimhan, R.P., Hartmann, G., Zarnegar, R., Michalopoulos, G.K., Birchmeier, W. and Comoglio, P.M. (1991b) Scatter factor and hepatoycte growth factor are indistinguishable ligands for the *met* receptor. *EMBO J.* 10: 2867–2878.

Niehrs, C., Steinbeisser, H. and De Robertis E.M. (1994) Mesodermal patterning by gradient of the vertebrate homeobox gene goosecoid. *Science* 263: 817–820.

Park, M., Dean, M., Cooper, C.S., Schmidt, M., O'Brien, S.J., Blair, D.G. and Vande, W.G. (1986) Mechanism of *met* oncogene activation. *Cell* 45: 895–904.

Park, M., Dean, M., Kaul, K., Braun, M.J., Gonda, M.A. and Vande, W.G. (1987) Sequence of *met* protooncogene cDNA has features characteristic of the tyrosine kinase family of growth-factor receptors. *Proc. Natl. Acad. Sci. USA* 84: 6379–6383.

Ponzetto, C., Giordano, S., Peverali, F., Della, V.G., Abate, M.L., Vaula, G. and Comoglio, P.M. (1991) c-*met* is amplified but not mutated in a cell line with an activated *met* tyrosine kinase. *Oncogene* 6: 553–559.

Prat, M., Narsimhan, R.P., Crepaldi, T., Nicotra, M.R., Natali, P.G. and Comoglio, P.M. (1991) The receptor encoded by the human c-*met* oncogene is expressed in hepatocytes, epithelial cells and sold tumors. *Int. J. Cancer* 49: 323–328.

Riedel, H., Massozlia, S., Schlessinger, J. and Ullrich, A. (1988) Ligand activation of overexpressed epidermal growth factor receptors transforms NIH/3T3 mouse fibroblasts. *Proc. Natl. Acad. Sci. USA* 85: 1477–1481.

Rodrigues, G.A. and Park, M. (1993) Dimerization mediated through a leucine zipper activates the oncogenic potential of the *met* receptor tyrosine kinase. *Mol. Cell Biol.* 13: 6711–6722.

Rong, S., Bodescot, M., Blair, D., Dunn, J., Nakamura, T., Mizuno, K., Park, M., Chan, A., Aaronson, S. and Vande Woude, G.F. (1992) Tumorigenicity of the *met* proto-oncogene and the gene for hepatocyte growth factor. *Mol. Cell Biol.* 12: 5152–5158.

Rong, S., Jeffers, M., Resau, J.H., Tsarfaty, I., Oskarsson, M. and Vande Woude, G.F. (1993a) *met* expression and sarcoma tumorigenicity. *Cancer Res.* 53: 5355–5360.

Rong, S., Oskarsson, M., Faletto, D.L., Tsarfaty, I., Resau, J., Nakamura, T., Rosen, E., Hopkins, R. and Vande Woude, G.F. (1993b) Tumorigenesis induced by co-expression of human hepatocyte growth factor and the human *met* protooncogene leads to high levels of expression of the ligand and receptor. *Cell Growth & Diff.* 4: 563–569.

Rong, S., Segal, S., Anver, M., Resau, J.H. and Vande Woude, G.F. (1994) Invasiveness and metastasis of NIH/3T3 cells induced by Met-HGF/SF autocrine stimulation. *Proc. Natl. Acad. Sci. USA* 91: 4731–4735.

Rong, S., Donehower, L.A., Hansen, M.F., Strong, L., Tainsky, M., Jeffers, M., Resau, J.H., Hudson, E., Tsarfaty, I. and Vande Woude, G.F. (1995) *met* protooncogene product is overexpressed in tumors of p53 deficient mice and tumors of LI-Fraumeni patients. *Cancer Res.* 99: 1943–1970.

Rosen, E.M. and Goldberg, I.D. (1989) Protein factors which regulate cell motility. *In Vitro Cell Dev. Biol.* 25: 1079–1087.

Rosen, E.M., Meromsky, L., Setter, E., Vinter, D.W. and Goldberg, I.D. (1990) Smooth muscle-derived factor stimulates mobility of human tumor cells. *Invasion Metastasis* 10: 49–64.

Rosen, E.M., Meromsky, L., Setter, E., Vinter, D.W. and Goldberg, I.D. (1990) Purified scatter factor stimulates epithelial and vascular endothelial cell migration. *Proc. Soc. Exp. Biol. Med.* 195: 34–43.

Rosen, E.M., Grant, D., Kleinman, H., Jaken, S., Donovan, M.A., Setter, E., Luckett, P.M., Carley, W., Bhargava, M. and Goldberg, I.D. (1991) Scatter factor stimulates migration of vascular endothelium and capillary-like tube formation. *In:* I.D. Goldberg (ed.): *Cell Motility Factors.* Birkhäuser Verlag, Basel, pp 76–88.

Roussel, M.F., Dull, T.J., Rettenmier, C.W., Ralph, P., Ullrich, A. and Sherr, C.J. (1987) Transforming potential of the c-*fms* protoocogene (CSF-1 receptor). *Nature* 325: 549–552.

Roussel, M.F., J.R. Downing, C.W. Rettenmier and C.J. Sherr (1988) A point mutation in the extracellar domain of the human CSF-1 receptor (c-*fms* protoocogene product) activates its transforming potential. *Cell* 55: 979–988.

Rubin, J.S., Chan, A.M., Bottaro, D.P., Burgess, W.H., Taylor, W.G., Cech, A.C., Hirschfield, D.W., Wong, J., Miki, T., Finch, P.W. and Aaronsen, S.A. (1991) A broad-spectrum human lung fibroblast-derived mitogen is a variant of hepatocyte growth factor. *Proc. Natl. Acad. Sci. USA* 88: 415–419.

Rupp, R.A.W., and Weintraub, H. (1991) Ubiquitous *MyoD* transcription at the midblastula transition precedes induction-dependent *MyoD* expression in presumptive mesoderm of *X. laevis. Cell* 65: 927–937.

Russo, I.M., Tewari, M. and Russo, J. (1989) Morphology and development of the rat mammary gland. *In:* T.C. Jones, U. Mohr and R.D. Hunt (eds): *Integument and Mammary Glands.* Vol. 7, Springer-Verlag, New York, pp 233–266.

Seslar, S.P., Nakamura, T. and Byers, S.W. (1993) Regulation of fibroblast hepatocyte growth factor/scatter factor expression by human breast carcinoma cell lines and peptide growth factors. *Cancer Res.* 53: 1233–1238.

Shima, N., Nagao, M., Ogaki, F., Tsuda, E., Murakami, A. and Higashio, K. (1991a) Tumor cytotoxic factor/hepatocyte growth factor from human fibroblasts: cloning of its cDNA, purification and characterization of recombinant protein. *Biochem. Biophys. Res. Commun.* 180: 1151–1158.

Shima, N., Itagaki, Y., Nagao, M., Yasuda, H., Morinaga, T. and Higashio, K. (1991b) A fibroblast-derived tumor cytotoxic factor/F-TCF (hepatocyte growth factor/HGF) has multiple functions *in vitro. Cell Biol. Int. Rep.* 15: 397–408.

Sporn, M.B. and Roberts, A.B. (1985) Autocrine growth factors and cancer. *Nature* 313: 745–747.

Srivastava, S., Wang, S., Tong, Y., Pirolla, K. and Chang, E.H. (1993) Several mutant p53 proteins detected in cancer-prone families with Li-Fraumeni Syndrome exhibit transdominant effects on the biochemical properties of the wild-type p53. *Oncogene* 8: 2449–2456.

Stoker, M. and Perryman, M. (1985) An epithelial scatter factor released by embryo fibroblasts. *J. Cell Sci.* 77: 209–223.

Stoker, M., Gherardi, E., Perryman, M. and Gray, J. (1987) Scatter factor is a fibroblast-derived modulator of epithelial cell mobility. *Nature* 327: 239–242.

Stoker, M. (1989) Effect of scatter factor on motility of epithelial cells and fibroblasts. *J. Cell Physiol.* 139: 565–569.

Strain, A.J., Ismail, T., Tsubouchi, H., Arakaki, N., Hishida, T., Kitamura, N., Daikuhara, Y. and McMaster, P. (1991) Native and recombinant human hepatocyte growth factors are highly potent promoters of DNA synthesis in both human and rat hepatocytes. *J. Clin. Invest.* 87: 1853–1857.

Tajima, H., Higuchi, O., Mizuno, K. and Nakamura, T. (1992) Tissue distribution of hepatocyte growth factor receptor and its exclusive down-regulation in a regenerating organ after injury. *J. Biochem.* 111: 401–406.

Tsao, M.-S., Zhu, H., Giaid, A., Viallet, I., Nakamura, T. and Park, M. (1993) Hepatocyte growth factor/scatter factor is an autocrine factor for human normal bronchial epithelial and lung carcinoma cells. *Cell Growth & Diff.* 4: 571–579.

Tsarfaty, I., Resau, J.H., Rulong, S., Keydar, I., Faletto, D.L. and Vande Woude, G. (1992) The *met* proto-oncogene in mesenchymal to epithelial cell conversion. *Science* 263: 98–101.

Tsarfaty, I., Rong, S., Resau, J.H., Rulong, S., Pindo da Silva, P. and Vande Woude, G.F. (1994) The *met* proto-oncogene in mesenchymal to epithelial cell conversion. *Science* 263: 98–101.

Ullrich, A. and Schlessinger, J. (1990) Signal transduction by receptors with tyroskine kinase activity. *Cell* 61: 203–212.

Vogelstein, B. and Kinzler, K.W. (1992) p53 function and dysfunction. *Cell* 70: 523–526.

Weidner, K.M., Behrens, J., Vandekerckhove, J. and Birchmeier, W. (1990) Scatter factor: molecular characteristics and effect on the invasiveness of epithelial cells. *J. Cell Biol.* 111: 2097–2108.

Weidner, K.M. and Voss, E.J. (1991) Immunological characterization of xenogenic anti-metatype antibodies. *J. Biol. Chem.* 266: 2513–2519.

Weidner, K.M., Sachs, M. and Birchmeier, W. (1993) The *met* receptor tyrosine kinase transduces motility, proliferation, and morphogenic signals of scatter factor/hepatocyte growth factor in epthelial cells. *J. Cell Biol.* 121: 145–154.

Yoshinaga, Y., Fujita, S., Gotoh, M., Nakamura, T., Kikuchi, M. and Hirohashi, S. (1992) Human lung cancer cell line producing hepatocyte growth factor/scatter factor. *Jpn. J. Cancer Res.* 83: 1257–1261.

Zarnegar, R. and Michalopoulos, G. (1989) Purification and biological characterization of human hepatopoietin A, a polypeptide growth factor for hepatocytes. *Cancer Res.* 49: 3314–3320.

Epithelial–Mesenchymal Interactions in Cancer
ed. by I.D. Goldberg & E.M. Rosen
© 1995 Birkhäuser Verlag Basel/Switzerland

Tumor cell interactions with the vascular endothelium and their role in cancer metastasis

G.L. Nicolson

Department of Tumor Biology (108), The University of Texas M.D. Anderson Cancer Center, 1515 Holcombe Blvd., Houston, Texas 77030, USA

Introduction

The prevention of cancer deaths is dependent upon understanding the mechanism of malignant tumor cell spread from primary tumor sites to distant organ sites. Cancer metastases form by way of a complex series of sequential steps involving a variety of tumor cell and host properties. These steps typically involve growth and invasion of malignant cells at primary sites, followed by their penetration into lymphatics and blood circulation and body cavities. Once in these compartments, the malignant cells can detach and be transported to distant sites. There they can implant and invade surrounding tissues, and establish suitable microenvironments for survival and growth. This chapter will deal with the importance of tumor cell–endothelial interactions in the metastatic process.

The most obvious role of endothelial cells in the formation of distant metastases is in the attachment of blood-borne malignant cells to endothelial cells during their transport in the circulation. These interactions have been described as nonspecific or the result of random mechanical lodgment of tumor cells in the microcirculation, or alternatively they have been described as specific or the result of specific molecular recognition by adhesion molecules on malignant cells and endothelial cells (Nicolson, 1988a,b; Auerbach, 1992). Although the location of most regional metastases can be explained on the basis of anatomical and/or mechanical factors (Sugarbaker, 1981), clinical and experimental observations indicate that many malignant tumors preferentially metastasize to particular distant organ sites that would not be predicted solely from mechanical or anatomic considerations (Hart, 1982; Kiernan and Longenecker, 1983; Nicolson, 1982a, 1988a,b, 1993a).

Historically, two arguments have been advanced to explain the organ preference of formation of distant metastases. Ewing (1928) argued that tumor cells mechanically lodge in the first capillary bed encountered during their blood-borne transit, and that the anatomic structure and haemodynamics of the vascular system determine the secondary locations of tumor metastasis. In contrast, the 'seed and soil' hypothesis of Paget (1889) proposed that the formation of distant metastases was a consequence of the successful interaction of the tumor cells or 'seeds' with the microenvironment of a particular target organ or 'soil', which resulted in specific organ or tissue implantation, survival and growth of the tumor cells. These two hypotheses are not mutually exclusive, and organ colonization is probably determined both by haemodynamic factors and tumor cell and host characteristics unique to particular tumor cells and host environments (Weiss, 1983; Weiss et al., 1988; Nicolson, 1988a,b, 1989, 1991, 1993a).

The vascular endothelium and its underlying basement membrane make up the primary barrier between blood or lymph and the extravascular tissues. Specific interactions of circulating tumor cells and normal host leukoctye cells with components of the blood vessel lumen and underlying basement membrane are thought to be important for cell targeting to different tissues and their exit from the circulation. For example, lymphocytes that circulate continuously between the bloodstream and lymphatics cross the vascular barrier at specialized regions known as high endothelial venules (HEVs) (Ford et al., 1976; Stamper and Woodroff, 1976). These specialized HEV cells are morphologically and antigenically distinct from other microvessel endothelial cells (Dujvestijn et al., 1987). HEVs from different lymphoid tissues express antigenic differences that can be recognized by circulating lymphocytes (Gallatin et al., 1983; Streeter et al., 1988a) and used to initiate interactions with and invasion of the capillary endothelium at these sites (Ford et al., 1976; deBono, 1977). Malignant tumor cells that preferentially colonize certain organ sites must also adhere to and invade through endothelial cells and basement membranes during blood-borne tumor spread (Nicolson, 1988a,b, Belloni and Tressler, 1990; Belloni and Nicolson, 1992). Alternatively, some blood-borne malignant cells can adhere to microvessel endothelial cells and proliferate intravascularly (Roos and Dingemans, 1979; Kawaguchi et al., 1983).

Organ heterogeneity of endothelial cells

The microcirculation has its own characteristic structure, and its length, diameter, wall thickness and ultrastructure varies considerably between endothelial cells in different tissues and organs (Majno, 1965; Rhoden, 1980). For example, the continuous endothelium can be either thick or thin depending on the type of endothelial cells. Continuous thick capillaries (endothelial cells $> 2\ \mu$m thick) are found in skeletal tissue, cardiac smooth muscle, testes and ovary tissues; whereas continuous thin capillaries (endothelial cells < 1 μm thick) are typical in tissues of the central nervous system, lung, dermis, vasa rectum of the kidney, spleen, thymus, bone marrow and bone. These latter endothelial cells are generally flat and hexagonal in shape in culture; both the apical and basolateral endothelial cell surfaces are rich in cytoplasmic and pinocytotic vesicles that are adjacent to the plasma membrane. Weibel–Palade bodies can be found in human arterial capillaries, but these are usually rare (Stemmerman, 1980).

The integrity and continuity of the endothelium is maintained by gap junctions and desmosomes, except in the central nervous system where tight junctions predominate. Endothelial cells also have multiple points of attachment to a basement membrane that contains laminin, various proteoglycans (particularly heparan sulfate proteoglycan), fibronectin, and collagen types III, IV, and V. Basement membranes are also sources of vascular heterogeneity, and variations exist in the distributions of matrix proteins and bound factors that are involved in growth regulation and haemostasis (Kefalides, 1975; Carley et al., 1988; Wang et al., 1985).

Although most endothelial cells are morphologically similar, there are distinctive regional differences in vascular endothelial cells. For example, in different organs and tissues there are differences in the sizes and thicknesses of cells (large vessel endothelial cells are thicker than microvascular endothelial cells) as well as variations in the numbers of specific organelles, types of junctions and cell surface receptors and surface charged molecules. Using cationic ferritin and lectin–peroxidase conjugates to probe the fenestrated endothelial surfaces of pancreas and intestinal mucosa, it was demonstrated (Simionescue et al., 1982, 1985) that the microvascular system is a complex network of differentiated cell surface microdomains characterized by differences in negative charge and glycoconjugate expression. Lectin colloidal gold complexes have been used to demonstrate differential expression of glycoconjugates on the luminal and abluminal surfaces of microvessels, for example, in the brain (Vorbrodt et al., 1986). Some differences were also noted in the distributions of lectin receptors found on the three types of brain microvessels (capillaries, arterioles and venules). For example, it was found (Ponder and Wilkinson, 1983) that the *Dolichos biflorus* agglutinin-binding sites were differentially expressed in different organs and also on different vessels within the same organ. Auerbach (1992) used lectin cytofluorigraphic analysis of various endothelial cells to demonstrate quantitative differences in lectin receptor expression.

Cell labeling techniques have been used to demonstrate differences in glycoconjugate expression on organ endothelial cells. *In situ* cell-surface labeling techniques combined with Western enzyme-linked lectin analysis were used by us to demonstrate that microvessel endothelia *in vivo* and isolated microvessel endothelial cells *in vitro* express organ-related differences in cell surface glycoproteins (Belloni and Nicolson, 1988). Certain lectins, such as concanavalin A, reacted with the microvascular glycoproteins from all organs, whereas other lectins reacted selectively with the microvascular glycoproteins from specific organs. For example, wheat germ agglutinin (WGA) was highly reactive with brain microvessel endothelial cell glycoproteins, and *Ricinus communis* agglutinin (RCA) bound preferentially to lung microvessel glycoproteins. Some of the glycoproteins appeared to be selectively expressed in a particular organ, such as the WGA-binding brain glycoproteins gp135 and gp90, whereas others, such as RCA-binding gp120, appeared to be expressed in several organ-derived microvessels. These studies strengthen the conclusion that endothelial cells from different tissues possess unique displays of cell-surface glycoconjugates (Belloni and Nicolson, 1992).

Antigens are differentially expressed on endothelial cells in certain organs and tissues (Auerbach, 1992). The antigenic heterogeneity of murine vascular endothelium was first demonstrated by Pressman (1964), who generated antibodies against vascular trees from different organs. Differences in the antigenic composition of various organ microvascular endothelia were demonstrated after intravenous injection of labeled polyclonal antibodies raised against isolated organ-derived vessels. For the most part, the labeled antibodies localized to the organ which was used as a source of immunogen. More recently, using panels of monoclonal antibodies produced by immunization with isolated microvascular cells or heterogeneous populations of endothelial cells, Auerbach et al. (1985) demonstrated that the luminal surfaces of organ capillaries contain organ-associated antigens. Organ-associated endothelial antigens identified by monoclonal antibodies have been found for brain capillaries (Ghandour et al., 1982) and lung capillary endothelial cells (Kennel et al., 1988).

Further evidence supporting the tissue specificity of the vascular system has come from investigations on the molecular basis of lymphocyte homing and the identification of vascular addressins involved in homing (Gallatin et al., 1983). Monoclonal antibodies have been generated against tissue specific endothelial cell surface antigens involved in lymphocyte homing to mucosal lymphoid tissue (Streeter et al., 1988b) or peripheral lymphoid tissue (Streeter et al., 1988a). Other endothelial cell molecules involved in lymphocyte adhesion, such as intercellular adhesion molecule-1 (ICAM-1) (Dustin et al., 1986) and endothelial–leukocyte adhesion molecule-1 (ELAM-1) (Bevilacqua et al., 1987) are inducible in many tissues and contribute to transient heterogeneity of the endothelium at sites of trauma or inflammation.

Functional differentiation of the vascular endothelium

The vascular endothelium is functionally differentiated and metabolically distinct in each organ. Various organ endothelial cells participate
in metabolic and haemostatic functions, such as: (a) maintenance of a
non-thrombogenic luminal surface via production of prostacyclins; (b)
regulation of the coagulation pathway, clot formation and fibrinolysis;
(c) presentation of antigens to lymphoid cells; (d) secretion of chemotactic monokines; (e) active participation in leukocyte adhesion and
extravasation; (f) secretion of factors that coordinately stimulate the
growth of smooth muscle cells, pericytes and other cells; (g) clearance
and metabolism of vasoactive substances, such as angiotensin I,
bradykinin, ADP, and thrombin; (h) control of vasomotion; (i) regulation of perfusion and permeability at the microvascular level; (j) responses to and secretion of autocoids such as PGI_2, epinephrine, and
vasopressin; and (k) regulation of macromolecular transport between
the plasma and interstitial fluids (Belloni and Nicolson, 1992). The
endothelial cell functions associated with haemostasis are important
topics that are beyond the scope of this review. Here the functional
diversity of organ endothelia will be limited to a few properties, and
other reviews should be consulted (Shepro and Dunham, 1986; Weiss et
al., 1988; Belloni and Tressler, 1990; Gerritsen, 1987).

The movement of solutes, such as hormones, drugs, sugars, amino
acids, lipids and electrolytes, across capillary walls is regulated by
endothelial cells. Transcapillary passage of various molecules is controlled by: (a) endothelial cell surface microdomains that attract or
exclude molecules based on their size and charge; (b) endothelial cell
surface receptor-specific membrane transport; (c) endothelial cell surface ectoenzymes that chemically alter solutes transferred into, out of,
or between endothelial cells; and (d) endothelial cell synthesis and
secretion of solutes that establish concentration gradients between
blood plasma and the interstitial tissues.

Some models of the microvascular endothelial cell membrane describe
it as a static pore system, where solute and fluid flux are regulated by
changes in pressure and surface area. This may be a suitable description
under basal conditions; however, it cannot explain the increase in
secretion that occurs during trauma or inflammatory reactions. The
concept of variable mediator-stimulated pathways in which microvascular membrane permeability is physiologically controlled has evolved
from a static pore system in which fluid and solute flux are only
regulated by changes in blood flow. Variable macromolecular transport
pathways have been demonstrated, but not in all vascular beds. For
example, lung microvessels are not responsive to most classical
inflammatory mediators that show tissue-selective effects (Grega et al.,
1981).

Endocytosis, transendothelial cell channels, vasoactive stimuli, and endothelial desquamation all contribute to the regulation of the transport of components through the vascular endothelium. Regional and local tissue variability also exist in this complicated regulatory process. Cultured endothelial cell monolayers derived from a number of different vascular sites and species have been used extensively to study transendothelial cell permeability (Albeda et al., 1988). From such studies we know that the endothelial barrier function is partially attributable to components in the subendothelial matrix or basement membrane as well as the presence of tight junctional complexes between endothelial cells (Del Vecchio et al., 1987).

Tumors can dramatically alter the permeability and transport properties of blood vessels. In brain tumors a common problem is vasogenic edema that can cause subsequent neurological symptoms (Klatzo, 1967). In this example, structural alterations are present in tumor-induced blood vessels and peritumoral brain microvessels, such as increased numbers of pinocytotic vesicles, large vacuoles, and open junctions, which may cause changes in permeability (Groothuis and Vick, 1982). Additional changes caused by tumors that may contribute to changes in cerebrovascular permeability include focal destruction of brain capillaries (Roy and Chitra, 1989), increased hydrostatic pressure (Groothuis and Vick, 1982) and release of cytokine mediators (Criscuolo et al., 1988).

Endothelial cells can be stimulated to produce a variety of proteins, enzymes, and biochemical mediators. For example, tissue factor or thromboplastin is a glycoprotein produced by endothelial cells that forms an enzymatically active complex at the cell surface with Factor VII to activate the extrinsic coagulation pathway (Nemerson, 1966). Immunohistochemical localization indicates that thromboplastin is not uniformly expressed on endothelial cell surfaces, and it is particularly abundant in brain, lung and placenta (Gonmori and Takeda, 1976). Disseminated intravascular coagulation, a state often associated with acute bacterial infection, causes increased expression of tissue factor/thromboplastin on endothelial cells exposed to endotoxin (Colucci et al., 1983). Endothelial cell thromboplastin expression can also be increased by interleukin-1 and tumor necrosis factor, suggesting a more general link between inflammation and coagulation (Bevilacqua et al., 1985a,b, 1986).

Factor VIII is another endothelial cell-secreted complex that is important in homeostasis and blood coagulation. It consists of two glycoproteins that have distinct functional, biochemical and immunological properties and that are separately regulated (Jaffe, 1984). One component, Factor VIII:C, a glycoprotein of $M_r \sim 250\,000 - 300\,000$ has antihemophilic procoagulant activity. Monoclonal antibodies against FVIII:C have been used to determine the primary sources of the antigens–liver sinusoidal endothelial cells, mononuclear cells in the lymph nodes and alveolar septa of the lung (Kwast et al., 1986). Factor VIII:C was not found in the vascular beds of other organs.

Factor VIII-related protein or von Willebrand factor (vWF) is the major component (> 90% by weight) of plasma factor VIII and is composed of several subunits of $M_r \sim 200\,000 - 225\,000$. vWF binds to the subendothelial matrix and can interact with the platelet membrane receptor Gpab to mediate platelet adhesion. After thrombin stimulation, platelet adhesion to vWF occurs in an RGD-dependent manner via the platelet integrin GpIIb/IIIa (Plow et al., 1985). vWF has been identified in the endothelial cells of arteries, veins, arterioles, venules and capillaries as well as in megakaryocytes and platelets (Hoyer, 1981). Although factor VIIIR antigen has been identified in cultured endothelial cells and blood vessels of various species (Zetter, 1981), its expression on endothelial cells during long-term culture is often reduced to undetectable amounts (Hormonia, 1982). Dejana et al. (1989) found that vWF is also involved in the adhesion and spreading of cultured endothelial cells via an RGD-dependent mechanism.

Vasoactive substances induce vascular and endothelial cell prostaglandin PGI_2 synthesis (ADP, angiotensin II, PDGF, thrombin, and histamine). The response appears to be both species- and site-specific. For example, Hong (1980) demonstrated that thrombin-inducible PGI_2 synthesis occurred in human umbilical cord endothelial cells, but this response was not detected in calf or pig aortic endothelial cells. Thrombin-stimulated synthesis of PGI_2 was found in sheep umbilical cord vein endothelial cells, but not in sheep aorta (Goldsmith and Kisker, 1982). Prostacyclin metabolism varies in different endothelial cell cultures. For example, cultured rat aortic endothelial cells degrade PGI_2 into keto-acids (Sun and Taylor, 1978), but this was not seen in canine pulmonary microvascular endothelial cells (Dusting et al., 1978).

Endothelial cells synthesize and secrete various molecules associated with the subendothelial matrix, such as collagens, laminin, fibronectin, proteoglycans and some of the factors discussed above (Factor VIII, vWF) (Kefalides, 1975, 1980; Sage, 1984). The secretion and deposition of extracellular matrix components differs in various types of endothelial cells. Differences in the types of matrix molecules produced by endothelial cells and their distributions have been reviewed elsewhere (Kefalides, 1980; Sage, 1984). Our analyses of brain, lung, and liver organ-derived microvascular endothelial cell lines isolated from various species indicate that endothelial cells from different organs differentially express various cell surface and matrix components (Belloni et al., 1992). For example, Factor VIII/vWF antigen was detected in all of the endothelial cell cultures, except in the liver-derived endothelial cells. The intensity of antibody reactivity to Factor VIII/vWF antigen, however, was variable within particular cultures, and antibody reactivity decreased with culture passage. The finding that liver-derived sinusoidal endothelial (HSE) cells were negative for Factor VIII/vWF antigen (Belloni et al., 1992) was consistent with our inability to stain mouse liver sinusoids with this antibody and with previous reports on the distribution of this antigen *in vivo* (Irving et al., 1984). HSE cells selectively internalized modified low density lipoprotein, as did bovine aortic, bovine pulmonary, and rat lung endothelial cells; however, both mouse and human brain-derived microvascular endothelial cells lacked this property or it was not detectable (Belloni et al., 1992). Pitas et al. (1985) reported a similar *in vivo* distribution of receptors for acetylated low density lipoprotein, and rat brain microvascular endothelial cell cultures also appeared to lack these receptors and variably expressed other endothelial cell markers (Rupnick et al., 1988). The differences found in organ-derived endothelial cell cultures, for the most part, were consistent with phenotypic variations that have been observed *in vivo*.

Organ-derived endothelial cells display distinct patterns of growth responsiveness to purified enzymes and growth factors, such as thrombin and endothelial cell growth factor (ECGF) (Belloni et al., 1992). Although microvascular endothelial cells established from various organs responded to ECGF, maximum growth stimulation of lung endothelial cells was achieved by thrombin alone. In contrast, brain endothelial cells required combinations of thrombin and ECGF, or thrombin and subendothelial extracellular matrix, for maximum growth response. Liver sinusoidal endothelial (HSE) cells were relatively unresponsive to thrombin, but were the most responsive cells to EGF. High affinity thrombin-binding sites were differentially expressed on the organ-derived endothelial cell cultures, and the growth responsiveness of these cells correlated with their expression of high-affinity receptors (Belloni et al., 1992). These data indicate that microvessel endothelial cells from different organs possess unique characteristics that are probably related to their different functions.

Metastatic tumor cell adhesive interactions with the vascular endothelium

The importance of tumor cell–host cell interactions in the organ preference of metastatic colonization has been demonstrated in several tumor systems. Greene and Harvey (1964) first suggested that the organ distribution patterns of metastatic foci were dependent on the formation of sufficient adhesive bonds between arrested tumor cells and endothelial cells, and they hypothesized that these interactions were similar to lymphocyte/endothelial cell interactions at sites of inflammation. They also suggested that endothelial cells in various organs must express different receptors for certain tumor cells. Evidence for specific organ cell recognition and adhesion in preferential organ metastasis was demonstrated in *in vitro* aggregation studies with single organ cell suspensions (Nicolson and Winkelhake, 1975). A similar correlation between organ-specific adhesion and metastatic properties was subsequently shown in other tumor systems (Schirrmacher et al., 1980; Auerbach et al., 1987; Nicolson et al., 1983, 1985, 1989). Tissue-specific tumor cell adhesion has also been demonstrated using cryostat tissue sections (Netland and Zetter, 1985).

Monolayers of cultured endothelial cells derived from bovine aorta or human unbilical veins have been used to investigate the adhesive properties of various tumor cell lines *in vitro*. Although differences in metastatic and nonmetastatic tumor cell adhesion to large vessel endothelial cells were usually not observed, the use of microvessel endothelial cells resulted in organ preferences in adhesion in many but not all metastatic systems (Nicolson, 1989; Belloni and Nicolson, 1992; Auerbach, 1992). Metastatic cells bound to endothelial cells and stimulated endothelial cell retraction at cell junctions adjacent to the adherent malignant cells. Eventually the malignant cells penetrated the endothelial cell monolayer, bound to the subendothelial matrix and destroyed this structure (Kramer and Nicolson, 1979; Menter et al., 1987; Lapis et al., 1988) (Fig. 1). This behavior was uncharacteristic of most of the noninvasive, non-neoplastic cells examined (Kramer and Nicolson, 1979).

Using microvessel endothelial cell monolayers to measure the adhesive interactions of metastatic and nonmetastatic tumor cells, the kinetics of tumor cell adhesion to the appropriate endothelial cell type usually correlated with their metastatic properties (Belloni and Nicolson, 1992; Pauli et al., 1992). For example, in analyzing the adhesive characteristics of 11 cloned cell lines derived from a rat rhandomyosarcoma, four of five clones displaying low lung-colonizing capacity showed low rates of attachment to endothelial cell monolayers, whereas all of the high lung-colonizing lines had high rates of attachment (Korach et al., 1986). In general, the subendothelial matrix is a better adhesive substrate for most tumor cells than the endothelial cell surfaces (Kramer et al., 1980; Nicolson et al., 1981). Moreover, in some metastatic systems differential adhesion of tumor cells to specific organ subendothelial matrix has been seen (Lichtner et al., 1989). In an interesting study, Pauli et al. (1988) found that by growing bovine aortic endothelial cells on biomatrix derived from different organs, the endothelial cells acquired organ-related adhesive properties. They found that lung metastatic tumor cells adhered better on the lung biomatrix-grown aortic endothelial cells than endothial cells grown on other organ biomatrix, whereas liver metastatic malignant cells bound at higher rates to liver biomatrix-grown endothelial cells, suggesting that the subendothelial matrix plays a role in determining the types of cell surface molecules expressed on endothelial cells.

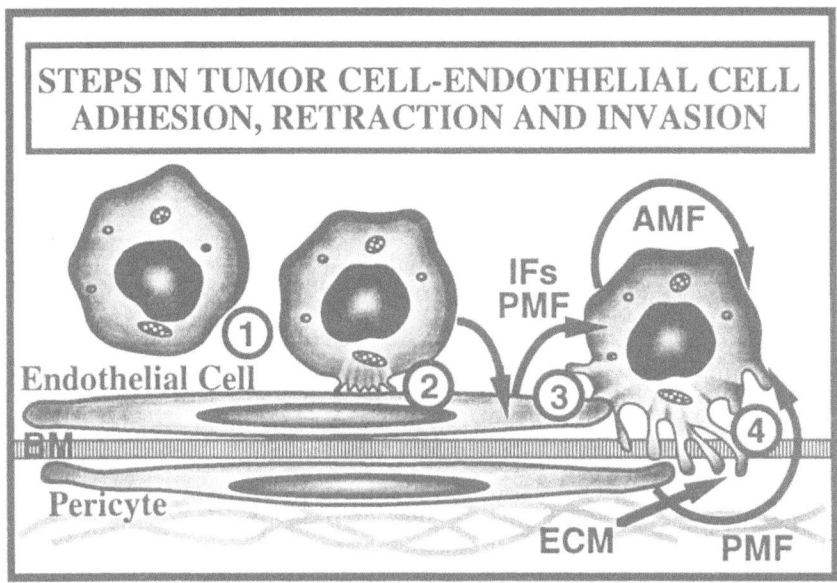

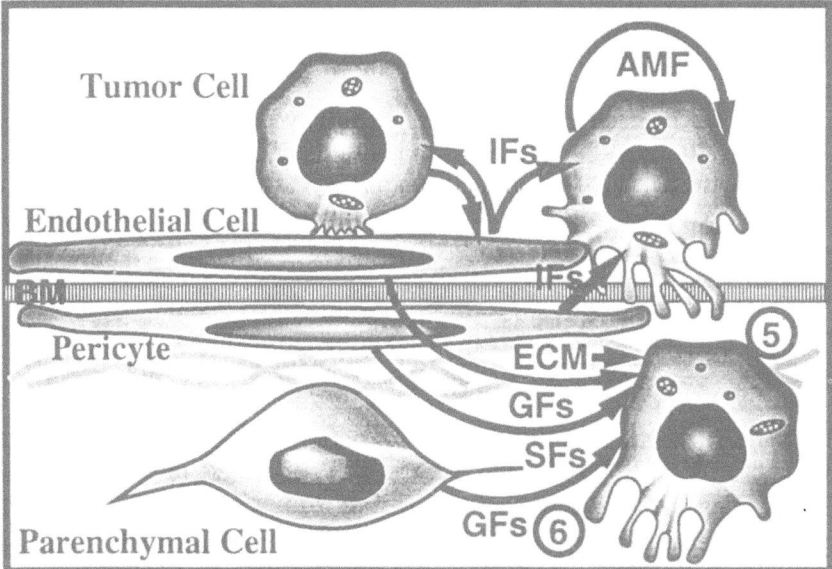

Figure 1. Sequence of events during blood-borne metastatic cell attachment and invasion of the vascular endothelium and underlying basement membrane. (1) tumor cell approaches endothelial cell surface in a microvessel; (2) tumor cell attachment to endothelial cell surface; (3) endothelial cell retraction and tumor cell migration to basement membrane stimulated by paracrine motility factors (PMFs) and autocrine motility factors (AMFs); (4) tumor cell attachment to and invasion of the basement membrane stimulated by invasion factors (IFs); (5) tumor cell invasion into parenchymal tissue; and (6) tumor cell response to survival factors (SFs) and paracrine growth factors (GFs).

Table 1. Examples of microvessel endothelial cell systems for the study of the organ preference of metastatic cell adhesion

Tumor system	Organ preference of adhesion	Reference
Brain-colonizing melanoma	Brain > Lung > Other	Nicolson, 1982b, 1988a
Liver-colonizing lymphoma	Liver > Lung > Other	Nicolson et al., 1989
Liver-colonizing hepatoma	Liver > Ovary	Auerbach et al., 1987
Ovary-colonizing teratoma	Ovary > Liver	Auerbach et al., 1987
Liver-colonizing large-cell lymphoma	Liver > Lung > Other	Nicolson et al., 1989
Lung-colonizing rhabdo-myosarcoma	Lung > Other	Korach et al., 1986
Lung-colonizing mammary adenocarcinoma	Lung > Liver > Other	Lichtner et al., 1989

The adhesion rates of various tumor cells to specific subendothelial matrix components, such as laminin, fibronectin and type IV collagen, differ (Liotta et al., 1983; Netland and Zetter, 1986; Chung et al., 1988; McCarthy et al., 1985). Although in many metastatic systems the rates of adhesion of tumor cells to specific extracellular matrix components correlated with their metastatic behaviors, it is difficult to generalize on the differences in adhesion properties of metastatic tumor cells to extracellular matrix components. Adhesion to the subendothelial matrix is probably important in stabilizing tumor cell interactions with the endothelium, and it initiates the events that lead to invasion of the basement membrane.

Organ-derived microvascular endothelial cell cultures have been developed to examine the organ adhesion preferences of metastatic cells. Many of these tumor systems show organ preferences in adhesion that correlate with their metastatic properties (Tab. 1). For example, brain-colonizing mouse melanoma cells adhered to brain-derived endothelial cells at faster rates than to endothelial cells of unrelated origin (Nicolson, 1982b; 1988a). Similar data have been obtained for brain-colonizing human melanoma sublines (Fig. 2). In this case the 70W subline of human MeWo melanoma line is capable of colonizing the brains of nude mice (Ishikawa et al., 1988), and we found that the 70W cells adhered at higher rates to brain-derived microvessel endothelial cells than did other MeWo sublines that fail to colonize brain (Fig. 2) (Nicolson et al., 1995). Auerbach et al. (1987) examined the adhesive interactions of metastatic cells with microvascular endothelial cells and correlated the rates of adhesion with their tissue origin or metastatic organ preference. Hepatoma and glioma cell lines showed preferential adhesion to the endothelial cells derived from the organ of tumor origin, whereas metastatic teratoma cells preferentially attached to endothelial cells derived from ovaries, the target organ for metastasis of these cells. Adhesion preference was not observed when large vessel endothelial cell

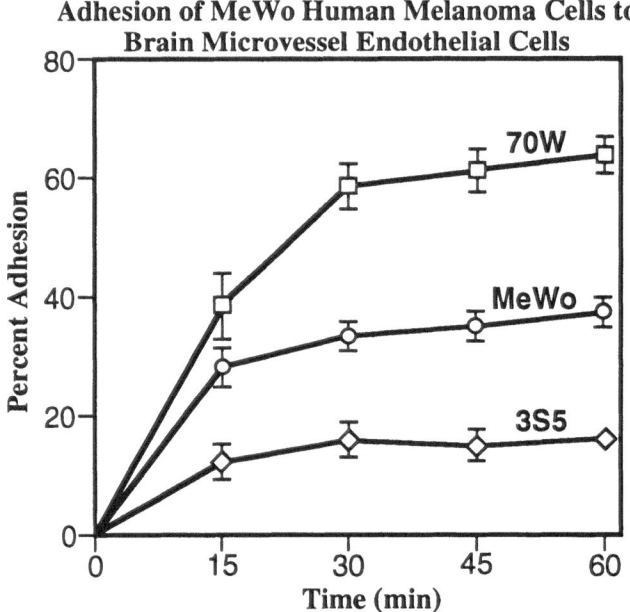

Figure 2. Adhesion of human melanoma cells to microvessel endothelial cell monolayers. MeWo melanoma variant lines were derived as described (Ishikawa et al., 1988). [^{35}S]labeled tumor cells (1×10^5/well) were incubated with brain endothelial cell monolayers, adherent cells harvested at various times, and the number of adherent cells determined by scintillation counting. 70W, brain-colonizing highly metastatic MeWo subline; MeWo, parental line; 3S5, poorly metastatic MeWo subline. The highly brain-metastatic 70W subline adheres at faster rates to the target organ brain microvessel endothelial cells.

monolayers were substituted for microvessel endothelial cells. Roos et al. (1985) have described high affinity binding between a murine lymphosarcoma line that colonizes mouse liver and freshly isolated rat liver endothelial cells or hepatocytes. Adhesion was inhibited by antibodies specific for tumor cell plasma membrane components. Using a murine large-cell lymphoma system it was found that the liver-colonizing sublines adhered at higher rates to murine liver sinusoidal endothelial cells, whereas only the lung-colonizing subline bound at high rates to lung microvessel endothelial cell monolayers (Nicolson et al., 1989). Thus different metastatic systems show variations in endothelial and subendothelial matrix adhesive properties, but in general, their adhesive properties can be related to their metastatic properties. In these and other studies, tissue-specific adhesion molecules were often expressed by organ endothelial and parenchymal cells, whereas other adhesion molecules may be cell type-specfic. The organ-expressed adhesion molecules used by metastatic cells did not appear to be the same for each malignant cell system (Roos et al., 1984). Tumor cell subpopulations may use additional adhesion mechanisms to attach to specific organ endothelial cells;

it is likely that several different adhesion systems are used by metastatic cells to adhere to organ endothelial cells.

Some of the organ-associated endothelial cell surface determinants involved in adhesion of metastatic tumor cells have been identified using affinity procedures. For example, to identify organ-associated recognition molecules expressed on endothelial cell surfaces, ^{125}I-labeled lysates of endothelial cells were adsorbed onto lightly glutaraldehyde-fixed tumor cells. Once bound, the components were solubilized in sodium dodecyl sulfate (SDS) detergent and analyzed by SDS-polyacrylamide gel electrophoresis–autoradiography. In the few examples where this approach has been used, the majority of the adsorbed radiolabeled proteins migrated in the molecular weight region known to contain integrins ($M_r \sim 120\,000$ to $180\,000$). But there were also many low M_r proteins that bound to the fixed tumor cells. Moreover, the individual protein profiles were unique for each endothelial cell type (Belloni and Tressler, 1990; Belloni and Nicolson, 1992). Zhu et al. (1992) have identified a Ca^{2+}-dependent lung endothelial cell adhesion molecule (LuECAM) that selectively promotes adhesion of lung-colonizing murine melanoma cells in a process that can be blocked by antibodies or soluble adhesion molecules. These results suggest that tumor cell adhesion may be directed to specific tissue-associated adhesion molecules expressed by the microvascular endothelial cells in different organs (Pauli and Lee, 1988).

Endothelial cell surface components involved in metastatic tumor cell adhesion have also been identified by the technique of tumor cell–Western adhesion. Using this technique, the cell-binding components were identified by adding viable biotinylated or radiolabeled tumor cells to Western transfers containing blots of detergent-solubilized endothelial cell surface membrane proteins separated by SDS-polyacrylamide gel electrophoresis. Using biotinylated murine large-cell lymphoma tumor cells, five endothelial cell-derived glycoproteins ($M_r \sim 48\,000$, $\sim 32\,000$, $\sim 30\,000$, $\sim 25\,000$, and $\sim 18\,000$) were identified with the Western transfers of liver or lung microvessel endothelial cell surface proteins (Fig. 3). More of the highly metastatic tumor cells bound to HSE Western transfers, and they also bound quantitatively on the basis of their organ preference of metastasis, suggesting that differential adhesion to organ-derived endothelial cells may be due to quantitative not qualitative differences in the expression of endothelial cell surface components. The five endothelial cell surface components also bound lymphokine-activated leukocytes, suggesting that their normal function is to bind normal hemopoietic cells. These five endothelial tumor cell-binding proteins have been called Endothelial Common Adhesion Molecules (ECAMs). Although they appear to be involved in tumor cell attachment to microvessel endothelial cells, it is unlikely that these molecules alone can confer endothelial cell adhesive specificities

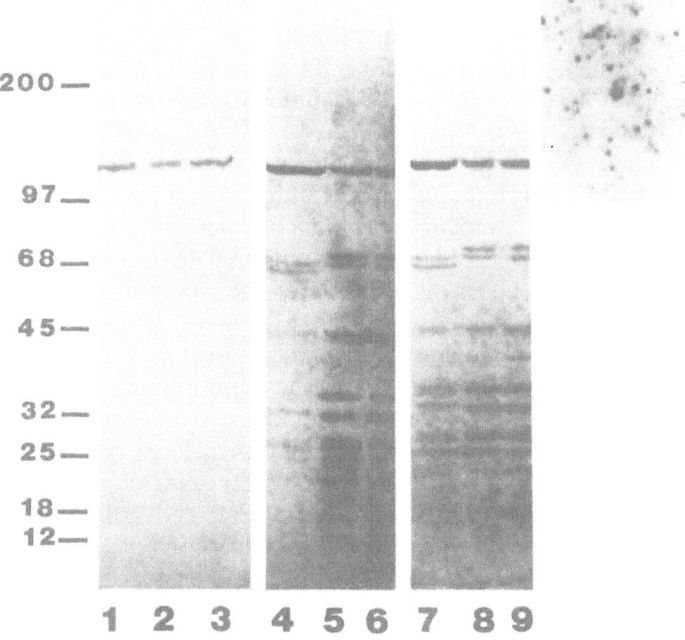

Figure 3. Identification of RAW117 tumor cell-binding glycoproteins on microvessel endothelial cells by tumor cell–Western labeling. Glycoproteins expressed by organ-derived endothelial cells, bovine aortic (lanes 1, 4 and 7); mouse hepatic sinusoidal (lanes 2, 5 and 8); mouse
lung microvessel (lanes 3, 6 and 9) endothelial cells were detergent solubilized, separated on
5–15% polyacrylamide gradient gels in SDS, transferred to nitrocellulose, and used for
Western cell adhesion by incubation of the blot with biotin-labeled RAW117 cells. Adhesion
of biotin-labeled RAW117 cells highly metastatic to liver (RAW117-H10), lung and other sites
(RAW117-L17) or poorly metastatic parental cells (RAW117-P) to endothelial cell glycoproteins was detected with strepavidin–peroxidase. RAW117-P (lanes 1–3); RAW117-H10
(lanes 4–6); RAW117-L17 (lanes 7–9). Inset shows high magnification of RAW117 cells
bound to glycoprotein band on blot.

to metastatic cells. Polyclonal antisera have been raised against three of
the individual endothelial cell-binding glycoproteins (gp32, gp25, and
gp18) and their normal tissue distributions have been determined using
immunohistochemistry. We found that gp32 and gp25 were expressed at
high levels on the apical surfaces of kidney microvessels, glomeruli and
liver sinusoids, and at lower levels on lung and brain microvessels. In
contrast, gp18 was localized in the subendothelial spaces of many blood
vessels, especially in the lung. Additionally, at high antibody dilutions
anti-gp18 selectively stained subendothelial components in lung,
whereas anti-gp25 specifically stained the endothelial cell surfaces of
hepatic sinusoids and bile ducts. Kidney glomeruli were selectively
stained by anti-gp32. At low antibody dilutions the three different
antibodies stained the microvessels in most tissues, indicating that these
endothelial cell adhesion-associated glycoprotiens are quantitatively not

qualitatively distributed in various organ endothelia. Adhesion of biotinylated metastatic large-cell lymphoma cells was only partially inhibited ($\sim 40\%$) by the three antibodies used together (there was no inhibition with any of the antibodies used alone), indicating that additional components (such as integrins and selectins) are probably involved in mediating the adhesion of tumor cells to microvascular endothelial cells. At least one of the endothelial cell surface components ($M_r \sim 32\,000 - 34\,000$) may be a galectin or galactose-specific lectin. Lotan et al. (1994) have described two galectins (galectins 1 and 3) that are expressed on liver sinusoidal endothelial cells. Antibodies against these galectins partially inhibited the adhesion of the highly metastatic large-cell lymphoma cells to liver sinusoidal endothelial cells, indicating their involvement in adhesion (Lotan et al., 1994). Another component of $M_r \sim 35\,000$ was isolated and sequenced and shown to be an annexin (annexin II) (Tressler et al., 1993). Antibodies against annexin II were localized at the cell surface of large-cell lymphoma cells (Yeatman et al., 1993), and in endothelial cell adhesion experiments anti-annexin II inhibited adhesion by approximately 40% (Tressler et al., 1993). Thus it appears that tumor cells use multiple adhesion systems (integrins, annexins, selectins) to adhere to microvessel endothelial cells.

It appears that many of the endothial cell surface adhesion components used by circulating tumor cells are similar to or the same as components used by normal leukocytes to adhere to endothelial cells (Yednock and Rosen, 1989). Butcher and Weissman (1984) have proposed that lymphocyte–endothelial cell adhesion components may vary with cell differentiation and antigenic stimulation, whereas the endothelial recognition/attachment components vary with anatomic site. Wood et al. (1988) described several endothelial–monocyte antigens expressed by immune-activated human endothelial cells that appear to be similar to the glycoproteins described above in the Western cell adhesion experiments. That multiple types of adhesion molecules are selectively expressed by various organ endothelia, including galectins, proteoglycans, integrins, annexins and other cell adhesion molecules (and their receptors) and are involved in site-specific tumor cell–endothelial cell adhesion appears to be quite likely. In addition, the types of molecules that are involved in leukocyte cell homing, such as intercellular adhesion molecule-1 (ICAM-1) (Dustin et al., 1986; Dustin and Springer, 1988), endothelial–leukocyte adhesion molecule-1 (ELAM-1) (Bevilacqua et al., 1987; 1989), CD44 (Stamenkovic, 1994), selectins (Zimmerman et al., 1992; Stamenkovic, 1994), integrins (Miller et al., 1986; Tressler et al., 1989), plus other molecules (Zhu et al., 1992; Zocchi et al., 1993) are, as mentioned above, collectively involved in tumor cell adhesion to endothelial cells.

Integrins are important cell adhesion and signalling molecules involved in tumor cell–endothelial cell and tumor cell–subendothelial matrix adhesion. Many of the integrin superfamily of adhesion receptors that are expressed on various cells recognize the tripeptide Arg-Gly-Asp (RGD) sequence as the primary site for cation-dependent binding, for example, to fibronectin, vitronectin, laminin, vWF, and ICAM-1 (Hynes, 1987). Integrin expression often changes during malignant progression (Juliano and Vasner, 1993). For example, there was a good correlation between increased expression of $\beta3$ integrins and malignancy of malignant melanomas (Albeda et al., 1990). Transfection of αv integrin subunit cDNA into melanoma cells that have been selected for low expression of $\alpha v \beta3$ restored expression of this integrin as well as tumorigenicity of the transfected melanoma cells (Felding-Haberman et al., 1992).

The involvement of integrins in metastasis has been shown primarily in experiments where *in vitro* invasion or *in vivo* dissemination of tumor cells could be inhibited by integrin-binding fibronectin, laminin, RGD-containing peptides, or proteins (McCarthy et al., 1985; Humphries et al., 1988). For example, using a membrane invasion culture assay, Welch et al. (1989) found that RGD-containing peptides inhibited the invasion of human melanoma cells. RGD-containing polymers have also been used to inhibit tumor cell–endothelial cell adhesion. Although the tetrapeptide RGDS at concentrations as high as 1 mg/ml, as well as antibodies directed against various known lymphoid cell integrins (LFA-1, Mac-1, gp150/95 or integrin $\beta1$ subunit), were unsuccessful in inhibiting large-cell lymphoma cell–endothelial cell adhesion, RGD polymers containing specific pentameric combinations of GRGDS, such as (GRGDS)$_4$ or (GRGES)$_4$ at low concentrations ($< 10 \, \mu g/ml$), were very effective in inhibiting the adhesion of highly metastatic cells but not poorly metastatic cells to murine liver sinusoidal cell monolayers (Tressler et al., 1989). Pierschbacher and Ruoslahti (1987) used cyclization of peptides to restrict their secondary structure and found that attachment to vitronectin could be inhibited using ten-fold lower concentrations than necessary when using the comparable uncyclized peptides. Danilov and Juliano (1989) reported that the valence and distribution of RGD sequences on a protein play an important role in determining the efficiency of integrin-mediated cell–substratum adhesion.

Peptide sequences other than those containing RGD appear to be important in cell–matrix adhesion. The YIGSR sequence derived from the $\beta1$ chain of laminin was identified as a cell-binding sequence, and it has been shown (Iwamoto et al., 1987) that peptides containing the YIGSR sequence can partially inhibit murine melanoma invasion *in vitro* and lung colony formation *in vivo*. In another metastatic system based on murine large-cell lymphoma sublines, tumor cells adhered to immobilized laminin, but YIGSR or its homopolymer (YIGSR)$_4$ had no effect on endothelial cell adhesive properties (Tressler et al., 1989).

Secretion of tumor cell motility factors by endothelial cells

An important property of organ-preferring metastatic cells is their ability to selectively invade secondary site tissues that are targets for metastatic colonization (Nicolson, 1988c; Nicolson et al., 1985, 1989). Tissue invasion requires both tumor cell adhesion (discussed above) and cell motility. Tumor cell motility was previously known to be an essential element of metastatic cell invasion (Varani, 1982; Sträuli and Haemmerli, 1984; Nicolson et al., 1989; Liotta et al., 1991). The ability of tumor cells to respond to organ-specific or organ-associated chemoattractants may be an important determinant in the selective migration of tumor cells into specific tissues.

Organ-derived chemotactic factors for tumor cells have been identified using conditioned medium from specific tissues, and in some cases the molecules involved have been partially purified. For example, Orr et al. (1980) identified a bone-derived chemotactic factor for cultured Walker carcinosarcoma cells, and Hujanen and Terranova (1985) found that tumor cells that metastasize to brain or liver migrate toward a concentration gradient of soluble components extracted from brain or liver, respectively. Hujanen and Terranova (1985) found that the chemotactic activity of their liver extract for liver-colonizing reticulum cell sarcoma cells was partially purified as a mixture of components of $M_r \sim 100\,000$ to $\sim 150\,000$ by gel filtration. Using a murine colon carcinoma metastatic system Bresalier et al. (1987) observed that liver-metastasizing cells preferentially migrated toward a concentration gradient of liver extract rather than to gradients made from extracts of lung or brain.

In addition to soluble factors, chemotactic factors present in extracellular matrix can stimulate the directed migration of tumor cells. Cerra and Nathanson (1989) noted that lung-colonizing melanoma cells bound to and migrated on lung extracellular matrix at higher rates than did poorly lung-colonizing or liver-colonizing cells. In contrast, liver extracellular matrix stimulated the migration of liver-colonizing melanoma and reticulum cell sarcoma cells at higher rates than lung-colonizing tumor cells (Cerra and Nathanson, 1991). The chemotactic activity present in mouse liver extracellular matrix reported by Cerra and Nathanson (1991) was eluted in four fractions of $M_r > 250\,000$, $\sim 245\,000$, $\sim 120\,000$ and $\sim 30\,000$ from gel fitration columns. The resorbing bone-derived chemotactic factor for bone-colonizing Walker carcinosarcoma cells was reported by Orr et al. (1980) to be of $M_r \sim 6000$. These reports suggest that organ-derived chemotactic factors present in secondary sites may play an important role in organ-selective tumor cell invasion after tumor cells adhere to endothelial cells and subendothelial matrix.

A likely source of organ-associated tumor cell motility factors is the organ microvessel endothelial cell. For example, we have examined the tumor cell motility factors secreted by target organ microvessel endothelial cells. Hepatic sinusoidal endothelial cells secreted tumor cell motility factors for both liver-metastatic lymphoid and epithelial tumor cells, and lung microvessel endothelial cells secreted motility factors for lung-metastatic cells (Hamada et al., 1993; Wakabayashi et al., 1994). The murine liver endothelial cell-secreted lymphoid motility-stimulating factor was purified by a five step procedure. The liver motility factor was a glycoprotein of $M_r > 200\,000$ that contained subunits of M_r $\sim 110\,000$ and $\sim 67\,000$. Sequence analysis of these subunits revealed that the motility factor was complement component C3b (Hamada et al., 1993). A quite different lymphoid tumor cell motility factor was found to be produced by murine lung microvessel endothelial cells that differentially stimulated the motility of lung metastatic lymphoid tumor cells (Wakabayashi et al., 1994). This motility factor has now been purified and sequenced, and shown to be monocyte chemotactic factor-1 (MCP-1), found previously to be made by endothelial cells (Sica et al., 1990). Thus, endothelial cells derived from lung and liver secrete quite different tumor cell paracrine motility factors that may contribute to organ preference of invasion of these sites by metastatic cells.

In some cases, organ endothelial cell-produced motility factors may act with other factors to promote organ-specific invasion. For example, brain endothelial cells produce potent motility factors for brain-metastatic cells (Hamada et al., 1994) that act in concert with brain trophic factors. Using brain-metastatic human melanoma cells we found that neurotrophins, such as nerve growth factor, stimulated invasion in the presence of brain endothelial cell motility factor (Herrmann et al., 1993). The neurotrophin appeared to upregulate the expression of basement degradative enzymes, whereas the brain endothelial cell motility factor increased the chemotactic migration of the brain-metastatic human melanoma cells (Herrmann et al., 1993; Marchetti et al., 1993; Nicolson et al., 1994).

In addition to paracrine motility factors, endothelial cells are known to produce a variety of known peptide growth factors and cytokines that possess tumor cell motility-stimulating activities (Mawatari et al., 1989; Børsum, 1991). Some of these factors, such as platelet-derived growth factor, fibroblast growth factors, granulocyte colony-stimulating factor, granulocyte–macrophage colony-stimulating factor, interleukin-6, monocyte chemoattractant protein-1 and interleukin-8, can stimulate the motility of various kinds of normal and malignant cells (Seppa et al., 1982; Bussolino et al., 1989; Matsushima et al., 1989; Wang et al., 1990; Stoker and Gherardi, 1991).

Cultured endothelial cells are also known to produce specific motility-stimulating extracellular matrix components, such as fibronectin, laminin, type IV collagen and thrombospondin (Kramer et al., 1985; McCarthy et al., 1985), that can stimulate tumor cell chemotaxis (Aznavoorian et al., 1990; Teraboletti et al., 1987). Collectively the release of motility-stimulating factors from endothelial cells and subendothelial matrix may cause differences in invasion of particular organ sites by metastatic cells.

Tumor cells can also produce their own autocrine motility factors. One of the important motility factors made by metastatic cells is autocrine motility factor (AMF). AMF is secreted by a wide variety of tumor cells, and AMF stimulates autocrine chemokinetic cell motility of melanoma, breast cancer and other cells (Liotta et al., 1986; Atnip et al., 1987; Silletti et al., 1991; Watanabe et al., 1991). Stracke et al. (1992) have described another autocrine motility factor, autotaxin, that is also made by malignant melanoma cells. A paracrine motility factor that appears to be important in malignant cell invasion is scatter factor (SF) or hepatocyte growth factor that induces the *in vitro* movement of epithelial cells and is produced by embryo fibroblasts and certain fibroblast cell lines (Gherardi et al., 1989). Selective tumor cell movement into various organs and tissues may thus be stimulated by combinations of autocrine and paracrine motility factors and extracellular matrix components and protein fragments. For example, we found that lung-metastatic large-

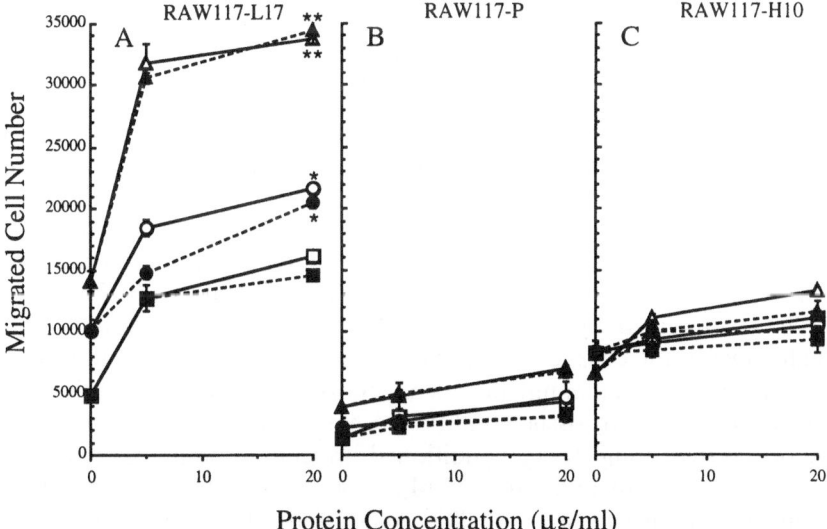

Figure 4. Migration of RAW117 tumor cell lines stimulated by mixtures of conditioned medium from microvessel endothelial cells or tumor cells. Each point and bar indicates migrated cell number ($\pm$ SD) of (A) RAW117-L17 (lung-metastatic), (B) RAW117-P (parental poorly metastatic) or (C) RAW117-H10 (liver-metastatic) at indicated protein concentration of the conditioned medium from murine lung microvessel endothelial cells (open symbols, solid lines) or murine lung fibroblasts (closed symbols, dotted lines) mixed with 0 μg (squares), 5 μg (circles), or 20 μg (triangles) of conditioned medium from (A) RAW117-L17 cells, (B) RAW117-P cells or (C) RAW117-H10 cells. P values were calculated according to Student's t-test: *, $P < 0.01$ compared with controls; **, $P < 0.001$ compared with controls.

cell lymphoma cells produce an autocrine motility factor and they respond to a lung microvessel endothelial cell motility factor (MCP-1) (Fig. 4) (Wakabayashi et al., 1994; 1995).

Secretion of metastatic cell growth factors by endothelial cells

The growth of malignant cells at particular organ sites is dependent on their responses to various concentrations of systemic and paracrine growth factors and inhibitors in individual organs. Differentially expressed paracrine growth and inhibitory factors at each site are particularly important as well as the ability of malignant cells to synthesize autocrine growth factors (Nicolson, 1993a,b,c). Using organ culture techniques various tissues have been examined for their abilities to produce paracrine growth factors for metastatic cells. For example, it was found (Horak et al., 1986; Naito et al., 1987) that lung-conditioned medium differentially stimulated the growth *in vitro* of lung-colonizing metastatic cells. Using poorly and highly metastatic mouse colon adenocarcinoma cell lines, Yamori et al. (1988) found that growth was stimulated by a lung extract; however, more metastatic cell lines were growth-stimulated at lower extract concentrations. Using lung- and ovary-colonizing melanoma sublines, we found that cell growth in serum-limited medium was differentially stimulated by nondialyzable factors from specific target organ tissues. For example, highly lung-colonizing melanoma cells were growth stimulated by lung-conditioned medium and highly ovary-colonizing melanoma cells were growth stimulated by ovary-conditioned medium more than organ-conditioned media generated from other tissues (Nicolson and Dulski, 1986). Using lung-metastasizing rat mammary adenocarcimona cell lines and organ-conditioned media we found that various metastatic cell lines were growth stimulated by both lung- and liver-conditioned medium. The highly metastatic cells, however, were growth stimulated at significantly higher rates than the nonmetastatic or poorly metastatic cells (Nicolson, 1988c). Using organ tissue-conditioned media Szaniawska et al. (1985) reported that lung tissue-conditioned medium contained both cell growth-stimulating molecules of $M_r \sim 50\,000-70\,000$ and growth-inhibitory molecules of $M_r \sim 12\,000-20\,000$ and 3000–5000. Collectively, these experiments established that each organ produces and secretes growth factors that differentially stimulate the proliferation of metastatic cells with organ preferences of metastasis.

One of the most potent metastatic cell growth-stimulating factors has now been purified to homogeneity from lung tissue-conditioned medium in our laboratory using a five-step purification procedure (Cavanaugh and Nicolson, 1989). This $M_r \sim 66\,000$ glycoprotein appears to be a transferrin, from its sequence and other data (Cavanaugh and Nicolson,

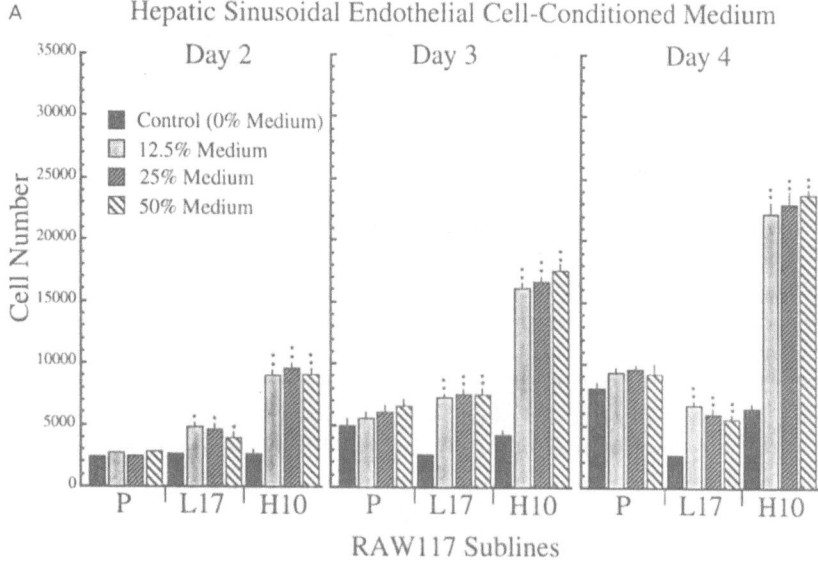

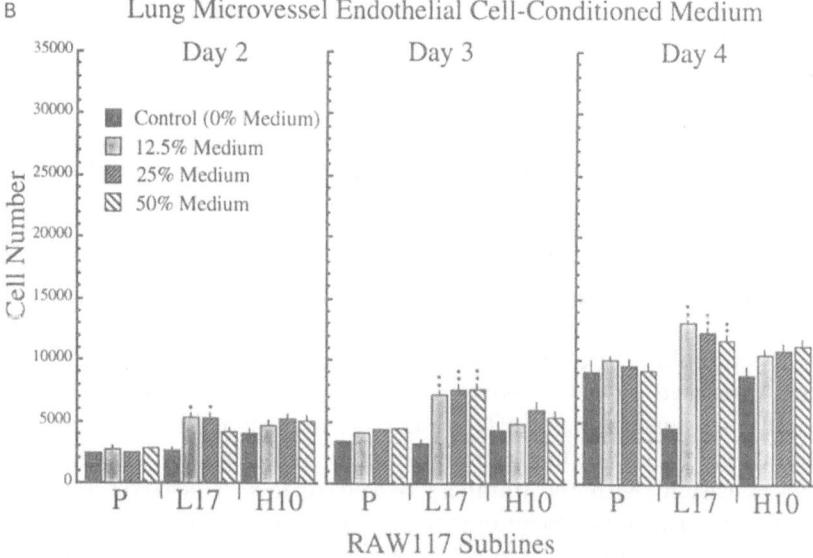

Figure 5. Growth of RAW117 tumor cell lines stimulated by conditioned medium from lung or liver microvessel endothelial cells. (A) Growth-stimulating activity of medium conditioned with hepatic sinusoidal endothelial cells for poorly metastatic parental RAW117-P, highly lung-metastatic RAW117-L17, and highly liver-metastatic RAW117-H10 at days 2, 3, and 4. Columns and error bars show means and standard deviations of triplicate samples, respectively. (B) Growth-stimulating activity of medium conditioned with lung microvessel endothelial cells for RAW117-P, -L17, and -H10 at days 2, 3, and 4. Cell growth assay was performed as described in Hamada et al. (1992). Columns and error bars show means and standard deviations of triplicate samples, respectively. P values were calculated according to Student's t-test: *, P < 0.01 compared with controls; **, P < 0.001 compared with controls.

1991). A different lung paracrine growth factor has been isolated and partially purified (Yamori et al., 1988). They found that murine lung tissue homogenates stimulated the growth of four highly metastatic cell clones of murine colon 26 carcinoma cells, but only one of four poorly metastatic cell clones were growth stimulated. The activity was trypsin- and heat-labile, and was partially purified by gel chromatography. The active component was eluted in the M_r range of $\sim 90\,000$ to $\sim 120\,000$.

Endothelial cells in various organs are likely to be the source of some of the organ paracrine growth factors that stimulate the growth of metastatic cells. To demonstrate this we measured the effects of endothelial-cell secreted growth factors and inhibitors on the proliferation of nonmetastatic and metastatic cells in cell culture in the presence of low concentrations of serum. Using organ-derived microvessel endothelial cells the effects of endothelial cell-derived growth factors on the proliferation of large-cell lymphoma cells was examined (Hamada et al., 1992). The patterns of stimulation/inhibition were almost exactly the same as found with organ-conditioned media. For example, liver-colonizing large-cell lymphoma cells were significantly more growth stimulated than the parental cells by conditioned medium from target hepatic sinusoidal endothelial cells, whereas lung-selected large-cell lymphoma cells that are capable of colonizing lung as well as liver were significantly more growth stimulated by lung microvessel endothelial cell-conditioned medium than by parental cells (Fig. 5). Using rat mammary adenocarcinoma sublines of different metastatic properties, similar results have been obtained with organ endothelial cells. The target organ endothelial cells produced the greatest growth stimulatory effects with the most metastatic cell lines (Nicolson, 1988c). Thus the organ growth properties of metastatic cells may be determined, in part, by the release of paracrine growth factors and inhibitors from organ microvessel endothelial cells.

Organ-derived endothelial cells differentially produce specific growth factors in culture. For example, transferrin and transferrin-like factors were produced by microvessel endothelial cells (Nicolson, 1993c). The relative amounts of the transferrin-like factors secreted from microvessel endothelial cells were in the order: brain > lung > liver. This result was also consistent with the numbers of cell surface transferrin receptors found on metastatic melanoma and mammary adenocarcinoma cells that were metastatic to these sites (brain-metastasizing > lung-metastasizing > liver-metastasizing) (Nicolson et al., 1990; 1992; Inoue et al., 1993). Thus the production of paracrine growth factors, such as the transferrin and transferrin-like factors by organ endothelial cells, may be important in determining the selective growth of metastatic cells at particular sites.

The physiologic roles of paracrine growth factors and inhibitors are unknown, but parenchymal cell- and endothelial cell-derived paracrine growth factors might be involved in normal local tissue regeneration and inflammation processes. When tissues are locally damaged due to invading tumor cells, paracrine growth factors may be released by parenchymal and endothelial cells to stimulate normal organ tissue repair.

Autocrine growth factors appear to be increasingly important as tumor cells progress to more malignant states. As they undergo tumor progression, highly malignant cells generally become less dependent on serum-derived growth factors for their growth *in vitro* (Chadwick and Lagarde, 1988; Rodeck et al., 1987). This is not, however, always found during tumor progression or metastasis. For example, the production and secretion of melanoma growth-stimulating activity by melanoma cells (Richmond et al., 1988) and bombesin-like (gastrin-releasing peptide) activity by small cell lung carcinoma cells (Cuttitta et al., 1985) were not related to stage or state of tumor progression. The general results obtained by a number of investigators indicate loss of growth factor requirements with tumor progression (Rodeck and Herlyn, 1991), and this is often accompanied by the secretion of autocrine motility factors.

The loss of certain growth factor and inhibitor responses in highly metastatic cancers and the ability of such cancers to colonize and grow at distinct secondary sites may be explainable by considering the growth properties of cancers at various stages of progression. When cancers progress to early metastatic stages, they often show restricted organ distributions of metastases or organ specificity of metastasis, whereas at the final stages of cancer progression near host death, cancers often colonize multiple organs and tissue sites. The most reasonable explanation for this is that cancer cells progress from mainly paracrine stimulatory and inhibitory growth mechanisms at the initial stages of metastatic progression to mainly autocrine stimulatory or even loss of all growth regulatory mechanisms at the final, terminal stages of progression (Nicolson, 1993a,b).

Secretion of metastatic cell growth inhibitors by endothelial cells

Some of the tissue-conditioned media in the tumor cell growth experiments described above were growth inhibitory, and it was natural to hypothesize that this was due to the release of inhibitory paracrine cytokines. For example, liver tissue factors and, at high concentrations, kidney tissue factors were inhibitory to all of the melanoma sublines examined by Nicolson and Dulski (1986), correlating with the low potential of these melanoma lines to colonize liver and kidney. When high liver-colonizing large-cell lymphoma cells were tested for their growth in serum-limited, liver-conditioned medium, they were significantly more growth stimulated than the poorly metastatic parental cells

(Nicolson, 1987). Lung-selected large-cell lymphoma cells that are capable of colonizing lung as well as liver were significantly more growth stimulated by lung- and liver-conditioned media than parental cells. Kidney-conditioned medium was only slightly stimulatory or inhibitory to the lung-selected cells, correlating with the lack of ability of these tumor cells to colonize.

Only a few organ-derived paracrine growth inhibitors for metastatic cells have been purified and identified. Using BSC-1 kidney cells, Holley et al. (1980) isolated a growth inhibitor of $M_r \sim 25\,000$ that was released into kidney cell-conditioned medium. This growth inhibitor was later identified as TGF-β (Tucker et al., 1984). A hepatic growth factor of $M_r \sim 26\,000$ has been purified from rat liver extracts, and this inhibitor reversibly blocked the proliferation of untransformed rat liver cells but not hepatoma cells (McMahon et al., 1982).

In addition to endothelial cell release of factors that stimulate growth of tumor cells, endothelial cells can inhibit the growth or possibly even kill metastatic cells. Endothelial cells can produce some of the cytokines described above; for example TGFβ and IL-6, potent growth inhibitors of some tumor cells (Sironi et al., 1989). Endothelial cells can also produce activated oxygen, such as oxygen free radicals as O_2^- which are rapidly converted to H_2O_2, $HO^{\cdot}$ and possibly singlet oxygen (Halliwell, 1989). To demonstrate that oxygen radicals might be involved in inhibiting tumor cells in the microvascular system, Glaves (1986) injected mice with superoxide dismutase to scavenge superoxide anions prior to intravenous injection of tumor cells. She found that the enzyme treatments resulted in increased tumor cell survival and experimental metastasis formation. Although inhibition of the activated oxygen released by activated leukocytes and NK cells could have accounted for the enhanced survival, Glaves (1986) suggested that endothelial cells were the main source of the activated oxygen radicals.

A different mechanism to explain the loss of tumor cells in the microcirculation has been advanced (Li et al., 1991). They found that endothelial cells could be activated by tumor necrosis factor-α and interferon-γ to become cytolytic to metastatic tumor cells. Although we have found that endothelial cells are cytostatic to tumor cells, we could not find evidence for cytolytic activities, even after cytokine activation. In the experiments of Li et al. (1991) the increase in tumor cell-endothelial cell adhesion induced by cytokine treatment (Bevilacqua et al., 1985b) was not taken into account, probably resulting in artifactually high values calculated for tumor cell cytolysis. We found that metastatic tumor cells became adherent to the cytokine-treated endothelial cell monolayers during the assay, and the endothelial cell-adherent tumor cells remained viable but incapable of proliferation. The endothelial cell-mediated tumor cll cytostasis was dependent on cell-contact and not affected by cytokine treatment.

Angiogenesis of metastatic tumors

An important tumor cell–endothelial cell interaction is angiogenesis that results in the neovascularization of tumors. This important topic will not be discussed in detail here; the reader is referred to several excellent reviews (Folkman, 1985; Folkman and Klagsbrun, 1987; D'Amore, 1988). Extravascular metastatic cells are generally growth-arrested after they have grown to form a metastasis of approximately 0.5 mm diameter, unless they receive a new blood supply (Folkman, 1985). Malignant tumors stimulate the sprouting, movement and proliferation of endothelial cells that are required for angiogenesis, and it appears that multiple molecules are involved in this process. Some of these molecules have been purified, sequenced and identified as known growth factors (for example, basic fibroblast growth factor [bFGF], vascular endothelial cell growth factor [VEGF], tumor growth factor-α, and angiogenin) (Folkman and Shing, 1992). Two of these angiogenic molecules, bFGF and VEGF, are produced by many types of malignant cells. The ability of tumor cells to stimulate angiogenesis has been correlated with metastasis. For example, the induction of blood vessels by invasive primary breast carcinomas may be an independent predictor of metastatic disease (Weidner et al., 1991).

Tumor cells do not have to be the sole source of angiogenic molecules. Tumor cells may recruit mast cells, macrophages and possibly other cells that can be stimulated to release angiogenic factors (Polverini and Leibovich, 1984). It seems clear that angiogenesis is an important step in the progression and growth of malignant tumors. Just as endothelial cells can secrete growth factors that act on tumor cells to stimulate motility and proliferation, tumor cells and tumor-associated macrophage and mast cells can secrete a variety of angiogenic factors that can act on endothelial cells in a reciprocal manner (Fig. 6) (Nicolson, 1993a).

Conclusions

Endothelial cells are importantly involved in tumor cell interactions during several steps of the metastatic process. They provide tumor cells with adhesive molecules to enhance arrest in the microcirculation, secrete motility factors to initiate cell extravasation, invasion factors to increase invasion of the subendothelial matrix, and they produce survival and growth factors to initiate and maintain growth of micrometastases. Together with parenchymal cells, fibroblasts, mast cells and other cell types, endothelial cells are essential to establishment of the appropriate microenvironmental conditions necessary for metastatic colonization of distant organ sites.

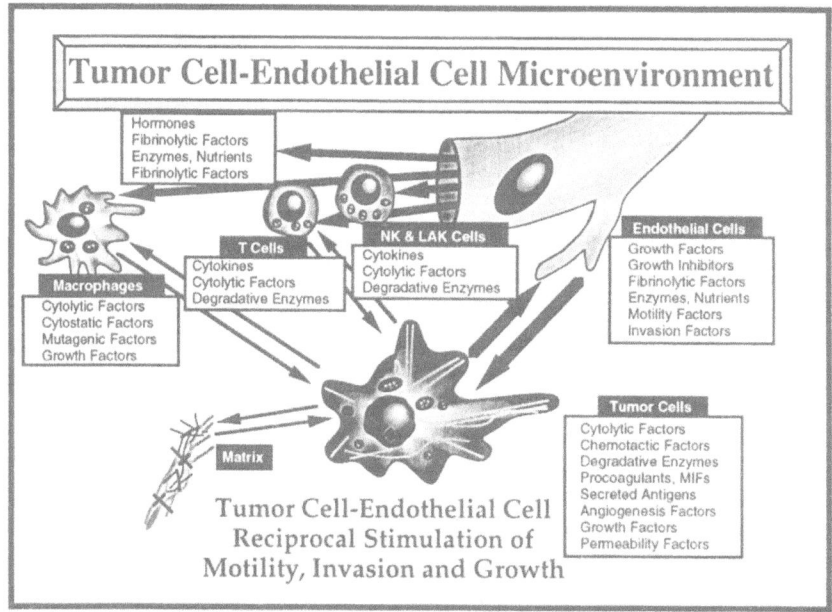

Figure 6. Bidirectional or reciprocal interactions between malignant cells and microvessel endothelial cells in their microenvironment. Tumor cells release cytokines and other factors that can affect microvessel endothelial cells and host tissue extracellular matrix. In turn, the endothelial cells release factors that stimulate or inhibit tumor cell proliferation.

Acknowledgments
The author's studies were supported by grants from the National Institutes of Health, National Cancer Institute (R35-CA44352 and RO1-CA63045). National Foundation for Cancer Research and the Nellie Connally Breast Cancer Research Fund.

References

Albeda, S.M., Sampson, P.M., Hastelton, F.R., McNiff, J.M., Mueller, S.N., Williams, S.K., Fishman, A.P. and Levine, E.M. (1988) Permeability characteristics of cultured endothelial cell monolayers. *J. Appl. Physiol.* 64: 308–322.

Albeda, S.M., Mette, S.A., Elder, D.E., Stewart, R.M., Damjanovic, L., Herlyn, M. and Buck, C.A. (1990) Integrin distribution in malignant melanoma: Association of the β_3 subunit with tumor progression. *Cancer Res.* 50: 6757–6764.

Atnip, K.D., Carter, L.M., Nicolson, G.L. and Dabbous, M.K. (1987) Chemotactic response of rat mammary adenocarcinoma cell clones to tumor-derived cytokines. *Biochem. Biophys. Res. Commun.* 146: 996–1002.

Auerbach, R., Alby, L., Morissey, L., Tu, M. and Joseph, J. (1985) Expression of organ-specific antigens on capillary endothelial cells. *Microvasc. Res.* 29: 401–406.

Auerbach, R., Lu, W., Pardon, E., Gumkowski, F., Kaminski, G. and Kaminski, M. (1987) Specificity of adhesion between murine tumor cells and capillary endothelium: An *in vitro* correlate of preferential metastasis *in vivo. Cancer Res.* 47: 1492–1496.

Auerbach, R. (1992) Endothelial cell heterogeneity: Its role as a determinant of selective metastasis. *In*: N. Simionescu and M. Simionescu (eds): *Endothelial Cell Dysfunctions.* Plenum Press, New York, pp 427–437.

Aznavoorian, S., Stracke, M.L., Krutzsch, H., Schiffmann, E. and Liotta, L.A. (1990) Signal transduction for chemotaxis and haptotaxis by matrix molecules in tumor cells. *J. Cell Biol.* 110: 1427–1438.

Belloni, P.N. and Nicolson, G.L. (1988) Differential expression of cell surface glycoproteins on various organ-derived microvascular endothelia and endothelial cell cultures. *J. Cell Physiol.* 136: 398–410.

Belloni, P.N. and Nicolson, G.L. (1992) The role of the vascular endothelium in cancer metastasis. *In*: N. Simionescu and M. Simionescu (eds): *Endothelial Cell Dysfunctions.* Plenum Press, New York, pp 395–425.

Belloni, P.N. and Tressler, R.J. (1990) Microvascular endothelial cell heterogeneity: interactions with leukocytes and tumor cells. *Cancer Metastatis Rev.* 8: 353–389.

Belloni, P.N., Carney, D.H. and Nicolson, G.L. (1992) Organ-derived endothelial cells exhibit differential rsponsiveness to thrombin and other growth factors. *Microvasc. Res.* 43: 20–45.

Bevilacqua, M.P., Pober, J.S., Wheeler, M.E., Cotran, R.S. and Gimbrone, M.A. (1985a) Interleukin-1 activation of vascular endothelium. Effects on procoagulant activity. *Am. J. Pathol.* 121: 393–403.

Bevilacqua, M.P., Pober, J.S., Wheeler, M.Z., Mendrick, D., Cotran, R.S. and Gimbrone, M.A. (1985b) Interleukin-1 acts on cultured human vascular endothelial cells to increase the adhesion of PMN, monocytes and related leukocyte cell lines. *J. Clin. Invest.* 76: 2003–2010.

Bevilacqua, M.P., Pober, J.S., Majeua, G.R., Fiers, W., Cotran, R.S. and Gimbrone, M.A. (1986) Recombinant TNF induce procagulant activity in cultured human endothelia: Characterization and comparison with the actions of IL-1. *Proc. Natl. Acad. Sci. USA* 83: 4533–4541.

Bevilacqua, M.P., Pober, J.S., Mendrick, D.L., Cotran, R.S. and Gimbrone, M.A. (1987) Identification of an inducible endothelial-leukocyte adhesion molecule. *Proc. Natl. Acad. Sci. USA* 84: 9238–9242.

Bevilacqua, M.P., Stengelin, S., Gimbrone, M.A. and Seed, B. (1989) Endothelial leukocyte adhesion molecule 1: Receptor for neutrophils related to complement regulatory proteins and lectins. *Science* 243: 1160–1165.

Børsum, T. (1991) Biochemical properties of vascular endothelial cells. *Virchows Archiv. B Cell Pathol.* 60: 279–286.

Bresalier, R.S., Hujanen, E.S., Raper, S.E., Roll, F.J., Itzkowtiz, S.H., Martin, G.R. and Kim, Y.S. (1987) An animal model for colon cancer metastasis: establishment and characterization of murine cell lines with enhanced liver-metastasizing ability. *Cancer Res.* 47: 1398–1406.

Bussolino, F., Wang, J.M., Defilippi, P., Turrini, F., Sanavio, F., Edgell, C-J.S., Aglietta, M., Arese, P. and Mantovani, A. (1989) Granulocyte- and granulocyte-macrophage colony-stimulating factors induce human endothelial cells to migrate and proliferate. *Nature* 337: 471–473.

Butcher, E.C. and Weissman, I. (1984) Lymphocytes, tissues and organs. *In*: W.E. Paul (ed.): *Fundamental Immunology*. Raven Press, New York, pp 109–127.

Carley, W.W., Milici, A.J. and Madri, J.A. (1988) Extracellular matrix specificity for the differentiation of capillary endothelial cells. *Exp. Cell Res.* 178: 426–434.

Cavanaugh, P.G. and Nicolson, G.L. (1989) Purification and some properties of lung-derived growth factor that differentially stimulates the growth of tumor cells metastatic to the lung. *Cancer Res.* 49: 3928–3933.

Cavanaugh, P.G. and Nicolson, G.L. (1991) Lung-derived growth factor for lung-metastasizing tumor cells: Identification as a transferrin. *J. Cell. Biochem.* 47: 261–267.

Cerra, R.F. and Nathanson, S.D. (1989) Organ-specific chemotactic factors present in lung extracellular matrix. *J. Surg. Res.* 46: 422–426.

Cerra, R.F. and Nathanson, S.D. (1991) Chemotactic activity present in liver extracellular matrix. *Clin. Exp. Metastasis* 9: 39–49.

Chadwick, D.E. and Lagarde, A.E. (1988) Coincidental acquisition of growth autonomy and metastatic potential during the malignant transformation of factor-dependent CCL 39 lung fibroblasts. *J. Natl. Cancer Inst.* 80: 318–325.

Chung, D.C., Zetter, B.R. and Brodt, N.P. (1988) Lewis lung carcinoma variants with different metastatic specificities adhere preferentially to differentially defined extracellular matrix molecules. *Invasion Metastasis* 8: 103–117.

Colucci, M., Balconi, R., Lorenzel, A., Pietra, D. and Semarara, N. (1983) Cultured endothelial cells generate tissue factor in response to endotoxin. *J. Clin. Invest.* 71: 1893–1901.

Criscuolo, G.R., Merrill, M.J. and Oldfield, E.H. (1988) Further characterization of malignant glioma-derived vascular permeability factor. *J. Neurosurg.* 69: 254–262.

Cuttitta, F., Carney, D.N., Mulshine, J., Moody, T.W., Fedorko, K., Fischler, A. and Minna, J.D. (1985) Bombesin-like peptides can function as autocrine growth factors in human small cell lung cancer. *Nature* 316: 823–825.

D'Amore, P.A. (1988) Antiangiogenesis as a strategy for antimetastasis. *Semin. Thromb. Hemostatasis* 14: 73–78.

Danilov, Y.N. and Juliano, R.L. (1989) (Arg-Gly-Asp)$_n$-Albumin conjugates as a model substratum for integrin-mediated cell adhesion. *Exp. Cell Res.* 182: 186–196.

deBono, D. (1977) Lymphocytes and the microcirculation. *Adv. Microcirc.* 7: 68–95.

Dejana, E., Lampugnani, M.G., Giorgio, M., Gaboli, M., DFederici, A.B., Ruggeri, Z.M. and Marchisio, P.C. (1989) Von Willebrand factor promotes endothelial cell adhesion via an Arg-Gly-Asp-dependent mechanism. *J. Cell Biol.* 109: 367–375.

Del Vecchio, P.J., Siflinger-Birnboim, A., Shepard, J.M., Bizios, R., Cooper, J.A. and Malik, A.B. (1987) Endothelial monolayer permeability to macromolecules. *Fed. Proc.* 46: 2511–2515.

Dujvestijn, A.M., Kerkhove, M., Bargatze, R.F. and Butcher, E.C. (1987) Lymphoid tissue and inflammation specific endothelial cell differentiation defined by monoclonal antibodies. *J. Immunol.* 138: 713–719.

Dustin, M.L., Rothlein, R., Bhan, A.K., Dinarello, C.A. and Springer, T.A. (1986) Induction by IL-1 and interferon-L tissue distribution, biochemistry, and function of a natural adherence molecule (ICAM-1). *J. Immunol.* 37: 245–254.

Dustin, M.L. and Springer, T.A. (1988) Lymphocyte function associated antigen-1 (LFA-1) interaction with intercellular adhesion molecule-1 (ICAM-1) is one of at least three mechanisms for lymphocyte adhesion to cultured endothelial cells. *J. Cell Biol.* 107: 321–331.

Dusting, G.J., Moncada, S. and Vane, J.R. (1978) Recirculation of prostacyclin (PGI$_2$) in the lung. *Br. J. Pharmacol.* 64: 315–319.

Ewing, J. (1928) *A treatise on tumors. Neoplastic Diseases*, Third Edition. Saunders, Philadelphia.

Felding-Haberman, B., Muelller, B.M., Romerdahl, C.A. and Cheresch, D.A. (1992) Involvement of integrin αV gene expression in human melanoma tumorigenicity. *J. Clin. Invest.* 89: 2018–2022.

Folkman, J. (1985) Toward an understanding of angiogenesis: Search and discovery. *Prospect. Biol. Med.* 29: 453–464.

Folkman, J. and Klagsbrun, M. (1987) Angiogenesis factors. *Science* 235: 442–447.

Folkman, J. and Shing, Y. (1992) Angiogenesis. *J. Biol. Chem.* 267: 10931–10934.

Ford, W.D., Sedgely, M. and Smith, M.E. (1976) The migration of lymphocytes across specialized endothelium. *Cell Tissue Kinet.* 9: 351–361.

Gallatin, W.M., Weissman, I.L. and Butcher, E.C. (1983) A cell surface molecule involved in organ-specific homing of lymphocytes. *Nature* 304: 30–34.

Gerritsen, M.E. (1987) Functional heterogeneity of vascular endothelial cells. *Biochem. Pharmacol.* 36: 2710–2721.

Ghandour, M.S., Langley, O.K., Gombos, G., Hirn, M., Hirsch, M.R. and Goridis, C.A. (1982) A surface marker for murine vascular endothelial cells defined by monoclonal antibody. *J. Histochem. Cytochem.* 30: 165–170.

Gherardi, E., Gray, J., Stoker, M., Perryman, M. and Furlong, R. (1989) Purification of scatter factor, a fibroblast-derived basic protein that modulates epithelial interactions and movement. *Proc. Natl. Acad. Sci. USA* 86: 5844–5848.

Glaves, D. (1986) Intravascular death of disseminated cancer cells mediated by superoxide anion. *Invasion Metastasis* 6: 101–111.

Goldsmith, J.C. and Kisker, C.T. (1982) Thrombin–endothelial cell interactions: Critical importance of vessel origin. *Thrombosis Res.* 25: 131–136.

Gommori, H. and Takeda, Y. (1976) Properties of human tissue thromboplastin from brain, lung and placenta. *Thrombos. Haemostas.* 36: 90–103.

Greene, H.S. and Harvey, E.K. (1964) The relationship between the dissemination of tumor cells and the distribution of metastases. *Cancer Res.* 24: 799–811.

Grega, G.J., Svesjo, E. and Haddy, F. (1981) Macromolecular permeability of the microvascular membrane: Physiological and pharmacological regulation. *Microcirculation* 1: 325–341.

Groothuis, D.R. and Vick, N.A. (1982) Brain tumors and the blood-brainbarrier. *Trends Neurosci.* 5: 232–235.

Halliwell, B. (1989) Auperoxide anion, vascular endothelium and reperfusion injury. *Free Radical Res. Commun.* 5: 315–318.

Hamada, J., Cavanaugh, P.G., Lotan, O. and Nicolson, G.L. (1992) Separable growth and migration factors for large-cell lymphoma cells secreted by microvascular endothelial cells derived from target organs for metastasis. *Br. J. Cancer* 66: 349–354.

Hamada, J.-I., Cavanaugh, P.G., Miki, K. and Nicolson, G.L. (1993) A metastatic tumor cell paracrine migration-stimulating factor secreted by mouse hepatic sinusoidal endothelial cells: Identification as complement component 3b. *Cancer Res.* 53: 4418–4423.

Hamada, J., Nicolson, G.L. Hosokawa, M. and Taaichi, N. (1994) Brain endothelial cell-derived motility factors and brain metastasis of mouse B16 melanoma cells. *Clin. Exp. Metastasis* 12(5): 53–54.

Hart, I.R. (1982) 'Seed and soil' revisited: mechanisms of site specific metastasis. *Cancer Metastasis Rev.* 1: 5–16.

Herrmann, J.L., Menter, D.G., Hamada, J.-I., Marchetti, D., Nakajima, M. and Nicolson, G.L. (1993) Mediation of NGF-stimulated extracellular matrix invasion by the human melanoma low-affinity p75 neurotrophin receptor: Melanoma p75 functions independent of trkA. *Mol. Biol. Cell.* 4: 1205–1216.

Holley, R.W., Bohlen, P., Fava, R., Baldwin, J.H., Kleeman, G. and Armour, R. (1980) Purification of kidney epithelial cell growth inhibitors. *Proc. Natl. Acad. Sci. USA* 77: 5989–5992.

Hong, S.L. (1980) Effect of bradykinin and thrombin on PGI synthesis in endothelial cells from calf and pig aorta and human umbilical cord vein. *Thrombosis Res.* 18: 787–795.

Horak, E., Darling, D.L. and Tarin, D. (1986) Analysis of organ-specific effects on metastatic tumor formation by studies *in vitro. J. Natl. Cancer Inst.* 75: 913–922.

Hormonia, M. (1982) Expression of factor VIII-related antigen and UEA lectin binding sites in endothelial cells during long-term culture. *Cell Biol. Int. Reports* 6: 1123–1134.

Hoyer, L.W. (1981) The factor VIII complex: Structure and function. *Blood* 58: 1–13.

Hujanen, E.S. and Terranova, V. P. (1985) Migration of tumor cells to organ-derived chemoattractants. *Cancer Res.* 45: 3517–3521.

Humphries, M.J., Yamada, K.M. and Olden, K. (1988) Investigation of the biological effects of anti-cell adhesive synthetic peptides that inhibit experimental metastasis of B16-F10 murine melanoma cells. *J. Clin. Invest.* 81: 782–790.

Hynes, R.O. (1987) Integrins: A family of cell surface receptors. *Cell* 48: 549–554.

Inoue, T., Cavanaugh, P.G., Steck, P., Brunner, N. and Nicolson, G.L. (1993) Differences in transferrin response and numbers of transferrin receptors on rat and human mammary carcinoma lines of different metastatic potentials. *J. Cell Physiol.* 156: 212–217.

Irving, M.G., Roll, F.J., Huang, S. and Montgomery-Bissel, D. (1984) Characterization and culture of sinusoidal endothelium from normal rat liver; lipoprotein uptake and collagen phenotype. *Gastroenterology* 82: 1233–1217.

Ishikawa, M., Dennis, J.W., Man, S. and Kerbel, R.S. (1988) Isolation and characterization of spontaneous wheat germ agglutinin-resistant human melanoma mutants displaying remarkably different metastatic profiles in nude mice. *Cancer Res.* 48: 665–670.

Iwamoto, F.A., Robey, J., Graf, M., Sasaki, S., Kleinman, H.K., Yamada, Y. and Martin, G.R. (1987) YIGSR, a synthetic laminin pentapeptide, inhibits experimental metastasis formation. *Science* 238: 1132–1134.

Jaffe, E. (1984) Synthesis of factor VIII by endothelial cells. *In*: E. Jaffe (ed.): *Biology of Endothelial Cells.* Martinus Nijhoff, MA, pp 209–214.

Juliano, R.L. and Varner, J.A. (1993) Adhesion molecules in cancer: The role of integrins. *Curr. Opin. Cell Biol.* 5: 819–826.

Kawaguchi, T., Kawaguchi, M., Miner, K.M., Lembo, T.M. and Nicolson, G.L. (1983) Brain meninges tumor formation by *in vivo*-selected metastatic B16 melanoma variants in mice. *Clin. Exp. Metastasis* 1: 247–259.

Kefalides, N.A. (1975) Basement membranes: Structural and biosynthetic considerations. *J. Invest. Dermatol.* 65: 85–92.

Kafalides, N.A. (1980) Chemistry of basement membranes: Structure and biosynthesis. *In*: B.M. Alttura, E. Davis and H. Harder (eds): *Vascular Endothelium and Basement Membranes*. Karger Press, Switzerland, pp 295–322.

Kennel, S.J., Lankford, T.K., Ullrich, R.L. and Jamashbi, R.J. (1988) Enhancement of lung tumor colony formation by treatment of mice with monoclonal antibodies to pulmonary capillary endothelial cells. *Cancer Res.* 48: 4964–4968.

Kiernan, M.W. and Longenecker, B.M. (1983) Organ specific metastasis with special reference to avian systems. *Cancer Metastasis Rev.* 2: 165–182.

Klatzo, I. (1967) Neuropathological aspects of brain edema. *J. Neuropathol. Exp. Neurol.* 26: 1–14.

Korach, S., Poupon, M.F., DuVillard, J.A. and Becker, M. (1986) Differential adhesiveness of rhabdomyosarcoma-derived cloned metastatic cell lines to vascular endothelial monolayers. *Cancer Res.* 46: 3624–3629.

Kramer, R.H. and Nicolson, G.L. (1979) Interactions of tumor cells with vascular endothelial cell monolayers: A model for metastatic invasion. *Proc. Natl. Acad. Sci. USA* 76: 5704–5708.

Kramer, R.H., Gonzalez, R. and Nicolson, G.L. (1980) Metastatic tumor cells adhere preferentially to the extracellular matrix underlying vascular endothelial cells. *Int. J. Cancer* 26: 639–645.

Kramer, R.H., Fuh, G.H., Bensch, K.G. and Karasek, M.A. (1985) Synthesis of extracellular matrix glycoproteins by cultured microvascular endothelial cells isolated from the dermis of neonatal and adult skin. *J. Cell. Physiol.* 123: 1–9.

Kwast, T.H., Stel, H.V., Cristen, E., Bertina, R.M. and Veerman, E.C. (1986) Localization of FVIII-procoagulant antigen. *Blood* 67: 222–227.

Lapis, K., Paku, S. and Liotta, L.A. (1988) Endothelialization of embolized tumor cells during metastasis formation. *Clin. Exp. Metastasis* 6: 73–89.

Li, L., Nicolson, G.L. and Fidler, I.J. (1991) Direct *in vitro* lysis of tumor cells by cytokine-activated murine vascular endothelial cells. *Cancer Res.* 51: 245–254.

Lichtner, R.B., Belloni, P.N. and Nicolson, G.L. (1989) Differential adhesion of metastatic rat mammary carcinoma cells to organ-derived microvessel endothelial cells and subendothelial matrix. *Exp. Cell Biol.* 57: 146–152.

Liotta, L.A., Rao, C.N. and Barsky, S.H. (1983) Tumor invasion and the extracellular matrix. *Lab. Invest.* 49: 636–649.

Liotta, L.A., Mandler, R., Murano, G., Katz. D.A., Gordon, R.K., Chiang, P.K. and Schiffmann, E. (1986) Tumor cell autocrine motility factor. *Proc. Natl. Acad. Sci. USA* 83: 3302–3306.

Liotta, L.A., Stracke, M.L., Aznavoorian, S.A., Beckner, M.E. and Schiffmann, E. (1991) Tumor cell motility. *Semin. Cancer Biol.* 2: 111–114.

Lotan, R., Belloni, P.N., Tressler, R.J., Lotan, D., Xu, X.-C. and Nicolson, G.L. (1994) Expression of galectins in endothelial cells and their involvement in tumor cell adhesion. *Glycoconjugate J.* 11: 462–468.

Majno, G. (1965) Ultrastructure of the vascular membrane. *In*: W.F. Hamilton and P. Dow (eds): *Handbook of Physiology*. Circulation Vol. III, Waverly Press, Baltimore, pp 2293–2375.

Marchetti, D., Menter, D., Jin, L., Nakajima, M. and Nicolson, G.L. (1993) Nerve growth factor effects on human and mouse melanoma cell invasion and heparanase production. *Int. J. Cancer* 55: 693–699.

Matsushima, K., Larsen, C.G., DuBois, G.C. and Oppenheim, J.J. (1989) Purification and characterization of a novel monocyte chemotactic and activating factor produced by a human myelomonocytic cell line. *J. Exp. Med.* 169: 1485–1490.

Mawatari, M., Kohno, K., Mizoguchi, H., Matsuda, T., Asoh, K., Damme, J.V., Welgus, H.G. and Kuwano, M. (1989) Effects of tumor necrosis factor and epidermal growth factor on cell morphology, cell surface receptors, and the production of tissue inhibitor of metalloproteinases and IL-6 in human microvascular endothelial cells. *J. Immunol.* 143: 1619–1627.

McCarthy, J.B., Basara, M.L., Palm, S.L., Sa, D.F. and Furcht, L.T. (1985) The role of cell adhesion proteins laminin and fibronectin in the movement of malignant and metastatic cells. *Cancer Metastasis Rev.* 4: 125–152.

McMahon, J.B., Farelly, J.G. and Iype, P.T. (1982) Purification and properties of a rat liver protein that specifically inhibits the proliferation of nonmalignant epithelial cells from rat liver. *Proc. Natl. Acad. Sci. USA* 79: 456–460.

Menter, D.G., Hatfield, J.S., Harkins, C., Sloane, B.F., Taylor, J.D., Crissman, J.J. and Honn, K.V. (1987) Tumor cell–platelet interactions *in vivo* and their relationship to *in vivo* arrest of hematogenously circulating tumor cells. *Clin. Exp. Metastasis* 5: 65–78.

Miller, L.J., Schwarting, R. and Springer, T.A. (1986) Regulated expression of the Mac-1, LFA-1, p150,95 glycoprotein family during leucocyte differentiation. *J. Immunol.* 137: 2891–2900.

Naito, S., Giavazzi, R. and Fidler, I.J. (1987) Correlation between the *in vitro* interaction of tumor cells with an organ environment and metastatic behavior *in vivo*. *Invasion Metastasis* 7: 16–29.

Nemerson, Y. (1966) The reaction between brain tissue factor and factor VII and X. *Biochem.* 5: 601–606.

Netland, P.A. and Zetter, B.R. (1985) Metastatic potential of B16 melanoma cells after *in vitro* selection for organ-specific adherence. *J. Cell Biol.* 101: 720–724.

Netland, P.A. and Zetter, B.R. (1986) Melanoma cell adhesion to defined extracellular matrix components. *Biophys. Biochem. Res. Commun.* 139: 515–522.

Nicolson, G.L. and Winkelhake, J.L. (1975) Organ specificity of blood-borne tumour metastasis determined by cell adhesion? *Nature* 255: 230–234.

Nicolson, G.L., Irimura, T., Gonzales, R. and Rouslahti, E. (1981) The role of fibronectin in adhesion of metastatic melanoma cells to endothelial cells and their basal lamina. *Exp. Cell Res.* 135: 461–465.

Nicolson, G.L. (1982a) Cancer metastasis: Organ colonization and the cell-surface properties of malignant cells. *Biochim. Biophys. Acta* 695: 113–176.

Nicolson, G.L. (1982b) Metastatic tumor cell attachment and invasion assay utilizing vascular endothelial cell monolayers. *J. Histochem. Cytochem.* 30: 214–220.

Nicolson, G.L., Muscali, J.J. and McGuire, E.J. (1983) Metastatic RAW117 lymphosarcoma as a model for malignant–normal cell interaction. Possible roles for cell surface antigens in determining the quantity and localization of secondary tumors. *Oncodevop. BioMed.* 4: 149–152.

Nicolson, G.L., Dulski, K.M., Basson, C. and Welch, D.R. (1985) Preferential organ attachment and invasion *in vitro* by B16 melanoma cells selected for different metastatic colonization and invasive properties. *Invasion Metastasis* 5: 144–158.

Nicolson, G.L. and Dulski, K.M. (1986) Organ specificity of metastatic tumor colonization is related to organ-selective growth properties of malignant cells. *Int. J. Cancer* 38: 289–294.

Nicolson, G.L. (1987) Differential growth properties of metastatic large cell lymphoma cells in target organ-conditioned medium. *Exp. Cell Res.* 168: 572–577.

Nicolson, G.L. (1988a) Organ specificity of tumor metastasis: Role of preferential adhesion, invasion, and growth of malignant cells at specific secondary sites. *Cancer Metastasis Rev.* 7: 143–188.

Nicolson, G.L. (1988b) Cancer metastasis: Tumor cell and host organ properties important in colonization of specific secondary sites. *Biochim. Biophys. Acta* 948: 175–224.

Nicolson, G.L. (1988c) Differential organ tissue adhesion, invasion, and growth properties of metastatic rat mammary adenocarcinoma cells. *Breast Cancer Res. Treat.* 12: 167–176.

Nicolson, G.L. (1989) Metastatic tumor cell interactions with endothelium, basement membrane and tissue. *Curr. Opinion Cell Biol.* 1: 1009–1019.

Nicolson, G.L., Belloni, P.N., Tressler, R.J., Dulski, K., Inoue, T. and Cavanaugh, P.G. (1989) Adhesive, invasive, and growth properties of selected metastatic variants of a murine large-cell lymphoma. *Invasion Metastasis* 9: 102–116.

Nicolson, G.L., Inoue, T., Van Pelt, C. and Cavanaugh, P.G. (1990) Differential expression of a M_r ~90,000 cell surface transferrin receptor-related glycoprotein on murine B16 metastatic melanoma sublines selected for enhanced brain or ovary colonization. *Cancer Res.* 50: 515–520.

Nicolson, G.L. (1991) Molecular mechanisms of cancer metastasis: Tumor and host properties and the role of oncogenes and suppressor genes. *Curr. Opinion Oncol.* 3: 75–92.

Nicolson, G.L., Cavanaugh, P.G. and Inoue, T. (1992) Differential stimulation of the growth of lung-metastasizing tumor cells by lung (paracrine) growth factors: Identification of transferrin-like mitogens in lung tissue-conditioned medium. *J. Natl. Cancer Inst. Monogr.* 13: 153–161.

Nicolson, G.L. (1993a) Cancer progression and growth: Relationship of paracrine and autocrine growth mechanisms to organ preference of metastasis. *Exp. Cell Res.* 204: 171–180.

Nicolson, G.L. (1993b) Paracrine/autocrine growth mechanisms in tumor metastasis. *Oncology Res.* 4: 389–399.

Nicolson, G.L. (1993c) Paracrine and autocrine growth mechanisms in tumor metastasis to specific sites with particular emphasis on brain and lung metastasis. *Cancer Metastasis Rev.* 12: 325–343.

Nicolson, G.L., Nakajima, M., Herrmann, J.L., Menter, D.G., Cavanaugh, P.G., Park, J.S. and Marchetti, D. (1994) Malignant melanoma metastasis to brain: Role of degradative enzymes and responses to paracrine growth factors. *J. Neuro-Oncol.* 18: 139–149.

Nicolson, G.L., Menter, D., Herrmann, J., Cavanaugh, P., Jia, L.-B., Hamada, J., Yun, Z. and Marchetti, D. (1995) Tumor metastasis to brain: Role of endothelial cells, neutrophins and paracrine growth factors. *Crit. Rev. Oncogenesis; in press.*

Orr, F.W., Varani, J., Gondek, M.D., Ward, P.A. and Mundy, G.R. (1980) Partial characterization of bone-derived chemotactic factors for tumor cells. *Am. J. Pathol.* 99: 43–52.

Paget, S. (1889) The distribution of secondary growths in cancer of the breast. *Lancet* 1: 571–573.

Pauli, B.V. and Lee, C.L. (1988) Organ preference of metastasis: The role of organ-specifically modulated endothelial cells. *Lab. Invest.* 58: 379–387.

Pauli, B.V. Johnson, R.C. and El-Sabban, M.E. (1992) Organotypic endothelial cell surface molecules mediate organ preference of metastasis. *In*: N. Simionescu and M. Simionescu, (eds): *Endothelial Cell Dysfunctions.* Plenum Press, New York, pp 439–454.

Pierschbacher, M.D. and Rouslahti, E. (1987) Influence of stereochemistry of the sequence Arg-Gly-Asp-Xaa on binding specificity of cell adhesion. *J. Biol. Chem.* 262: 17294–17298.

Pitas, R.E., Boyles, J., Mahley, R.W. and Montgomery-Bissell, D. (1985) Uptake of chemically modified low density lipoproteins *in vivo* is mediated by specific endothelial cells. *J. Cell Biol.* 100: 103–117.

Plow, E.F., Piersbacher, M.D., Rouslahti, E., Marguerie, G.A. and Ginsberg, M.H. (1985) The effect of Arg-Gly-Asp-containing peptides on fibrinogen and VonWillebrand factor binding to platelets. *Proc. Natl. Acad. Sci. USA* 82: 8057–8062.

Polverini, P.J. and Leibovich, J.S. (1984) Induction of neovascularization *in vivo* and endothelial proliferation *in vitro* by tumor-associated macrophages. *Lab. Invest.* 51: 635–642.

Ponder, B.A. and Wilkinson, M.M. (1983) Organ related differences of dolichos biflorus agglutinin to vascular endothelium. *Dev. Biol.* 96: 535–544.

Pressman, D. (1964) Certain aspects of tissue specific antigens. *Canadian Cancer Conf.* 5: 363–376.

Rhoden, L. (1980) Structure and metabolism of connective tissue proteoglycans. *In*: W.J. Lennarz (ed.): *The Biochemistry of Proteoglycans.* Plenum Press, New York, pp 180–221.

Richmond, A., Balantien, E., Thomas, H.G., Flaggs, G., Barton, D.E., Spiess, J., Bordoni, R., Franke, U. and Derynck, R. (1988) Molecular characterization and chromosomal mapping of melanoma growth stimulatory activity, a growth factor structurally related to beta-thromboglobulin. *EMBO J.* 7: 2025–2033.

Rodeck, U., Herlyn, M., Menssen, H.D., Furlanetto, R.W. and Koprowski, H. (1987) Metastatic but not primary melanoma cell lines grow *in vitro* independently of exogenous growth factors. *Int. J. Cancer* 40: 687–690.

Rodeck, U. and Herlyn, M. (1991) Growth factors in melanoma. *Cancer Metastasis Rev.* 10: 89–101.

Roos, E. and Dingemans, K.P. (1979) Mechanisms of metastasis. *Biochim. Biophys. Acta* 560: 135–166.

Roos, E., Middlekoop, O.P. and Van de Pavert, I.V. (1984) Adhesion of tumor cells to hepatocytes: Different mechanisms for mammary carcinoma compared with lymphosarcoma cells. *J. Natl. Cancer Inst.* 73: 963–969.

Roos, E., Tulp, A., Middlekoop, O.P. and Van de Pavert, I.V. (1985) Interaction between lymhphoid tumor cells and isolated liver endothelial cells. *J. Natl. Cancer Inst.* 72: 1173–1180.

Roy, S. and Chitra, S. (1989) Ultrastructural study of microblood vessels in human brain tumors and peritumoral tissue. *J. Neuro. Oncol.* 7: 283–294.

Rupnick, M.A., Carey, A. and Williams, S.K. (1988) Phenotypic diversity in cultured cerebral microvascular endothelial cells. *In Vitro* 24: 435–444.

Sage, H. (1984) Collagen synthesis by endothelial cells in culture. *In*: E. Jaffe (ed.): *Biology of Endothelial Cells*. Martimus Nijhoff, MA, pp 161–177.

Schirrmacher, V., Chiensong-Popov, R. and Arnheiter, H. (1980) Hepatocyte tumor cell interactions *in vitro*. *J. Exp. Med.* 151: 984–989.

Seppa, H., Grotendorst, G., Seppa, S., Schiffmann, E. and Martin, G.R. (1982) Platelet-derived growth factor is chemotactic for fibroblasts. *J. Cell Biol.* 92: 584–588.

Shepro, D. and Dunham, B. (1986) Endothelial cell metabolism of biogenic amines. *Ann. Rev. Physiol.* 48: 335–368.

Sica, A., Wang, J.M., Colotta, F., Dejana, E., Mantovani, A., Oppenheim, J.J., Larsen, C.G., Zachariae, C.O.C. and Matsushima, K. (1990) Monocyte chemotactic and activating factor gene expression induced in endothelial cells by IL-1 and tumor necrosis factor. *J. Immunol.* 144: 3034–3038.

Silletti, S., Watanabe, H., Hogan, V., Nabi, I.R. and Raz, A. (1991) Purifications of B16-F1 melanoma autocrine motility factor and its receptor. *Cancer Res.* 51: 3507–3511.

Simionescue, M., Simionescue, N. and Palade, G.E. (1982) Preferential distribution of anionic sites on the basement membranes and abluminal aspect of the endothelium in fenestrated capillaries. *J. Cell Biol.* 95: 425–434.

Simionescue, M., Simionescue, N., Santoro, F. and Palade, G. (1985) Differentiated microdomains of the luminal plasmalemma of murine muscle capillaries: Segmental variations in young and old. *J. Cell Biol.* 100: 1296–1407.

Sironi, M., Breviario, F., Proserpio, P., Biondi, A., Vecchi, A., Damme, J., Dejana, E. and Mantovani, A. (1989) IL-1 stimulates IL-6 production in endothelial cells. *J. Immunol.* 142: 549–553.

Stamenkovic, I. (1994) Leukocyte interactions with endothelium and extracellular matrix: Role of selectins and CD44. *Adv. Mol. Cell Biol.* 9: 11–27.

Stamper, H.B. and Woodruff, J.J. (1976) Lymphocyte homing into lymph nodes: *In vitro* demonstration of the selective affinity of recirculating lymphocytes for HEV. *J. Exp. Med.* 144: 828–841.

Stemmerman, M.B. (1980) General properties of blood vessels: Vascular endothelia. *In*: D. Abramson and P. Dobrin (eds): *Blood Vessels and Lymphatics in Organ Systems*. Academic Press, New York, pp 25–31.

Stoker, M. and Gherardi, E. (1991) Regulation of cell movement: The motogenic cytokines. *Biochim. Biophys. Acta* 1072: 81–102.

Stracke, M.L., Krutzsch, H.C., Unsworth, E.J., Arested, A., Cioce, V., Schiffman, E. and Liotta, L.A. (1992) Identification, purification and partial sequence of autotaxin, a novel motility-stimulating protein. *J. Biol. Chem.* 267: 2524–2529.

Sträuli, P. and Haemmerli, G. (1984) The role of cancer cell motility in invasion. *Cancer Metastasis Rev.* 3: 127–141.

Streeter, P.R., Rouse, B.T. and Butcher, E.C. (1988a) Immunohistologic and functional characterization of vascular addressin involved in lymphocyte homing into peripheral lymph nodes. *J. Cell Biol.* 107: 1853–1862.

Streeter, P.R., Lakey-Berg, E., Rouse, B., Bargatz, R. and Butcher, E.C. (1988b) A tissue specific endothelial cell molecule involved in lymphocyte homing. *Nature* 331: 41–46.

Sugarbaker, E.V. (1981) Patterns of metastasis. *Cancer Biol. Rev.* 2: 235–278.

Sun, F.F. and Taylor, B.M. (1978) Metabolism of prostacyclin in rat. *Biochemistry* 17: 4096–5000.

Szanuiawska, B., Majewski, S., Maninski, M.J., Noremberg, K., Swierz, M. and Janik, P. (1985) Stimulatory and inhibitory activities of lung-conditioned medium on the growth of normal and neoplastic cells *in vitro*. *J. Nat. Cancer Inst.* 75: 303–316.

Teraboletti, G., Roberts, D.D. and Liotta, L.A. (1987) Thrombospondin-induced tumor cell migration: Haptotaxis and chemotaxis are mediated by different molecular domains. *J. Cell Biol.* 105: 2409–2415.

Tressler, R.J., Belloni, P.N. and Nicolson, G.L. (1989) Correlation of inhibition of tumor cell–endothelial cell adhesion by RGD-containing peptide polymers and metastatic potential: Role of integrin-dependent and -independent adhesion mechanisms. *Cancer Commun.* 1: 55–63.

Tressler, R.J., Updyke, T.V., Yeatman, T. and Nicolson, G.L. (1993) Extracellular annexin II is associated with divalent cation-dependent tumor cell-endothelial cell adhesion of metastatic RAW117 large-cell lymphoma cells. *J. Cell Biochem.* 53: 265–276.

Tucker, R.F., Shipley, G.D., Moses, H.L. and Holley, R.W. (1984) Growth inhibitor from
    BSC-1 cells closely related to platelet type beta-transforming growth factor. *Science* 226:
    705–707.
Varani, J. (1982) Chemotaxis of metastatic tumor cells. *Cancer Metastasis Rev.* 1: 17–28.
Vorbrodt, A.W., Dobrogowska, D.H., Lossinsky, A.S. and Wisniewski, H.M. (1986) Ultra-
    structural localization of lectin receptors on the luminal and abluminal aspects of brain
    micro-blood vessels. *J. Histochem. Cytochem.* 34: 251–261.
Wakabayashi, H., Cavanaugh, P.G. and Nicolson, G.L. (1994) Responses to paracrine
    chemotactic factor and autocrine chemokinetic factor correlates with lung-specific
    metastatic capability of mouse RAW117 large-cell lymphoma cells. *Br. J. Cancer* 70:
    1089–1094
Wakabayashi, H., Cavanaugh, P.G. and Nicolson, G.L. (1995) Purification and identification
    of mouse lung microvessel endothelial cell-derived chemoattractant for lung metastasizing
    murine RAW117 large-cell lymphoma cells: identification as mouse monocyte chemotactic
    protein-1. *Cancer Res.*; in press.
Wang, Z.W., Irimura, T., Nakajima, M., Belloni, P.N., Nicolson, G.L. (1985) Characteriza-
    tion of the ECM-associated GAG produced by untransformed and transformed bovine
    corneal endothelial cells in culture. *Eur. J. Biochem.* 153: 125–130.
Wang, J.M., Taraboletti, G., Matsushima, K., Damme, J.V. and Mantovani, A. (1990)
    Induction of haptotactic migration of melanoma cells by neutrophil activating protein/inter-
    leukin-8. *Biochem. Biophys. Res. Commun.* 169: 165–170.
Watanabe, H., Carmi, P., Hogan, V., Raz, T., Silletti, S., Nabi, I.R. and Raz, A. (1991)
    Purification of human tumor cell autocrine motility factor and molecular cloning of its
    receptor. *J. Biol. Chem.* 266: 13442–13448.
Weidner, N., Semple, J.P., Welch, W.R. and Folkman, J. (1991) Tumor angiogenesis and
    metastasis-correlation in invasive breast carcinoma. *New Eng. J. Med.* 324: 1–8.
Weiss, L. (1983) Random and nonrandom processes in metastasis, and metastatic efficiency.
    Invasion Metastasis 3: 193–208.
Weiss, L., Orr, F.W. and Honn, K.V. (1988) Interaction of cancer cells with the microvascu-
    lature during metastasis. *FASEB J.* 2: 12–21.
Welch, D.R., Lobl, T.J., Seftor, E.A., Wack, P.I.J., Aeed, P.A., Yohem, K.H., Seftor, R.E.B.
    and Hendrix, M.J. (1989) Use of the membrane invasion culture system (MICS) as a screen
    for anti-invasive agents. *Int. J. Cancer* 43: 449 457.
Wood, N.L., Schook, L.B., Stoker, E.J. and Mohanakumar, T. (1988) Biochemical character-
    ization of human vascular endothelial cell monolayer antigens defined by monoclonal
    antibodies. *Transplantation* 45: 787–792.
Yamori, T., Iida, H., Tsukagoshi, S. and Tsuruo, T. (1988) Growth stimulating activity of
    lung extract on lung-colonizing colon 26 clones and its partial characterization. *Clin. Exp.
    Metastasis* 6: 131–139.
Yeatman, T.J., Updyke, T.V., Dedman, J.R. and Nicolson, G.L. (1993) Expression of
    annexins on the surfaces of nonmetastatic and metastatic human and rodent tumor cells.
    Clin. Exp. Metastasis 11: 37–44.
Yednock, T.A. and Rosen, S.D. (1989) Lymphocyte homing. *Adv. Immunol.* 44: 313–378.
Zetter, B.R. (1981) The endothelial cells of large and small blood vessels. *Diabetes* 30: 24–28.
Zetter, B.R. (1988) Endothelial heterogeneity: Influence of vessel size, organ location, and
    species specificity. *In*: U. Ryan (ed.): *Biology of Vascular Endothelial Cells.* CRC Press,
    Boca Raton, Florida, pp 63–80.
Zhu, D., Cheng, C.F. and Pauli, B.U. (1992) Blocking of lung endothelial cell adhesion
    molecule-1 Lu-ECAM-1 inhibits murine melanoma lung metastasis. *J. Clin. Invest.* 89:
    1718–1724.
Zimmerman, G.A., Prescott, S.M. and McIntyre, T.M. (1992) Endothelial cell interactions
    with granulocytes: Tethering and signaling molecules. *Immunol. Today* 13: 93–100.
Zocchi, M.F., Vidal, M. and Poggi, A. (1993) Involvement of the CD56/N-CAM molecule in
    the adhesion of human solid tumor cell lines to endothelial cells. *Exp. Cell Res.* 204:
    130–135.

Epithelial–Mesenchymal Interactions in Cancer
ed. by I.D. Goldberg & E.M. Rosen
© 1995 Birkhäuser Verlag Basel/Switzerland

Stimulation and regulation of tumor cell motility in invasion and metastasis

M.D. Levine[1], L.A. Liotta[2] and M.L. Stracke[2]

[1]Harvard College, Cambridge, MA 02138, USA
[2]Laboratory of Pathology, National Cancer Institute, National Institutes of Health, Bethesda, MD 20892, USA

Summary. In this review, the role of extracellular factors in the stimulation and regulation of tumor cell motility are discussed. Tumor cells respond in a motile fashion to a variety of external ligands including autocrine motility factors, growth factors, and components of the extracellular matrix. Since tumor cell motility is a necessary component of tumor invasion and metastasis, we speculate that these protein factors could play important regulatory roles in tumor motility at different stages of the metastatic cascade.

Introduction

The transition from *in situ* tumor growth to invasive disease is defined by the ability of the tumor cells at the primary site to invade local tissues and to cross tissue barriers. This process is marked by loss of epithelial differentiation and development of a more fibroblastic or mesenchymal appearance (Birchmeier et al., 1993; Boyer et al., 1993). The tumor cell loses its attachment to the primary tumor mass, creates a pathway through surrounding stroma or basement membrane, and then moves through the pathway it has created. Thus, a defining characteristic of invasive cells as compared to *in situ* carcinoma cells, is their motility (Fig. 1) (Liotta and Stracke, 1988; Straüli and Haemmerli, 1984).

It has now been well established that active cellular locomotion is a necessary component of metastasis. Under normal physiological conditions, such as wound healing or inflammatory response, cell motility is tightly controlled. Much progress has been made in understanding the internal cellular changes that accompany this cellular motility, including changes in cytoplasmic fluidity and alterations in the microfilamentous framework (reviewed in Condeelis, 1993; Janney, 1994; Stossel, 1993, 1994). The internal cellular machinery of this motility is probably similar for tumor cells. However, since tumor cell motility appears to be aberrantly regulated or even autoregulated, the question of what initiates and maintains this locomotory response is highly relevant.

Tumor cells can respond in a motile fashion to a variety of agents, including extracellular matrix components, host-derived motility and

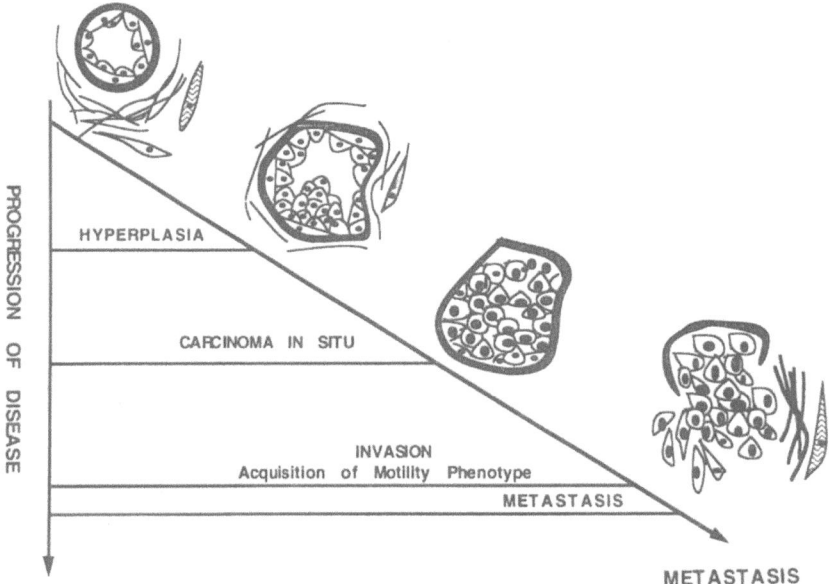

Figure 1. Progression of disease in intraductal breast carcinoma. The progression of an intraductal lesion toward carcinoma involves a transition from an epithelial to a more mesenchymal phenotype. In order for an *in situ* lesion to become invasive, this transition must include the capacity to locomote across the basement membrane.

growth factors, and tumor-secreted or autocrine factors. Although some of these same agents stimulate motility in embryonic or other physiologically motile cells, the tumor cells respond in a motile fashion to multiple diverse stimuli. This allows the cells to adapt a variety of microenvironments toward the stimulation and sustenance of a motile phenotype. Thus, the tumor cell appears to redirect normal signals for growth or adherence into signals for locomotion.

Autocrine motility factors

In early studies, Hayashi and colleagues described a 70 kDa chemotactic factor derived from whole cell extracts of rat hepatoma tumors grown subcutaneously in Donryu rats (Hayashi et al., 1970). This factor was chemotactic for several tumor cell lines including the cells of origin. Later, it was observed that cultured human melanoma cells (the A2058 cell line) secreted an attractant material into serum-free media (Liotta et al., 1986). This factor stimulated a motility response that was both chemokinetic (random) and chemotactic (directed toward a positive concentration gradient) in nature. Because this property accorded with the early dissemination of single cells from the primary tumor, the

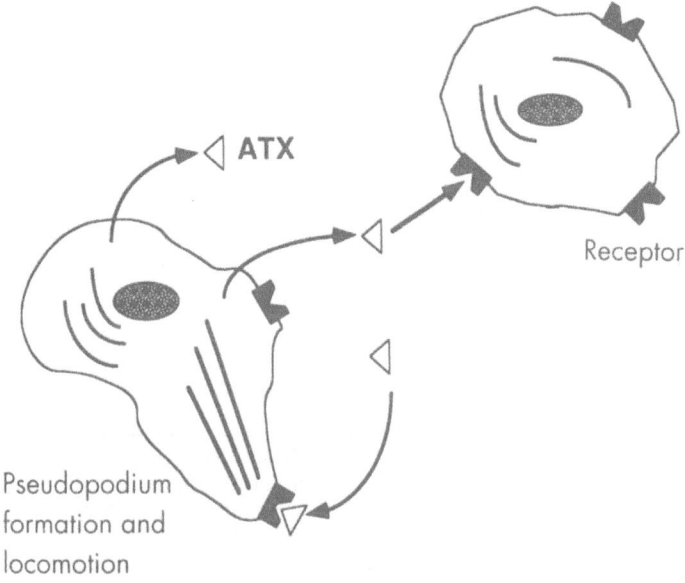

Figure 2. The autocrine motility factor hypothesis. Certain tumor cells synthesize and secrete a motility factor which can then activate the same or other tumor cells, presumably through a cell surface receptor.

autocrine motility factor (AMF) hypothesis was proposed (Fig. 2). Previous work (Todaro et al., 1980; Anzano et al., 1983) had demonstrated the presence of specific autocrine growth factors for tumor cells, thereby setting a precedent for autocrine control of neoplastic cell behavior.

A family of tumor cell-derived motility-inducing cytokines has now been reported (Tab. 1). Only the two growth factors, hepatocyte growth factor/scatter factor (HGF/SF) and insulin-like growth factor II (IGF-II), and the newly cloned 125 kDa glycoprotein from A2058 melanoma

Table 1. Autocrine motility factors

Cell of origin	Factor	References
Human melanoma	60 kDa	(Liotta et al., 1986)
Rat mammary carcinoma	53 kDa	(Atnip et al., 1987)
Rat/human malignant glioma	>10 kDa	(Ohnishi et al., 1990)
Human rhabdomyosarcoma	IGF-II	(El-Badry et al., 1990)
Murine melanoma	64 kDa	(Siletti et al., 1991)
Rat prostatic adenocarcinoma	<30 kDa	(Evans et al., 1991)
ras-transfected NIH 3T3	150–200 kDa	(Seiki et al., 1991)
Human melanoma	Autotaxin	(Stracke et al., 1992)
Murine fibrosarcoma	55&65 kDa	(Watanabe et al., 1994)
Lung and bladder carcinoma	SF/HGF	(Bellusci et al., 1994; Tsao et al., 1993)

cells (Murata et al., 1994) have been purified to homogeneity and cloned. In addition, the glycosaminoglycan, hyaluronic acid (Turley et al., 1991) may be secreted by certain tumor cells to induce a motile response in the same cells, functioning as an autocrine motility factor.

As a group, the autocrine motility factors are not specific for a given type of cancer cell and have a wide spectrum of activity on many types of cancer cells (Evans et al., 1991; Hayashi et al., 1970; Kohn et al., 1990; Seiki et al., 1991). However, in general, these cytokines have little effect on leukocytes (Liotta et al., 1986; Ozaki et al., 1971) and less stimulatory effect on NIH-3T3 cells than PDGF (Liotta et al., 1986).

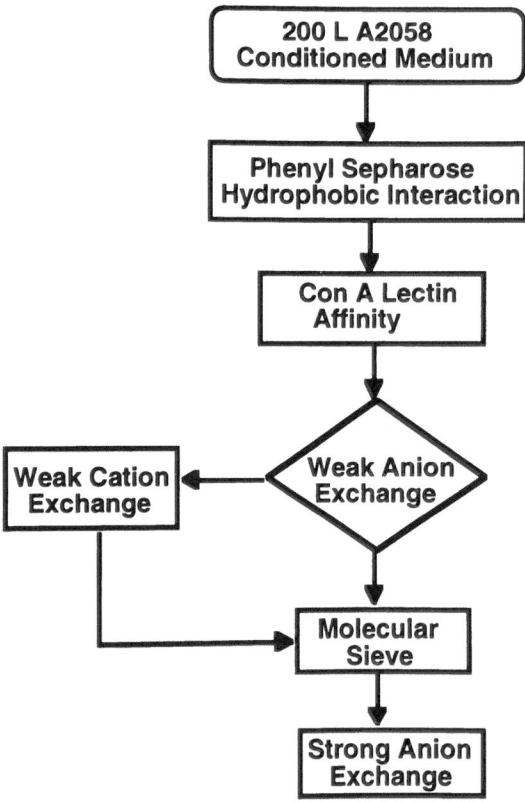

Figure 3. Schemata of ATX purification. Because ATX is secreted in relatively small amounts, 200 L of A2058 conditioned medium was collected under serum-free conditions but with supplements of (10 μg/ml) insulin to maintain cell growth and (5 μg/ml) bovine serum albumin to act as a carrier protein. The conditioned medium was concentrated with low protein-binding ultrafiltration membranes and then separated via a series of liquid chromatographic columns, as shown. Since the motility-stimulating activity was tested after each chromatographic procedure to select positive fractions, chromatographic media were limited to those which separate proteins under relatively mild conditions, maintaining activity.

The 60 kDa cytokine produced by the A2058 cell line is a protein that stimulates pseudopodial extrusion to initiate cell locomotion (Guirguis et al., 1987). This response is sensitive to pertussis toxin (Stracke et al., 1987), to inhibitors of methylation (Liotta et al., 1986), and to pharmacologic agents that act on microtubules or microfilaments, such as *cis*-tubulozole, taxol, and cytochalasin B or D (Stracke et al., 1993). Although a G protein might be involved in the signal transduction, a

(A)

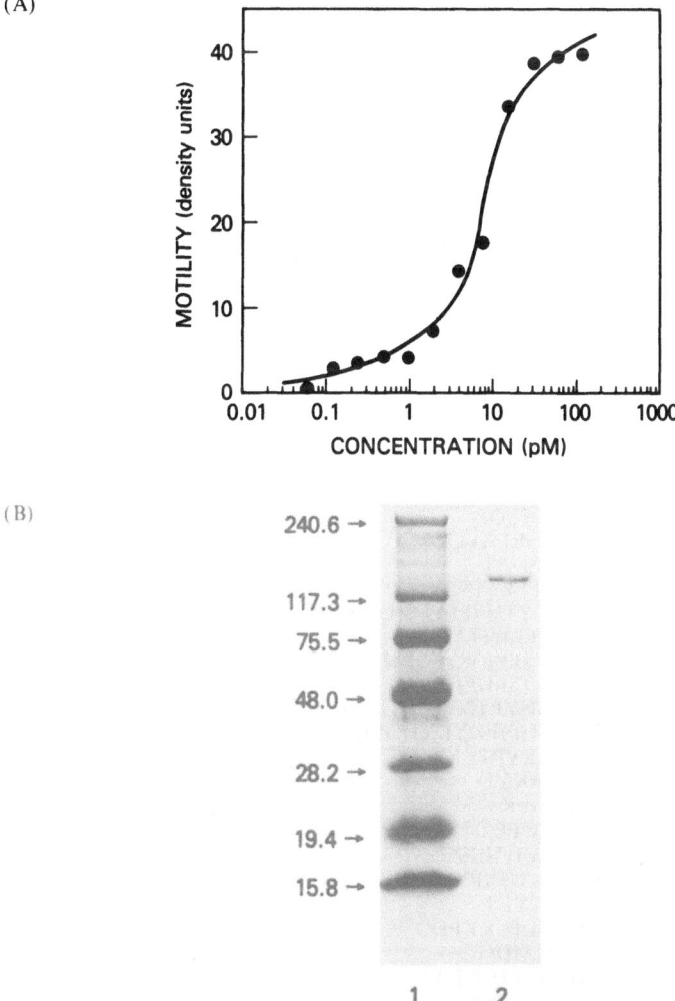

(B)

Figure 4. Characterization of ATX activities. (A) Dilution curve of ATX. Purified ATX was serially diluted and tested for motility-stimulating activity. This result, with unstimulated background motility subtracted out, shows that activity is half-maximal at ~ 500 pM ATX. (B) SDS-polyacrylamide gel electrophoresis of purified ATX, shown in lane 2. (Lane 1 contains molecular weight standards.)

variety of agents that affect adenylate cyclase have no effect on AMF-
stimulated motility, indicating that cAMP is not the necessary second
messenger (Stracke et al., 1987). Recently, a possible cell surface receptor
to a murine melanoma autocrine motility factor has been isolated (Siletti
et al., 1991). Antibodies produced against this 78 kDa glycoprotein
stimulated motility (Nabi et al., 1990) and enhanced metastatic ability in
high metastatic variants (Watanabe et al., 1991).

Autotaxin, a novel autocrine motility factor

We have recently isolated a potent new cytokine (molecular mass
120 kDa) from the conditioned medium from A2058 cells. Utilizing
sequential chromatographic methods (Fig. 3), we purified this factor to
homogeneity (Stracke et al., 1992 (Fig. 4B)). This new cytokine, termed
autotaxin (ATX), is a basic glycoprotein with pI $\sim$ 7.8. ATX is active in
the high picomolar to low nanomolar range (Fig. 4A) stimulating both

Table 2. Peptide sequences for autotaxin

Peptide	Amino acid sequence
ATX-18	WHVAXN
ATX-19	PXLDVYK
ATX-20	YPAFK
ATX-29	PEEVTRPNYL
ATX-34B	RVWNYFQR
ATX-41	HLLYGRPAVLY
ATX-47	YDVPFDAT
ATX-48	(S/V)PPFENINLY
ATX-59	TFPNLYT(F/L)ATG(G/L)YW
ATX-100	XGGQPLWITATK
ATX-101/223A	VNSMQTVFVGYGPTFK
ATX-102	DIEHLTSLDFFR
ATX-103	TEFLSNYLTNVDDITLVPGTLGR
ATX-104	VNVISGPIFDYDYDGLHDTEDK
ATX-204	MHTARVRD
ATX-205	FSNNAKYD
ATX-209	VMPNIEK
ATX-210	TARGWECT
ATX-212	XDSPWT(N)ISGS
ATX-214	LRSCGTHXPYM
ATX-215/34A	TYLHTYES
ATX-213/217A	AIIANLTCKKPDQ
ATX-216	IVGQLMDG
ATX-218/44	TSRSYPEILT(P/L)
ATX-223B/24	QAEVSSVPD
ATX-224	RCFELQEAGPPD(D/C)
ATX-229	SYTSCCHDFDEL
ATX-230	XFNHQWGGQQP
ATX-239	AAEC(V/P)A
ATX-244/53	QMSYGFLFPPYLSSSP

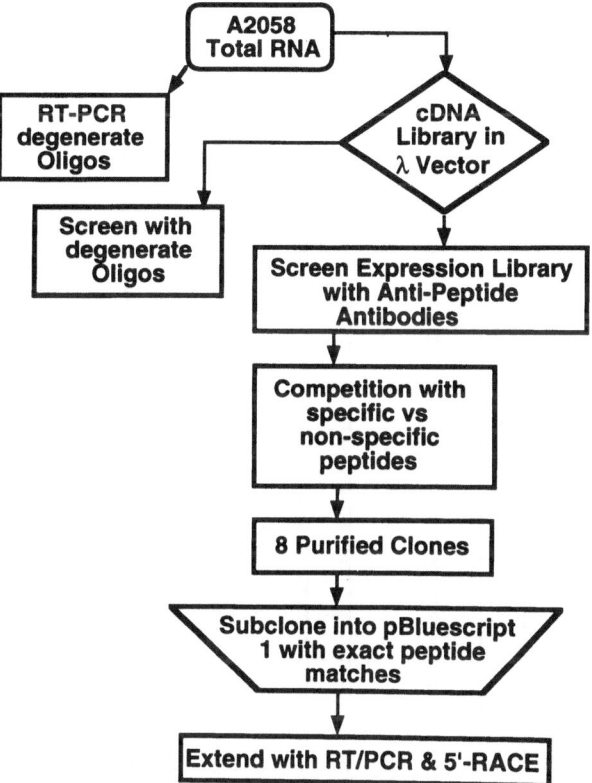

Figure 5. Schemata of ATX cloning. Several strategies were tried to obtain the ATX cDNA clone. Attempts to utilize degenerate oligos deduced from known peptide sequences were unsuccessful – whether the oligos were employed for reverse transcription/polymerase chain reaction (RT/PCR) amplifications or for library screening. However, when an A2058 expression library was screened with affinity-purified anti-peptide antibodies, a number of positive clones were obtained. These were confirmed by antibody competition with specific peptides but not unrelated peptides. Sequence analysis of these clones revealed a single partial cDNA clone of the autotaxin gene containing the 3′ terminus of the protein. Utilizing known sequence from this partial clone, we extended the cDNA clone toward the 5′ end with RT/PCR amplifications and rapid amplification of cDNA ends (5′ RACE).

chemotactic and chemokinetic responses in the same A2058 cells. When cells are pretreated with pertussis toxin, this motile response is abolished. Sequence information, obtained by Edman degradation on 30 purified enzymatically-digested peptides, confirmed that the protein is unique with no significant homology to growth factors or previously described motility factors (Tab. 2). Several of these peptides were synthesized, conjugated to bovine serum albumin, and injected into rabbits to produce antipeptide antibodies. After affinity purification, these antipeptide antibodies recognized ATX on immunoblots, but were unable to neutralize activity or immunoprecipitate protein.

```
ATX   MARRSSFQSCQIISLFTPAVGVSICLGFTAHRIKFAEGWEEGPPTVLSDSPWTNISGSCKGRCFELQEAGPPDCRCDNLCKSYTSCCHDF
         ||                          |        |         |||||          |||||        ||   |
PC-1  MDVGEEPLEKAAARATAKDPNTYKVLSLVLSVCVLTTIL.......QCIFG....LKPSCAKEVK.SCKGRCF...ERTFGNCRCDAACVELGNCCLDY

ATX   DELCLKTARGWECTKDRCGEVRNEENACHCSEDCLARGDCCTNYQVVCKGESHWVDDDCEEIKAAECPAGFVRPLIIFSVDGFRASYMKKGSKVMPNIE
       | ||   | |||  | | ||  |  | || | | |||  |||||| ||||  | |||||   |  ||||||  || |||||| |||||||| |  | |
PC-1  QETCIEPEHIWTCNKFRCGEKRLTRSLCACSDDCKDKGDCCINYSSVCQGEKSWVEEPCESINEPQCPAGFETPPTLLFSLDGFRAEYLHTWGGLLPVIS

ATX   KLRSCGTHSPYNRPVVPTKTFPNLYTLATGLYPESHGIVGNSMYDPVFDATPHLRGREKFMHRWWGGQPLWITATKQGVKAGTFWS..........
      || ||| | | |||||||||||| | |  |||||||||| | | ||| || |     ||  |    || | |  | | ||| ||
PC-1  KLKKCGTYTKNNRPVVPTKTFPNHYISIVTGLYPESHGIIDNKMYDPKMNASFSLKSKEKFNPEWYKGEPIWVTAKYQGLKSGTFWPGSDVEINGIFPDI

ATX   .....VVIPHERRILTILRWLTLPDHERPSVYAFYSEQDFSGHLYGPFGPESSYGSPFTPAKRPKRVAPKRRQERPVAPPKRRRRKIHRMDHYAAET
           | |||||| || || |||||||  |  |   ||| |||| || ||| | || |  ||   ||
PC-1  YKMYNGSVPFEERILAVLQWLQLPKDERPHFYTLYLEEFDSSGHSYGPVSSE..........

ATX   RQDKWTNPLREIDKIVGQLMDGLKQLLKLRRCVNVIFVGDHGMEDVTCDRTEFLSNYLTNVDDITLVPGTLGRIR.SKFSNN.AKYDPKAIIANLTCKKPD
         |          | |||||||  || | | ||   ||||||  ||| | | |  | ||   || | |  ||      |         | | ||| |
PC-1  ...VIKALQRVDGWVGMLMDGLKMLHRCLNLILISDHGMEQGGSCKKYIYLNKYLGDVKNIKVIYQPAARLRPSDVPDKYYSFNYEGIARNLSCREPN

ATX   OHFKPYLKQHLPKRLHYANNRRIEDIHLLVERRWHVARKPLDVYLKDSGKCFFQGDHGFDNKVNSMQTVFVGYGPTFKYKTKVPPENIELYNVMCDLLG
      ||||||||  |||||| |   ||| |  |           |||  | | | ||  ||        |  ||||||||| |  |   |||| | || |
PC-1  OHFKPYLKHFLPKRLHFAKSDRIEPLTFYLDPQWQLALNFSE..EKYCGSGF....HGSDNVFSNMQALFVGYGPGFKHGIEADTFENIEVYNLMCDLLN

ATX   LKPAPNNGTHGSLNHLLRTWTFRPTWPESVTRPNYPGLIMYLQSDFDLGCTCDDKVEPKNKLD.ELNKRLHTKGSTEERHLLYGRPAVLYRTR.YDILYHT
      | |||||||||||||| |                 |  |        | |  ||      |        |  |            ||||  |
PC-1  LTPAPNNGTHGSLNHLLKNPVYTPKHPKEV.HFLVQCPFTRNFRDNLQCSCNFSILFIEDFQTQFNLTVAEEKIIKHETLFYGRPRVLQKENTICLLSQH

ATX   DFESGYSEIFLMLLWTSYTVSKQAEVSSVPDHLTSCVRPDVRVSPSFSQNCLAYKNDKQMSYQFLFPYLSSSPEAKY.DAFLVTNWVPMYPAFKRVWNY
       |   |||   ||||||||  |  | ||  |     |  |  | | |  |  |  |  | |||  ||||||  |    |  |      ||  | |
PC-1  QFMSGYSQDIIMPLWTSYTVDRNDSFS..TEDFSNCLYQDFRIPLSPVHKCSFYKNNTKVSYGFLSPPQLMKQNSSQIYSEALLTNIVPMYQSFQVIWRY

ATX   FQRVLVKKYASERNGVNVISGPIFDYDYDGLGHDTEDKIKQ...YVEGSSIFVPTHYYSIITSCLDFTQPADKCDGPLSVSSFILPHRPDNEESCNSSEDE
       | | | ||| |||||||| |||| |||||  ||  ||       ||||| | |||| |||||||  |||| || || ||| ||||| ||  ||| |  |
PC-1  FHDTLLRKYAEERNGVNVVSGPVFDFDYDGRCDSLENLRQKRAVIRNMQEILLPTHFFVLTSCKDFSQTPLECKN.LDTLAFILPHRTDNSESCVHGKHD

ATX   SKWVEELMKMHTARVRDIEHLTSLDFFRKTSRSRYSETILTLKTYLETYESEI
      | |||||   | |  |  ||| | | |          |   | | |
PC-1  SSWVEELLLHRARITDVEHITGLSFYQQRKEFVSDILKLKTHLFTFSQED
```

Initial attempts to clone ATX utilized oligonucleotides deduced from the known peptide sequences (Murata et al., 1994). However, whether these oligonucleotides were used directly as radiolabeled probes, or as primers in reverse transcriptase/polymerase chain reaction amplifications fron A2058 mRNA, these procedures were unsuccessful (Fig. 5). Consequently, the affinity-purified antipeptide antibody (anti-ATX-102), which gave the strongest signal on immunoblots, was used to screen an A2058 cDNA expression library in λgt11. One clone, designated 4C11, appeared to contain the 3' terminus of ATX. The 4C11 clone contained 1084 base pairs, including the polyadenylated tail and the AATAAA polyadenylation signal motif. The open reading frame region was 628 bp long and coded for 209 amino acids with matches for 8 previously identified ATX peptides. Database analysis of the 4C11 clone revealed a homology with PC-1, found on the surface of activated B cells and plasma cells. This homology was utilized to estimate the relative positions of several of the larger ATX peptides. The 4C11 clone was then extended using reverse transcriptase/polymerase chain reaction and 5' rapid amplication of cDNA ends (5' RACE) methodology. The total length of the resultant cDNA was 3251 bp with a 2745 bp open reading frame that encoded a 915 amino acid-deduced peptide sequence. This deduced sequence has been shown to match all 30 of the previously sequenced peptides, providing strong evidence that we sequenced the correct cDNA for purified autotaxin.

Searches of protein databases revealed that ATX is a new protein and confirmed that the homology between ATX and PC-1 (Buckley et al., 1990; Funakoshi et al., 1992) was present throughout the length of the extracellular portion of the molecules with a 45% amino acid identity and a 57% nucleotide identity between the 2 sequences (Fig. 6). PC-1 is an N-glycosylated membrane protein found on differentiated plasma cells and a few non-lymphoid tissues (Harahap and Goding, 1988). Its function is unknown but it is an ectoenzyme with several defined enzymatic activities, including type I alkaline phosphodiesterase, nucleotide pyrophosphatase, and threonine-specific kinase (Oda et al., 1991; Rebbe et al., 1991). When the domain structures of the two proteins are compared, a number of interesting similarities become apparent (Fig. 7). Both have adjacent somatomedin B domains near the amino terminus of their extracellular portions. Somatomedin B, derived from the amino terminus of vitronectin, is a cysteine-rich region that forms the presumed binding site for activated type 1 plasminogen activator

---

Figure 6. Comparison of amino acid sequences of ATX and PC-1. The amino acid sequences of ATX and PC-1 are compared. Amino acid identity is indicated by a vertical line between the sequences.

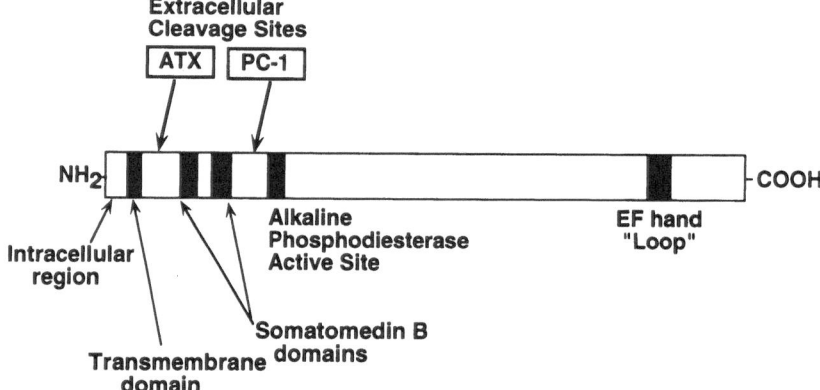

Figure 7. Domain structure of ATX and PC-1. Putative domains are indicated for the two homologous proteins, ATX and PC-1. The presumed cleavage site for each protein is also indicated.

inhibitor (PAI) in extracellular matrix and in plasma (Seiffert et al., 1994; Seiffert and Loskutoff, 1991). This homology suggests a kinship with extracellular matrix proteins. In addition, both ATX and PC-1 have regions homologous to the active site of bovine type I phosphodiesterase (Culp et al., 1985). ATX and PC-1 also both contain the loop region of an EF hand, which is a calcium binding domain that structurally forms a helix-loop-helix (Kretsinger, 1987). The loop region is the actual calcium binding site. Other proteins which lack one or both helical regions are variably capable of binding calcium; however, this binding has not been demonstrated for either PC-1 or ATX.

Despite the similarities, there are also a number of important differences between ATX and PC-1. First, the intracellular region of ATX is only 11 amino acids long and is different from the 24 intracellular amino acids found in PC-1 (Buckley et al., 1990). Likewise, the transmembrane domains are dissimilar. Although PC-1 normally exists in the membrane as a dimer, soluble forms have been described in normal mouse serum, transfected mouse L cells, and supernatants of plasmacytoma cells (Belli et al., 1993). When the soluble form of PC-1 is generated, the presumed cleavage site is between the second somatomedin B domain and the phosphodiesterase active site. In contrast, ATX appears to be cleaved proximal to both somatomedin B domains. The soluble forms of the two molecules appear therefore to be substantially different.

It remains to be seen what relationship this complex domain structure has to the motility-stimulating capacity of ATX. If ATX is an extracellular kinase/phosphodiesterase like PC-1, its capacity to modulate protein phosphorylation could effect both extracellular matrix and cell surface proteins. ATX could, therefore, play both a regulatory role in the interaction of the migrating cell with its microenvironment as well as a direct role in the stimulation of motility.

The role of extracellular matrix in the induction and regulation of tumor cell motility

The extracellular matrix (ECM) is a network of macromolecules that surrounds cells and tissues to form a significant portion of the cellular microenvironment. The ECM is made up of structural molecules such as the collagens and elastin; adhesive molecules, including fibronectin, laminin, vitronectin, and thrombospondin; and a variety of proteoglycans and glycosaminoglycans (Kleinman et al., 1993; Timpl, 1993; Venstrom and Reichardt, 1993). These molecules of the ECM crosslink extensively with each other and bind to multiple cell surface receptors. Tumor cell interactions with the matrix throughout the process of invasion include attachment, proteolysis of matrix components, and migration through the matrix defect (Liotta et al., 1983). None of these functions is unique to tumor cell behavior. Attachment, proteolysis, and migration are steps of wound healing, trophoblast implantation, mammary gland involution, embryonic morphogenesis, and tissue remodeling (Lola and Graham, 1990). The difference between normal physiological processes and the pathogenic process of tumor cell invasion appears to be one of regulation.

The capacity to migrate in response to certain ECM proteins has been found to correlate positively with *in vivo* invasive and metastatic capacity of melanoma and carcinoma cell lines (Lester and McCarthy, 1992; Yabkowitz et al., 1993). Several components of the extracellular matrix have been found to stimulate locomotion in tumor cells. These include vitronectin (Basara et al., 1985), fibronectin (Aznavoorian et al., 1990b; Makabe et al., 1990; McCarthy and Furcht, 1984; McCarthy et al., 1986; Mensing et al., 1984), laminin (Aresu et al., 1991; Aznavoorian et al., 1990a,b; McCarthy and Furcht, 1984; McCarthy et al., 1983; Situ et al., 1984; Tashiro et al., 1991; Wewer et al., 1987), type I collagen (Faassen et al., 1992; Mooradian et al., 1992; Tchao, 1982), type IV collagen (Aznavoorian et al., 1990b; Chelberg et al., 1989), and thrombospondin (Taraboletti et al., 1987). For fibronectin (McCarthy et al., 1986), type IV collagen (Chelberg et al., 1990; Chelberg et al., 1989), and thrombospondin (Taraboletti et al., 1987; Yabkowitz et al., 1993), distinct domains within these molecules have been identified which promote motility of tumor cells, whereas other domains promote only adhesion. Certain fragments of these multidomain proteins have been shown to inhibit the formation of metastases when experimentally co-injected with tumor cells into mice (Humphries et al., 1986; Iwamoto et al., 1987; McCarthy et al., 1988; Nomizu et al., 1993). One peptide from the laminin A chain has been shown to induce invasive but not metastatic phenotype when cultured with tumorigenic, non-invasive murine melanoma cells (Royce et al., 1992).

A few of these extracellular matrix proteins induce tumor cell motility or invasion through integrin receptors (Leavesly et al., 1992; Tashiro et al., 1991). For example, anti-$\beta 1$ monoclonal antibody inhibits both human and murine tumor cell migration as well as *in vitro* (Matrigel assay) invasiveness at concentrations as low as 1 $\mu g/ml$ (Fujita et al., 1992; Kramer et al., 1989; Yamada et al., 1990). In murine melanoma, the $\alpha_{IIb}\beta_3$ integrin has been correlated with matrix adherence, platelet aggregation, and lung colony formation (Honn et al., 1992). In human melanomas, upregulation of the expression of certain integrins, including $\alpha_2\beta_1$, $\alpha_3\beta_1$, and $\alpha_6\beta_1$, correlates with metastatic potential (Danen et al., 1993) or tumor cell motility (Danen et al., 1993; Etoh et al., 1992; Yoshinaga et al., 1993). Evidence is also accumulating that expression of the vitronectin receptor, $\alpha_v\beta_3$, plays a role in the invasiveness of malignant melanoma (Albelda et al., 1990; Gehlsen et al., 1992; Marshall et al., 1991; Nip et al., 1992; Seftor et al., 1992). On hairy cell leukemia cells (Burthem et al., 1994) and pancreatic carcinoma cells (Leavesly et al., 1992), $\alpha_v\beta_3$ appears to be the major integrin responsible for the motility response. Pancreatic carcinoma cells, which normally express $\alpha_v\beta_5$ but not $\alpha_v\beta_3$, acquire the capacity to migrate in response to vitronectin and fibrinogen when transfected with cDNA for β_3. In hairy cell leukemia cells, anti-β_3 antibody strongly inhibits the motile response.

The human melanoma cell line, A2058, has been shown to respond in a locomotory fashion to multiple extracellular matrix proteins: laminin, fibronectin, type IV collagen and thrombospondin (Aznavoorian et al., 1990b; Taraboletti et al., 1987). These ECM proteins stimulate chemotaxis when they are in solution and haptotaxis when they are insoluble or substratum-bound. However, chemotactic and haptotactic stimulation by ECM proteins appear to act through different cell surface receptors and post-receptor signal transduction pathways (Aznavoorian et al., 1990). When cells are pretreated with pertussis toxin, the chemotactic response to laminin is diminished and the response to type IV collagen is abolished. In contrast, haptotaxis to these same proteins is insensitive to pertussis toxin. A likely explanation for these data is that chemotaxis and haptotaxis to the same ECM protein are mediated by distinct cell surface receptors, which recognize different domains of these large, multi-domain matrix proteins. This hypothesis is further supported by work of Taraboletti et al. (1987) which showed that the chemotaxis and haptotaxis promoting domains of thrombospondin are on opposite ends of the molecule. The carboxy-teminal region of thrombospondin appeared to stimulate haptotaxis and was inhibited by a specific monoclonal antibody (C6.7) as well as by the synthetic peptide

Table 3. Growth factors that effect tumor cell motility

Factor	Cell types	References
Bombesin	Small cell lung carcinoma	(Ruff et al., 1987)
acidic FGF	Bladder carcinoma	(Vallés et al., 1990)
basic FGF	Prostatic carcinoma	(Pienta et al., 1991)
	Teratocarcinoma	(Schofield et al., 1992)
HGF/SF	Carcinomas	(Stoker and Gherardi, 1989; Weidner et al., 1990)
	Lung carcinoma	(Tsao et al., 1993)
	Renal carcinoma	(Bellusci et al., 1994)
Histamine	Melanoma/carcinoma	(Tilly et al., 1990)
IGF-I	Melanoma	(Stracke et al., 1988)
IGF-II	Rhabdomyosarcoma	(El-Badry et al., 1990)
IL-1	Breast carcinoma	(Verhasselt et al., 1992)
IL-6	Ductal breast carcinoma	(Tamm et al., 1989)
IL-8	Melanoma	(Wang et al., 1990)
NGF	Embryonal carcinoma	(Kahan and Kramp, 1987)
PDGF	Teratocarcinoma	(Liapi et al., 1990)
TGFβ1	Adenocarcinoma of lung	(Mooradian et al., 1992)

GRGDS. The amino-terminal heparin-binding domain of thrombospondin appeared to stimulate chemotaxis and was inhibited by a specific monoclonal antibody (A2.5) as well as by the sulfatides heparin and fucoidan.

Perhaps, during the initial phases of metastasis, the insoluble matrix proteins provide tumor cells with a pathway of activation, allowing the tumor cells to track through stromal tissues and along basement membranes. Proteolytic enzymes secreted either by the tumor cells or by the host, including type IV (Liotta et al., 1979; Liotta et al., 1982) and interstitial collagenases (Woolley, 1984), cathepsin B (Sloane and Honn, 1984), and plasminogen activator (Mignatti et al., 1986) could then result in localized pools of soluble, partially degraded matrix proteins. These soluble pools would then provide an additional chemotactic stimulus to motility (Nabeshima et al., 1986).

Growth factors and other host-derived cytokines

Formation of a successful metastatic nidus requires that the tumor cells find a microenvironment capable of supporting cell growth. Certain highly metastatic cell lines produce their own necessary autocrine growth factors (Anzano et al., 1983; Cai et al., 1994; El-Badry et al., 1990; Halaban et al., 1988; Huff et al., 1986; Rodeck et al., 1987; Todaro et al., 1980; Williams et al., 1992). However, host-secreted, growth-supporting factors may still be advantageous to establishing colonies of metastatic cells. Several growth factors stimulate chemotactic motility in tumor cells. These growth factors appear to be somewhat specific for cellular origin (Tab. 3). Many of these cytokines also appear to act as mitogens for the same tumor cells that they stimulate to migrate (El-Badry et al., 1990; Pienta et al., 1991; Schofield et al., 1992; Tilly et al., 1990; Vallés et al., 1990).

The insulin-like growth factors (IGF) and insulin stimulate a pertussis toxin insensitive chemotactic response in A2058 cells (Stracke et al., 1988). This response is strongest to IGF-I and appears to activate the cells through a type I IGF receptor since a monoclonal antibody, specific for the type I IGF receptor (Jacobs et al., 1986), inhibits both ^{125}I-labelled IGF-I binding and IGF-I-induced motility in A2058 cells (Stracke et al., 1989). In addition, cross-linking experiments with ^{125}I-labelled IGF-I reveal a typical heterotetrameric receptor with IGF-I bound to the larger α subunit. Both insulin and IGF-I have been implicated as necessary growth factors for culture of primary human melanoma cells (Rodeck et al., 1987). In similar experiments, IGF-II has been found to stimulate motility through a type II IGF/mannose-6-phosphate receptor in human rhabdomyosarcoma cells (Minniti et al., 1992). IGF-II stimulates mitogenesis in these same cells through a type I IGF receptor.

Scatter factor was first isolated as a paracrine factor made by fibroblasts that stimulated loss of cell–cell adhesion and scattering in epithelial cells (Stoker et al., 1987). Partial peptide analysis of purified scatter factor revealed that this factor was identical to the previously identified and sequenced hepatocyte growth factor (Weidner et al., 1991). The cell surface receptor for HGF/SF has been demonstrated to be the proto-oncogene *met*, a heterodimeric tyrosine kinase (Bottaro et al., 1991; Naldini et al., 1991) which transduces both the motility and proliferative cellular responses (Weidner et al., 1993). Recently, HGF/SF has been found to stimulate motility in a variety of tumors and to behave as an autocrine motility factor in a few tumor cell lines, such as lung (Tsao et al., 1993) and bladder carcinomas (Bellusci et al., 1994). HGF/SF also enhances tumor invasiveness in breast (Rosen et al., 1994), small intestinal (Sunitha et al., 1994), bladder (Bellusci et al., 1994; Weidner et al., 1990), pancreatic and lung carcinomas (Weidner et al., 1990). Presence of the c-*met* receptor on melanoma cells has been associated with invasiveness (Natali et al., 1993); transformation of NIH 3T3 cells, which produce HGF/SF endogenously, with the murine *met* oncogene induces an invasive phenotype (Rong et al., 1994).

These data suggest that growth factors, acting through 'normal' receptor mechanisms, may stimulate a locomotory response in tumor cells. They may serve as 'homing' factors for tumor cells which have reached the vasculature, directing the tumor cells to extravasate into a secondary site which provides a suitable microenvironment for growth.

Protein factors that inhibit tumor cell invasiveness

Regulation of tumor cell motility and invasion could potentially involve host-secreted inhibitory factors as well as stimulatory factors. Two peptide factors that inhibit tumor cell invasion of a mesothelial cell monolayer were purified from acid-ethanol extracts of bovine liver (Isoai et al., 1990). These factors, termed tumor invasion-inhibiting factors (IIF)-1 and -2, had $M_r \sim 5000$ and 2000, respectively. The primary structure has been determined for IIF-2 (Isoai et al., 1992). It is a twenty-one amino acid peptide, found to be identical to the carboxyl-terminal region of a DNA-binding non-histone protein. IIF-2 inhibits chemotactic migration of B16 murine melanoma cells to laminin and fibronectin, as well as invasion through Matrigel (Isoai et al., 1992). In addition, bovine serum albumin-conjugated IIF-2, which remained in the circulation longer than unconjugated peptide, inhibits formation of pulmonary metastases by murine melanoma, colon adenocarcinoma, squamous cell carcinoma, and human fibrosarcoma cell lines (Isoai et al., 1994). The IIF-2 peptide is now thought to inhibit metastasis by preventing extravasation of blood-borne tumor cells and by directly inhibiting tumor cell motility.

Tumor-secreted and tumor-induced factors that stimulate host cell motility

Tumor cells interact with the host in various ways that can influence their metastatic capability. For example, tumor cells can produce chemotactic factors that affect host cell motility or induce the synthesis of chemotactic factors by the host cells themselves.

Several breast carcinoma cell lines have been shown to secrete a protein factor that stimulates fibroblast motility (Gleiber and Schiffmann, 1984) and may influence the fibrosis seen around tumors. The skin metastases of human breast carcinomas also synthesize a factor chemotactic for melanocytes (Konomi et al., 1992). Several different tumor cell lines secrete monocyte chemotactic protein-1 (MCP-1) (Bottazzi et al., 1990; Mantovani et al., 1986; Zachariae et al., 1990), and human melanoma cells have been shown to produce IL-8 (Zachariae et al., 1991). These cytokines are members of the superfamily of basic, heparin-binding chemotactic proteins termed chemokines that act as mediators in the inflammatory response (Oppenheim et al., 1991; Strieter et al., 1994). Production of these leukocyte chemotactic factors by the tumor cells may influence such diverse processes as angiogenesis and production of needed growth factors by host cells.

In addition, fetal and cancer patient fibroblasts have been shown to produce a migration-stimulating factor which is not made by normal adult fibroblasts (Schor et al., 1988). This 119 kDa protein is an

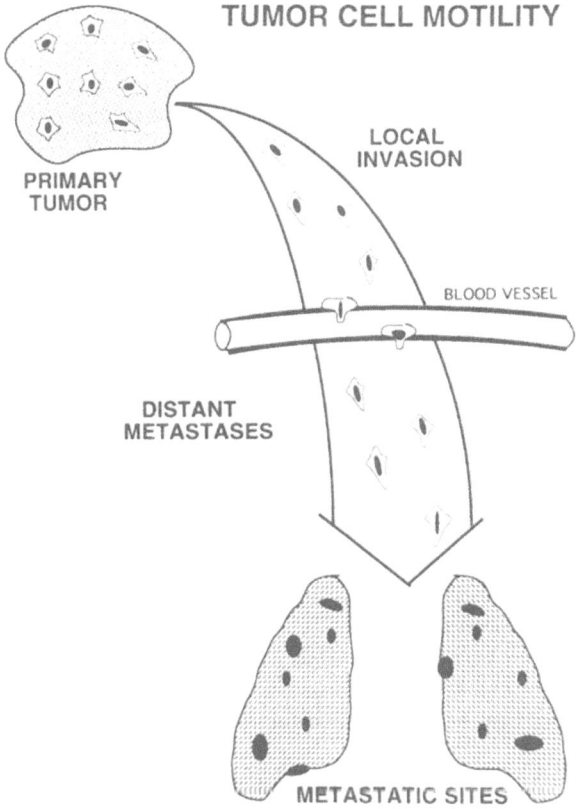

Figure 8. Tumor cell motility. Tumor cells are capable of responding to a variety of stimuli in a motile fashion. These different stimuli may take on greater or lesser significance during different stages of the metastatic cascade. The initial motile stimulus in or near the primary tumor may be an autocrine factor such as autotaxin. As the tumor cell moves into the stroma and invades local blood vessels and lymphatics, extracellular matrix proteins, either insoluble within the matrix or partially degraded in solution, may play a part in directing this motility. Once inside the vascular tree, the 'homing' of the tumor cell to an appropriate metastatic site may depend on cytokines such as IGF-1 which stimulate growth as well as a chemotactic response.

autocrine motility factor for fibroblasts which acts by stimulating hyaluronic acid secretion (Grey et al., 1989). Partial sequence analysis of the amino terminus of the migration-stimulating factor revealed a homology with the gelatin-binding domain of fibronectin (Schor et al., 1993). Since production of this factor is associated with fetal-like fibroblasts, the factor may be associated with a disruption of normal epithelial-mesenchymal interactions (Schor et al., 1993).

Conclusion

In summary, tumor cells have proved capable of responding in a motile fashion to a wide variety of stimuli, including factors which normally stimulate cell adhesion or growth (Fig. 8). These different stimuli could take on greater or lesser importance during different stages of tumor cell invasion and progression, influencing both the time course of each step and the ultimate site of the metastatic nidus. In addition, this flexibility of cellular activation indicates a likely mechanism by which tumor cells adapt to the different microenvironments that they encounter during the metastatic cascade.

References

Albelda, S.M., Mette, S.A., Elder, D.E., Stewart, R., Damjanovich, L., Herlyn, M. and Buck, C.A. (1990) Integrin distribution in malignant melanoma: Association of the β_3 subunit with tumor progress. *Cancer Res.* 50: 6757–6764.

Anzano, M.A., Roberts, A.B., Smith, J.M., Sporn, M.B. and DeLarco, J.E. (1983) Sarcoma growth factor from conditioned medium of virally transformed cells is composed of both type α and type β transforming growth factors. *Proc. Natl. Acad. Sci. USA* 80: 6264–6268.

Aresu, O., Nicolò, G., Allavena, G., Melchiori, A., Schmidt, J., Kopp, J.B., d'Amore, E., Chader, G.D. and Albini, A. (1991) Invasive activity, spreading and chemotactic response to laminin are properties of high but not low metastatic mouse osteosarcoma cells. *Invasion Metastasis* 11: 2–13.

Atnip, K.D., Carter, L.M., Nicolson, G.L. and Dabbous, M.K. (1987) Chemotactic response of rat mammary adenocarcinoma cell clones to tumor-derived cytokines. *Biochem. Biophys. Res. Commun.* 146: 996–1002.

Aznavoorian, S., Liotta, L.A. and Kupchik, H.Z. (1990a) Characteristics of invasive and noninvasive human colorectal adenocarcinoma cells. *J. Natl. Cancer Inst.* 82: 1485–1492.

Aznavoorian, S., Stracke, M.L., Krutzsch, H.C., Schiffmann, E. and Liotta, L.A. (1990b) Signal transduction for chemotaxis and haptotaxis by matrix molecules in tumor cells. *J. Cell Biol.* 110: 1427–1438.

Basara, M.L., McCarthy, J.B., Barnes, D.W. and Furcht, L.T. (1985) Stimulation of haptotaxis and migration of tumor cells by serum spreading factor. *Cancer Res.* 45: 2487–2494.

Belli, S.I., van Driel, I.R. and Goding, J.W. (1993) Identification and characterization of a soluble form of the plasma cell membrane glycoprotein PC-1 (5'-nucleotide phosphodiesterase). *Eur. J. Biochem.* 217: 421–428.

Bellusci, S., Moens, G., Gaudino, G., Comoglio, P., Nakamura, T., Thiery, J.-P. and Jouanneau, J. (1994) Creation of an hepatocyte growth factor/scatter factor autocrine loop in carcinoma cells induces invasive properties associated with increased tumorigenicity. *Oncogene* 9: 1091–1099.

Birchmeier, W., Weidner, K.M. and Behrens, J. (1993) Molecular mechanisms leading to loss of differentiation and gain of invasiveness in epithelial cells. *J. Cell Sci.* 17: 159–164.

Bottaro, D.P., Rubin, J.S., Faletto, D.L., Chan, A.M.-L., Kmiecik, T.E., Vande Woude, G.F. and Aaronson, S.A. (1991) Identification of the hepatocyte growth factor as the c-*met* proto-oncogene product. *Science* 251: 802–804.

Bottazzi, B., Colotta, F., Sica, A., Nobili, N. and Mantovani, A. (1990) A chemoattractant expressed in human sarcoma cells (tumor-derived chemotactic factor, TDCF) is identical to monocyte chemoattractant protein-1/monocyte chemotactic and activating factor (MCP-1/MCAF). *Int. J. Cancer* 45: 795–797.

Boyer, B., Vallés, A.-M., Tucker, G.C., Delouvée, A. and Thiery, J.P. (1993) Involvement of cell motility in tumor progression. *In:* G. Jones, C. Wigley and R. Warn (eds): *Cell Behaviour: Adhesion and Motility.* The Company of Biologists Ltd., Cambridge, UK, pp 183–195.

Buckley, M.F., Loveland, K.A., McKinstry, W.J., Garson, O.M. and Goding, J.W. (1990) Plasma cell membrane glycoprotein PC-1: cDNA cloning of the human molecule, amino acid sequence, and chromosomal location. *J. Biol. Chem.* 265: 17506–17511.

Burthem, J., Baker, P.K., Hunt, J.A. and Cawley, J.C. (1994) hairy cell interactions with extracellular matrix: Expression of specific integrin receptors and their role in the cell's response to specific adhesive proteins. *Blood* 84: 873–882.

Cai, J., Gill, P.S., Masood, R., Chandrasoma, P., Jung, B., Law, R.E. and Radka, S.F. (1994) Oncostatin-M is an autocrine growth factor in kaposi's sarcoma. *Am. J. Pathol.* 145: 74–79.

Chelberg, M.K., Tsilibary, E.C., Hauser, A.R. and McCarthy, J.B. (1989) Type IV collagen-mediated melanoma cell adhesion and migration: involvement of multiple, distinct domains of the collagen molecule. *Cancer Res.* 49: 4796–4802.

Chelberg, M.K., McCarthy, J.B., Skubitz, A.P.N., Furcht, L.T. and Tsilibary, E.C. (1990) Characterization of a synthetic peptide from type IV collagen that promotes melanoma cell adhesion, spreading, and motility. *J. Cell Biol.* 111: 262–270.

Condeelis, J. (1993) Life at the leading edge: the formation of cell protrusions. *Ann. Rev. Physiol.* 9: 411–444.

Culp, J.S., Blytt, H.J., Hermodson, M. and Butler, L.G. (1985) Amino acid sequence of the active site peptide of bovine intestinal 5'-nucleotide phosphodiesterase and identification of the active site residue as threonine. *J. Biol. Chem.* 260: 8320–8324.

Danen, E.H.J., van Muijen, G.N.P., van de Wiel-van Kemenade, E., Jansen, K.F.J., Ruiter, D.J. and Figdor, C.G. (1993) Regulation of integrin-mediated adhesion to laminin and collagen in human melanocytes and in non-metastatic and highly metastatic human melanoma cells. *Int. J. Cancer* 54: 315–321.

El-Badry, O.M., Minniti, C., Kohn, E.C., Houghton, P.J., Daughaday, W.H. and Helman, L.J. (1990) Insulin-like growth factor II acts as an autocrine growth and motility factor in human rhabdomyosarcoma tumors. *Cell Growth & Diff.* 1: 325–331.

Etoh, T., Byers, H.R. and Mihm, M.C., Jr. (1992) Integrin expression in malignant melanoma and their role in cell attachment and migration on extracellular matrix proteins. *J. Dermatol.* 19: 841–846.

Evans, C.P., Walsh, D.S and Kohn, E.C. (1991) An autocrine motility factor secreted by the Dunning R-3327 rat prostatic adenocarcinoma cell subtype AT2.1 *Int. J. Cancer* 49: 109–113.

Faassen, A.E., Schrager, J.A., Klein, D.J., Oegema, T.R., Couchman, J.R. and McCarthy, J.B. (1992) A cell surface chondroitin sulfate proteoglycan, immunologically related to CD44, is involved in Type I collagen-mediated melanoma cell motility and invasion. *J. Cell Biol.* 116: 521–531.

Fujita, S., Suzuki, H., Kinoshita, M. and Hirohashi, S. (1992) Inhibition of cell attachment, invasion and metastasis of human carcinoma cells by anti-integrin β_1 subunit antibody. *Jpn. J. Cancer Res.* 3: 1317–1326.

Funakoshi, I., Kato, H., Horie, K., Yano, T., Kobayashi, H., Inoue, T., Suzuki, H., Fukui, S., Tsukahara, M., Kajii, T. and Yamashina, I. (1992) Molecular cloning of cDNAs for human fibroblast nucleotide pyrophosphatase. *Arch. Biochem. Biophys.* 295: 180–187.

Gehlsen, K.R., Davis, G.E. and Sriramarao, P. (1992) Integrin expression in human melanoma cells with differing invasive and metastatic properties. *Clin. Exp. Metastasis* 10: 111–120.

Gleiber, W.E. and Schiffmann, E. (1984) Identification of a chemoattractant for fibroblasts produced by breast carcinoma cell lines. *Cancer Res.* 44: 3398–3402.

Grey, A.-M., Schor, A.M., Rushton, G., Ellis, I. and Schor, S.L. (1989) Purification of the migration stimulating factor produced by fetal and breast cancer patient fibroblasts. *Proc. Natl. Acad. Sci. USA* 86: 2438–2442.

Guirguis, R., Margulies, I., Taraboletti, G. and Liotta, L. (1987) Cytokine-induced pseudopodial protrusion is coupled to tumour cell migration. *Nature* 329: 261–263.

Halaban, R., Kwon, B.S., Ghosh, S., Bovis, P.D. and Baird, A. (1988) bFGF as an autocrine growth factor for human melanomas. *Oncogene Res.* 3: 177–186.

Harahap, A.R. and Goding, J.W. (1988) Distribution of the murine plasma cell antigen PC-1 in non-lymphoid tissues. *J. Immunol.* 141: 2317–2320.

Hayashi, H., Yoshida, K., Ozaki, T. and Ushijima, K. (1970) Chemotactic factor associated with invasion of cancer cells. *Nature* 226: 174–175.

Honn, K.V., Chen, Y.Q., Timar, J., Onoda, J.M., Hatfield, J.S., Fligiel, S.E.G., Steinert, B.W., Diglio, C.A., Grossi, I.M., Nelson, K.K. and Taylor, J.D. (1992) $\alpha_{IIb}\beta_3$ integrin expression and function in subpopulations of murine tumors. *Exp. Cell Res.* 201: 23–32.

Huff, K.K., Kaufman, D., Gabbay, K.H., Spencer, E.M., Lippman, M.E. and Dickson, R.B. (1986) Secretion of an insulin-like growth factor-I-related protein by human breast cancer cells. *Cancer Res.* 46: 4613–4619.

Humphries, M.J., Olden, K. and Yamada, K.M. (1986) A synthetic peptide from fibronectin inhibits experimental metastasis of murine melanoma cells. *Science* 233: 467–470.

Isoai, A., Giga-Hama, Y., Shinkai, K., Mukai, M., Akedo, H. and Kumagai, H. (1990) Purification and characterization of tumor invasion-inhibiting factors. *Jpn. J. Cancer Res.* 81: 909–914.

Isoai, A., Giga-Hama, Y., Shinkai, K., Mukai, M., Akedo, H. and Kumagai, H. (1992) Tumor invasion-inhibiting factor 2: Primary structure and inhibiting effect on invasion *in vitro* and pulmonary metastasis of tumor cells. *Cancer Res.* 52: 1422–1426.

Isoai, A., Goto-Tsukamoto, H., Yamori, T., Oh-hara, T., Tsuruo, T., Siletti, S., Raz, A., Watanabe, H., Akedo, H. and Kumagai, H. (1994) Inhibitory effects of tumor invasion-inhibiting factor 2 and its conjugate on disseminating tumor cells. *Cancer Res.* 54: 1264–1270.

Iwamoto, Y., Robey, F.A., Graf, J., Sasaki, M., Kleinman, H.K., Yamada, Y. and Martin, G.R. (1987) YIGSR, a synthetic laminin pentapeptide, inhibits experimental metastasis formation. *Science* 238: 1132–1134.

Jacobs, S., Cook, S., Svoboda, M.E. and Van Wyk, J.J. (1986) Interaction of the monoclonal antibodies alpha IR-1 and alpha IR-3 with insulin and somatomedin-C receptors. *Endocrinol.* 118: 223–226.

Janney, P.A. (1994) Phosphoinositides and calcium as regulators of cellular actin assembly and disassembly. *Ann. Rev. Physiol.* 56: 169–191.

Kahan, B.W. and Kramp, D.C. (1987) Nerve growth factor stimulation of mouse embryonal cell migration. *Cancer Res.* 47: 6324–6328.

Kleinman, H.K., Weeks, B.S., Schnaper, H.W., Kibbey, M.C., Yamamura, K. and Grant, D.S. (1993) The laminins: A family of basement membrane glycoproteins important in cell differentiation and tumor metastasis. *Vit. Hormones* 47: 161–186.

Kohn, E.C., Francis, E.A., Liotta, L.A. and Schiffmann, E. (1990) Heterogeneity of the motility response in malignant tumor cells: a biological basis for the diversity and homing of metastatic cells. *Int. J. Cancer* 46: 287–292.

Konomi, K., Imayama, S., Nagae, S., Terasaka, R., Chijiwa, K. and Yashima, Y. (1992) Melanocyte chemotactic factor produced by skin metastases of a breast carcinoma. *J. Surg. Oncol.* 50: 62–66.

Kramer, R.H., McDonald, K.A., Crowley, E., Ramos, D.M. and Damsky, C.H. (1989) Melanoma cell adhesion to basement membrane mediated by integrin-related complexes. *Cancer Res.* 49: 393–402.

Kretsinger, R.H. (1987) Calcium coordination and the calmodulin fold: Divergent versus convergent evolution. *Cold Spring Harbor Symp. Quant. Biol.* 52: 499–510.

Leavesly, D.I., Ferguson, G.D., Wayner, E.A. and Cheresh, D.A. (1992) Requirement of the integrin β_3 receptor for carcinoma cell spreading or migration on vitronectin and fibronectin. *J. Cell Biol.* 117: 1101–1107.

Lester, B.R. and McCarthy, J.B. (1992) Tumor cell adhesion to the extracellular matrix and signal transduction mechanisms implicated in tumor cell motility, invasion, and metastasis. *Cancer Metastasis Rev.* 11: 31–44.

Liapi, C., Raynaud, F., Anderson, W.B. and Evain-Brion, D. (1990) High chemotactic response to platelet-derived growth factor of a teratocarcinoma differentiated mesodermal cell line. *In Vitro Cell. Dev. Biol.* 26: 388–392.

Liotta, L.A., Abe, S., Robey, P. and Martin, G. (1979) Preferential digestion of basement membrane collagen by an enzyme derived from a metastatic murine tumor. *Proc. Natl. Acad. Sci. USA* 76: 2268–2272.

Liotta, L.A., Thorgeirsson, U.P. and Garbisa, S. (1982) Role of collagenases in tumor cell invasion. *Cancer Metastasis Rev.* 1: 277–288.

Liotta, L.A., Rao, C.N. and Barsky, S.H. (1983) Tumor invasion and the extracellular matrix. *Lab. Investigation* 49: 636–649.

Liotta, L.A., Mandler, R., Murano, G., Katz, D.A., Gordon, R.K., Chiang, P.K. and Schiffmann, E. (1986) Tumor cell autocrine motility factor. *Proc. Natl. Acad. Sci. USA* 83: 3302–3306.

Liotta, L.A. and Stracke, M.L. (1988) Tumor invasion and metastases: biochemical mechanisms. *In:* M.E. Lippmen and R.B. Dickson (eds): *Breast Cancer: Cellular and Molecular Biology.* Kluwer Academic Publishers, Boston, pp 223–238.

Lola, P.K. and Graham, C.H. (1990) Mechanisms of trophoblast invasiveness and their control: The role of proteases and protease inhibitors. *Cancer Metastasis Rev.* 9: 369–380.

Makabe, T., Saiki, I., Murata, J., Ohdate, Y., Kawase, Y., Taguchi, Y., Shimojo, T., Kumizuka, F., Kato, I. and Azuma, I. (1990) Modulation of haptotactic migration of metastatic melanoma cells by the interaction between heparin and heparin-binding domain of fibronectin. *J. Biol. Chem.* 265: 14270–14276.

Mantovani, A., Ming, W.J., Balotta, C., Abdeljalil, B. and Bottazzi, B. (1986) Origin and regulation of tumor-associated macrophages: the role of tumor-derived chemotactic factor. *Biochim. Biophys. Acta* 865: 59–67.

Marshall, J.F., Nesbitt, S.A., Helfrich, M.H., Horton, M.A., Polakova, K. and Hart, I.R. (1991) Integrin expression in human melanoma cell lines: Heterogeneity of vitronectin receptor composition and function. *Int. J. Cancer* 49: 924–931.

McCarthy, J.B., Palm, S.L. and Furcht, L.T. (1983) Migration by haptotaxis of a Schwann cell tumor line to the basement membrane glycoprotein laminin. *J. Cell Biol.* 110: 1427–1438.

McCarthy, J.B. and Furcht, L.T. (1984) Laminin and fibronectin promote the haptotactic migration of B16 melanoma cell *in vitro. J. Cell Biol.* 98: 1474–1480.

McCarthy, J.B., Hagen, S.T. and Furcht, L.T. (1986) Human fibronectin contains distinct adhesion- and motility-promoting domains for metastatic melanoma cells. *J. Cell Biol.* 102: 179–188.

McCarthy, J.B., Skubitz, A.P.N., Palm, S.L. and Furcht, L.T. (1988) Metastasis inhibition of different tumor types by purified laminin fragments and a heparin-binding fragment of fibronectin. *J. Natl. Cancer Inst.* 80: 108–115.

Mensing, H., Albini, A., Krieg, T., Pontz, B.F. and Müller, P.K. (1984) Enhanced chemotaxis of tumor-derived and virus-transformed cells to fibronectin and fibroblast-conditioned medium. *Int. J. Cancer* 33: 43–48.

Mignatti, P., Robbins, E. and Rifkin, D.B. (1986) Tumor invasion through the human amniotic membrane: Requirement for a proteinase cascade. *Cell* 47: 487–498.

Minniti, C.P., Kohn, E.C., Grubb, J.H., Sly, W.S., Oh, Y., Müller, H.L., Rosenfeld, R.G and helman, L.J. (1992) The insulin-like growth factor-II (IGF-II)/mannose 6-phosphate receptor mediates IGF-II-induced motility in human rhabdomyosarcoma cells. *J. Biol. Chem.* 267: 9000–9004.

Mooradian, D.L., McCarthy, J.B., Komanduri, K.V. and Furcht, L.T. (1992) Effects of transforming growth factor-$\beta 1$ on human pulmonary adenocarcinoma cell adhesion, motility, and invasion *in vitro. J. Natl. Cancer Inst.* 84: 523–527.

Murata, J., Lee, H.Y., Clair, T., Krutzsch, H.C., Årestad, A.A., Sobel, M.E., Liotta, L.A. and Stracke, M.L. (1994) cDNA cloning of the human tumor motility-stimulating protein, autotaxin, reveals a homology with phosphodiesterase. *J. Biol. Chem.* 269: 30479–30484.

Nabeshima, K., Kataoka, H. and Koono, M. (1986) Enhanced migration of tumor cells in response to collagen degradation products and tumor cell collagenolytic activity. *Invasion Metastasis* 6: 270–286.

Nabi, I.R., Watanabe, H. and Raz, A. (1990) Identification of B16-F1 melanoma autocrine motility-like receptor. *Cancer Res.* 50: 409–414.

Naldini, L., Vigna, E., Narsimhan, R.P., Gaudino, G., Zarnegar, R., Michalopoulos, G.K. and Comoglio, P.M. (1991) hepatocyte growth factor (HGF) stimulates the tyrosine kinase activity of the receptor encoded by the proto-oncogene c-*met. Oncogene* 6: 501–504.

Natali, P.G., Nicotra, M.R., Di Renzo, M.F., Prat, M., Bigotti, A., Cavaliere, R. and Comoglio, P.M. (1993) Expression of the c-*met*/HGF receptor in human melanocyte neoplasms: Demonstration of the relationship to malignant melanoma tumour progression. *Br. J. Cancer* 68: 746–750.

Nip, J., Shibata, H., Loskutoff, D.J., Cheresh, D.A. and Brodt, P. (1992) Human melanoma cells derived from lymphatic metastases use integrin $\alpha_v\beta_3$ to adhere to lymph node vitronectin. *J. Clin. Invest.* 90: 1406–1413.

Nomizu, M., Yamamura, K., Kleinman, H.K. and Yamada, Y. (1993) Multimeric forms of Tyr-Ile-Gly-Ser-Arg (YIGSR) peptide enhance the inhibition of tumor growth and metastasis. *Cancer Res.* 53: 3459–3461.

Oda, Y., Kuo, M.-D., Huang, S.S. and Huang, J.S. (1991) The plasma cell membrane glycoprotein, PC-1, is a threonine-specific protein kinase stimulated by acidic fibroblast growth factor. *J. Biol. Chem.* 266: 16791–16795.

Ohnishi, T., Arita, N., Hayakawa, T., Izumoto, S., Taki, T. and Yamamoto, H. (1990) Motility factor produced by malignant glioma cells: Role in tumor invasion. *J. Neurosurg.* 73: 881–888.

Oppenheim, J.J., Zachariae, C.O.C., Mukaida, N. and Matsushima, K. (1991) Properties of the novel proinflammatory supergene 'intercrine' cytokine family. *Annu. Rev. Immunol.* 9: 617–648.

Ozaki, T., Yoshida, K., Ushijima, K. and Hayashi, H. (1971) Studies on the mechanisms of invasion in cancer. II. *In vivo* effects of a factor chemotactic for cancer cells. *Int. J. Cancer* 7: 93–100.

Pienta, K.J., Isaacs, W.B., Vindivich, D. and Coffey, D.S. (1991) The effects of basic fibroblast growth factor and suramin on cell motility and growth of rat prostate cancer cells. *J. Urol.* 145: 199–202.

Rebbe, N.F., Tong, B.D., Finley, E.M. and Hickman, S. (1991) Identification of nucleotide pyrophosphatase/phosphodiesterase I activity associated with the mouse plasma cell differentiation antigen PC-1. *Proc. Natl. Acad. Sci. USA* 88: 5192–5196.

Rodeck, U., Herlyn, M., Menssen, H.D., Furlanetto, R.W. and Koprowski, H. (1987) Metastatic but not primary melanoma cell lines grow *in vitro* independently of exogenous growth factors. *Int. J. Cancer* 40: 687–690.

Rong, S., Segal, S., Anver, M., Resau, J.H. and Vande Woude, G.F. (1994) Invasiveness and metastasis of NIH 3T3 cells induced by Met-hepatocyte growth factor/scatter factor autocrine stimulation. *Proc. Natl. Acad. Sci. USA* 91: 4731–4735.

Rosen, E.M., Knesel, J., Goldberg, I.D., Jin, L., Bhargava, M., Joseph, A., Zitnik, R., Wines, J., Kelley, M. and Rockwell, S. (1994) Scatter factor modulates the metastatic phenotype of the EMT6 mouse mammary tumor. *Int. J. Cancer* 57: 706–714.

Royce, L.S., Martin, G.R. and Kleinman, H.K. (1992) Induction of an invasive phenotype in benign tumor cells with a laminin-A chain synthetic peptide. *Invasion Metastasis* 12: 149–155.

Ruff, M., Schiffmann, E., Terranova, V. and Pert, C.B. (1987) Neuropeptides are chemoattractants for human tumor cells and monocytes: A possible mechanism for metastasis. *Clin. Immunol. Immunopath.* 37: 387–396.

Schofield, P.N., Granerus, M., Lee, A., Ekström, T.J. and Engström, W. (1992) Concentration-dependent modulation of basic fibroblast growth factor action on multiplication and locomotion of human teratocarcinoma cells. *FEBS Lett.* 298: 154–158.

Schor, S.L., Schor, A.M., Grey, A.M. and Rushton, G. (1988) Fetal and cancer patient fibroblasts produce an autocrine migration-stimulating factor not made by normal adult cells. *J. Cell Sci.* 90: 391–399.

Schor, S.L., Grey, A.M., Ellis, I., Schor, A.M., Coles, B. and Murphy, R. (1993) Migration stimulating factor (MSF): Its structure, mode of action and possible function in health and disease. *In:* G. Jones, C. Wigley and R. Warn (eds): *Cell Behaviour: Adhesion and Motility.* The Company of Biologists Ltd., Cambridge, UK, pp 235–251.

Seftor, R.E.B., Seftor, E.A., Gehlsen, K.R., Stetler-Stevenson, W.G., Brown, P.D., Ruoslahti, E. and Hendrix, M.J.C. (1992) Role of the $\alpha_v\beta_3$ integrin in human melanoma cell invasion. *Proc. Natl. Acad. Sci. USA* 89: 1557–1561.

Seiffert, D. and Loskutoff, D.J. (1991) Evidence that type 1 plasminogen activator inhibitor binds to the somatomedin B domain of vitronectin *J. Biol. Chem.* 266: 2824–2830.

Seiffert, D., Ciambrone, G., Wagner, N.V., Binders, B.R. and Loskutoff, D.J. (1994) The somatomedin B domain of vitronectin: structural requirements for the binding and stabilization of active type 1 plasminogen activator inhibitor. *J. Biol. Chem.* 269: 2659–2666.

Seiki, M., Sato, H., Liotta, L.A. and Schiffmann, E. (1991) Comparison of autocrine mechanisms promoting motility in two metastatic cell lines: human melanoma and *ras*-transfected NIH3T3 cells. *Int. J. Cancer* 49: 717–720.

Siletti, S., Watanabe, H., Hogan, V., Nabi, I.R. and Raz, A. (1991) Purification of B16-F1 melanoma autocrine motility factor and its receptor. *Cancer Res.* 51: 3507–3511.

Situ, R., Lee, E.C., McCoy, J.P., Jr. and Varani, J. (1984) Stimulation of murine tumour cell motility by laminin. *J. Cell Sci.* 70: 167–176.

Sloane, B.R. and Honn, K.V. (1984) Cysteine proteinases and metastasis. *Cancer Metastasis Rev.* 3: 249–263.

Stoker, M., Gherardi, E., Perryman, M. and Gray, J. (1987) Scatter factor is a fibroblast-derived modulator of epithelial cell mobility. *Nature* 327: 239–242.

Stoker, M. and Gherardi, E. (1989) Scatter factor and other regulators of cell mobility. *Brit. Med. Bull.* 45: 481–491.

Stossel, T.P. (1993) On the crawling of animal cells. *Science* 260: 1086–1094.

Stossel, T.P. (1994) The machinery of cell crawling. *Scientific American* 271: 54–63.

Stracke, M.L., Guirguis, R., Liotta, L.A. and Schiffmann, E. (1987) Pertussis toxin inhibits stimulated motility independently of the adenylate cyclase pathway in human melanoma cells. *Biochem. Biophys. Res. Comm.* 146: 339–345.

Stracke, M.L., Kohn, E.C., Aznavoorian, S., Wilson, L.L., Salomon, D., Krutzsch, H.C., Liotta, L.A. and Schiffmann, E. (1988) Insulin-like growth factors stimulate chemotaxis in human melanoma cells. *Biochem. Biophys. Res. Comm.* 153: 1076–1083.

Stracke, M.L., Engel, J.D., Wilson, L.L., Rechler, M.M., Liotta, L.A. and Schiffmann, E. (1989) The type I insulin-like growth factor receptor is a motility receptor in human melanoma cells. *J. Biol. Chem.* 264: 21544–21549.

Stracke, M.L., Krutzsch, H.C., Unsworth, E.J., Årestad, A., Cioce, V., Schiffmann, E. and Liotta, L.A. (1992) Identification, purification, and partial sequence analysis of autotaxin, a novel motility-stimulating protein. *J. Biol. Chem.* 267: 2524–2529.

Stracke, M.L., Soroush, M., Liotta, L.A. and Schiffmann, E. (1993) Cytoskeletal agents inhibit motility and adherence of human tumor cells. *Kidney International* 43: 151–157.

Straüli, P. and Haemmerli, G. (1984) The role of cancer cell motility in invasion. *Cancer Metastasis Rev.* 3: 127–141.

Strieter, R.M., Koch, A.E., Antony, V.B., Fick, R.B. Jr., Standiford, T.J. and Kunkel, S.L. (1994) The immunopathology of chemotactic cytokines: The role of interleukin-8 and monocyte chemoattractant protein-1. *J. Lab. Clin. Med.* 123: 183–197.

Sunitha, I., Meighen, D.L., Hartman, D.-P., Thompson, E.W., Byers, S.W. and Avigan, M.I. (1994) Hepatocyte growth factor stimulates invasion across reconstituted basement membranes by a new human small intestinal cell line. *Clin. Exp. Metastasis* 12: 143–154.

Tamm, I., Cardinale, I., Krueger, J., Murphy, J.S., May, L.T. and Sehgal, P.B. (1989) Interleukin 6 decreases cell–cell association and increases motility of ductal breast carcinoma cells. *J. Exp. Med.* 170: 1649–1669.

Taraboletti, G., Roberts, D.D. and Liotta, L.A. (1987) Thrombospondin-induced tumor cell migration: haptotaxis and chemotaxis are mediated by different domains. *J. Cell Biol.* 105: 2409–2415.

Tashiro, K.-I., Sephel, G.C., Greatorex, D., Sasaki, M., Shirashi, N., Martin, G.R., Kleinman, H.K. and Yamada, Y. (1991) The RGD-containing site of the mouse laminin A chain is active for cell attachment, spreading, migration and neurite outgrowth. *J. Cell. Physiol.* 146: 451–459.

Tchao, R. (1982) Novel forms of epithelial cell motility on collagen and on glass surfaces. *Cell Motil.* 4: 333–341.

Tilly, B.C., Tertoolen, L.G.J., Remorie, R., Ladoux, A., Verlaan, I., de Laat, S.W. and Moolenaar, W.H. (1990) Histamine as a growth factor and chemoattractant for human carcinoma and melanoma cells: action through Ca^{2+}-mobilizing H_1 receptors. *J. Cell Biol.* 110: 1211–1215.

Timpl, R. (1993) Proteoglycans of basement membranes. *Experientia* 49(5): 417–428.

Todaro, G.J., Fryling, C. and DeLarco, J.E. (1980) Transforming growth factors produced by certain human tumor cells: polypeptides that interact with epidermal growth factor receptors. *Proc. Natl. Acad. Sci. USA* 77: 5258–5262.

Tsao, M.-S., Zhu, H., Giaid, A., Viallet, J., Nakamura, T. and Park, M. (1993) Hepatocyte growth factor/scatter factor is an autocrine factor for normal bronchial epithelial and lung carcinoma cells. *Cell Growth & Diff.* 4: 571–579.

Turley, E.A., Vandeligt, K. and Clary, C. (1991) Hyaluronan and a cell-associated hyaluronan binding protein regulate the locomotion of *ras*-transformed cells. *J. Cell Biol.* 112: 1041–1047.

Vallés, A.M., Boyer, B., Badet, J., Tucker, G.C., Barritault, D. and Thiery, J.P. (1990) Acidic fibroblast growth factor is a modulator of epithelial plasticity in a rat bladder carcinoma cell line. *Proc. Natl. Acad. Sci. USA* 87: 1124–1128.

Venstrom, K.A. and Reichardt, L.F. (1993) Extracellular matrix 2: Role of extracellular matrix molecules and their receptors in the nervous system. *FASEB J.* 7: 996–1003.

Verhasselt, B., Van Damme, J., van Larebeke, N., Put, W., Bracke, M., De Potter, C. and Mareel, M. (1992) Interleukin-1 is a motility factor for human breast carcinoma cells *in vitro*: Additive effect with interleukin-6. *Eur. J. Cell Biol.* 59: 449–457.

Wang, J.M., Taraboletti, G., Matsushima, K., Van Damme, J. and Montovani, A. (1990) Induction of haptotactic migration of melanoma cells by neutrophil activating protein/interleukin-8. *Biochem. Biophys. Res. Comm.* 169: 165–170.

Watanabe, H., Nabi, I.R. and Raz, A. (1991) The relationship between motility factor receptor internalization and the lung colonization of murine melanoma cells. *Cancer Res.* 51: 2699–2705.

Watanabe, H., Kanbe, K. and Chigira, M. (1994) Differential purification of autocrine motility factor derived from a murine protein-free fibrosarcoma. *Clin. Exp. Metastasis* 12: 155–163.

Weidner, K.M., Behrens, J., Vandekerckhove, J. and Birchmeier, W. (1990) Scatter factor: molecular characteristics and effect on the invasiveness of epithelial cells. *J. Cell Biol.* 111: 2097–2108.

Weidner, K.M., Arakaki, N., Hartmann, G., Vandekerckhove, J., Weingart, S., Rieder, H., Fonatsch, C., Tsubouchi, H., Hishida, T., Daikuhara, Y. and Birchmeier, W. (1991) Evidence for the identity of human scatter factor and human hepatocyte growth factor. *Proc. Natl. Acad. Sci. USA* 88: 7001–7005.

Weidner, K.M., Hartmann, G., Naldini, L., Comoglio, P.M., Sachs, M., Fonatsch, C., Rieder, H. and Birchmeier, W. (1993) Molecular characteristics of HGF-SF and its role in cell motility and invasion. *In:* I.D. Goldberg and E.M. Rosen (eds): *Hepatocyte Growth Factor-Scatter Factor (HGF-SF) and the C-Met Receptor.* Birkhäuser Verlag, Basel, pp 311–328.

Wewer, U.M., Taraboletti, G., Sobel, M.E., Albrechtsen, R. and Liotta, L.A. (1987) Role of laminin receptor in tumor cell migration. *Cancer Res.* 47: 5691–5698.

Williams, N.N., Györfi, T., Iliopoulos, D., Herlyn, D., Greenstein, D., Linnenbach, A.J., Daly, J.M., Jensen, P., Rodeck, U. and Herlyn, M. (1992) Growth factor independence and invasive properties of colorectal carcinoma cells. *Int. J. Cancer* 50: 274–280.

Woolley, D.E. (1984) Collagenolytic mechanisms in tumor cell invasion. *Cancer Metastasis Rev.* 3: 361–372.

Yabkowitz, R., Mansfield, P.J., Dixit, V.M. and Suchard, S.J. (1993) Motility of human carcinoma cells in response to thrombospondin: Relationship to metastatic potential and thrombospondin structural domains. *Cancer Res.* 53: 378–387.

Yamada, K.M., Kennedy, D.W., Yamada, S.S., Gralnick, H., Chen, W.-T. and Akiyama, S.K. (1990) Monoclonal antibody and synthetic peptide inhibitors of human tumor cell migration. *Cancer Res.* 50: 4485–4496.

Yoshinaga, I.G., Vink, J., Dekker, S.K., Mihm, M.C., Jr. and Byers, H.R. (1993) Role of $\alpha_3\beta_1$ and $\alpha_2\beta_1$ integrins in melanoma cell migration. *Melanoma Res.* 3: 435–441.

Zachariae, C.O.C., Anderson, A.O., Thompson, H.L., Appella, E., Mantovani, A., Oppenheim, J.J. and Matsushima, K. (1990) Properties of monocyte chemotactic and activating factor (MCAF) purified from a human fibrosarcoma cell line. *J. Exp. Med.* 171: 2177–2182.

Zachariae, C.O.C., Thestrup-Pedersen, K. and Matsushima, K. (1991) Expression and secretion of leukocyte chemotactic cytokines by normal human melanocytes and melanoma cells. *J. Invest. Dermatol.* 97: 593–599.

Epithelial-Mesenchymal Interactions in Cancer
ed. by I.D. Goldberg & E.M. Rosen
© 1995 Birkhäuser Verlag Basel/Switzerland

Tumor-stromal cytokine interactions in ovarian neoplasms

B.M. Kacinski

Therapeutic Radiology, Obstetrics & Gynecology and Dermatology, Yale University School of Medicine, New Haven, CT 06520-8040, USA

Overview

Ovarian surface epithelium – normal physiology

During the normal monthly ovulatory cycle, the surface epithelium of the ovary undergoes a set of changes which includes (1) epithelial and stromal cell destruction at the site of ovulation, (2) epithelial and stromal cell proliferation and migration into the ruptured follicle, and (3) remodeling and re-epithelialization of the folicular 'wound' (Nicosia et al., 1989; Dietl and Marzusch, 1993). These changes in ovarian epithelial and stromal (fibroblast and vascular) structure are accompanied by infiltrates of inflammatory cells (including macrophages, granulocytes, and T-cells) which dispose of cellular debris and secrete a variety of soluble factors which alter the motility, proliferation, and differentiated phenotypes of inflammatory cells, surface epithelium, fibroblasts, and vascular endothelium.

This monthly cycle of cell destruction, proliferation, and regeneration is regulated by many juxta- and paracrine interactions involving a variety of peptide and steroid hormones and their receptors. Point mutation, rearrangement, or deletions of chromosomal DNA sequences resulting from spontaneous (or mutagen-induced) errors of DNA replication alter the coding or regulatory sequences of genes encoding any of these hormones, their receptors, or of 'regulatory' genes which control their levels of expression. These might sufficiently perturb normal ovarian epithelial cell proliferation and differentiation to stimulate the development of benign, borderline, and invasively-malignant neoplasms from normal surface epithelial precursors.

At least some of these hormones and hormone receptors are the normal cellular counterparts of so-called 'dominant' oncogenes and/or tumor suppressor genes whose altered expression and/or function have been implicated in the pathogenesis of a wide variety of human malignancies, including those of the ovary.

Oncogenes and tumor suppressors

Several well-known protooncogenes (*sis; erbB*1; *erbB*2; *fms; met*, etc.) encode cytokines or cytokine receptors. The ultimate result of the functional activation (by *mutation*) and/or over-expression of wild-type cytokine receptors is the increased proliferation, invasive differentiation, and abnormal interactions with stromal elements (increased *angiogenesis, immune cell infiltration, fibroblast proliferation*) characteristic of neoplastic cells of ovarian carcinomas and other malignancies.

Similar characteristics are observed in such non-neoplastic *invasive* cells as activated hematopoietic and immune cells (eg., 'activated' macrophages, granulocytes, and T-cells) and implanting trophoblast. Certain of these shared phenotypes may permit tumor cells to *evade* host immune defenses (much as does the implanting placenta), resist killing by otherwise toxic biological (eg., *TNF-α, Γ-interferon*) and chemotherapeutic agents (eg., alkylating agents, CDDP, anthracyclines), stimulate neovascularization (directly or via angiogenic cytokines produced by activated immune cells) and be stimulated to proliferate or differentiate by some of those very cytokines and lymphokines which have been considered for therapeutic use as immunoadjuvants or myelostimulants in immuno- or myelosuppressed cancer patients.

Peptide hormone growth factors

A variety of peptide hormones and their receptors have been implicated in growth regulation of ovarian carcinomas. Several recent reports demonstrate that a variety of growth factors, including EGF, TGF-α, FGF, IL-1, TNF-α and IGF-1 stimulate the *in vitro* proliferation of at least some ovarian carcinoma cell lines whose proliferation is often inhibited by TGF-β and Γ-interferon (Berchuck et al., 1990b; Naylor et al., 1990; Wimalasena et al., 1993; Mobus et al., 1993; Wu et al., 1993, 1994). Ovarian carcinoma-derived cell lines can also respond to other cytokines such as TNF-α and *hepatocyte growth factor* (the c-*met* ligand) by increases in cell motility, protease production, and 'invasiveness' (Naylor et al., 1990; Malik et al., 1990; Rosen et al., 1991; Marth et al., 1990a). Ovarian carcinoma-derived cells or tumor-derived lines also synthesize many potent cytokines including TGF-β (Berchuck et al., 1990b), a potent modulator of fibroblast and immune cell function; CSF-1 (Kacinski et al., 1990a; Kacinski et al., 1990b) and IL-6 (Watson et al., 1990) which are potent stimulators of macrophage proliferation and/or differentiation; PDGF-A (Henriksen et al., 1993), and FGFs (DiBlasio et al., 1993) which stimulate the proliferation of fibroblasts and endothelial cells; TNF-α (Naylor et al., 1990; Wu et al., 1993,

1994); TGF-α (Kommos et al., 1990); and, very likely, many other uncharacterized factors which may function as autocrine cytokines for ovarian carcinoma cells and/or paracrine modifiers of host stromal and immune cell function.

Ascitic fluid factors

While ascitic fluid from ovarian carcinoma patients does not appear to contain significant TGF-α or EGF activity (Wilson et al., 1991), it does contain many potent cytokines active on malignant epithelial cells and host immune cells including CSF-1 (Price et al., 1993), IL-6 (Watson et al., 1990), TGF-β (Wilson et al., 1991), TNF-α, (Moradi et al., 1993) coagulation factors (Wilson et al., 1991), and an incompletely character- ized activity which facilitates the intraperitoneal growth of certain human ovarian carcinoma cell lines in nude mice (Mills et al., 1990). *In vivo*, CSF-1 and IL-6 (Watson et al., 1990), produced abundantly by many ovarian carcinomas and activated macrophages and present at high levels in malignant ascites may act as autocrine growth and differentiation factors for malignant epithelial and host immune cells. In particular, CSF-1 (and possibly IL-6) might adversely modify host-immune anti-tu- mor responses by decreasing the high levels of macrophage *MHC class II* antigen expression (Willman et al., 1989) required for effective presentation of novel tumor antigens to CD4 + T-helper cells and by stimulating macrophages to synthesize such cytokines as IL-1, IL-6, TGF-α, & TGF-β which could stimulate tumor cell proliferation and/or dissemination. TGF-β, synthesized by some ovarian carcinoma-derived cell lines and active upon others as a negative regulator of epithelial cell growth, is also a potent immunosuppressant able to completely block T cell-mediated cytolytic responses to even very strongly antigenic tumors (Willman et al., 1989; Torre-Amione et al., 1990).

Growth factor receptor expression – biology and prognostic significance

As suggested by the above studies, ovarian carcinoma cells also express a range of different growth factor receptors which mediate responses to a wide variety of peptide hormones synthesized by tumor cells them- selves or by host stromal and immune cells.

Epidermal growth factor receptor (EGF-R)

Not unexpectedly, normal ovarian surface epithelial cells express EGF- Rs (Rodriguez et al., 1991; Bauknecht et al., 1990) and respond to EGF when cultured *in vitro*. Ovarian carcinomas often also express EGF-Rs with or without the co-expression of an activating ligand such as TGF-α (Kommos et al., 1990; Bauknecht et al., 1990; Berchuck et al., 1991; Foekens et al., 1990).

In two recently published studies, ovarian carcinoma cell expression of EGF-R correlated with adverse prognosis (Berchuck et al., 1991; Foekens et al., 1990). However, in contrast, another study (Bauknecht et al., 1990) has suggested that EGF-R-positive carcinomas may actually respond somewhat better to combination chemotherapy, a result which may relate to the *in vitro* observation (Christen et al., 1990) that EGF renders some EGF-R-positive ovarian carcinoma cell lines more sensitive to CDDP, an essential component of most modern ovarian carcinoma chemotherapeutic regimens.

Macrophage colony-stimulating factor (CSF-1), interleukin 6, and their receptors

Ovarian carcinoma cell expression of macrophage colony-stimulating factor, interleukin 6 (IL-6) and their receptors (CSF-1R = *fms*, IL-6R) has been described by several groups of investigators (Kacinski et al., 1990a; Baiocchi et al., 1991; Kommoss et al., 1994).

CSF-1R (*fms*), initially characterized as the *transforming oncogene* of several *feline retroviruses*, was later shown to encode a 150 kDa *tyrosine kinase* cytokine receptor for which the soluble and membrane bound forms of the *macrophage, colony-stimulating factor, CSF-1* are only known ligands (Coussens et al., 1986; Browning et al., 1986; Yarden et al., 1986; Yarden et al., 1987). This receptor is structurally related to B-type receptor for PDGF which maps immediately upstream of the first exon of the CSF-1R gene. Both the PDGF-R and CSF-1R genes may have been generated by *gene duplication* and subsequent divergence of a common ancestral growth factor receptor locus. CSF-1R, PDGF-R, and several other closely-related tyrosine kinase receptors (FGF-Rs, c-*kit*) all contain a unique 'spacer' sequence which interrupts the internal tyrosine kinase domain. In c-*fms*, this unique spacer is quite antigenic and has proved a useful epitope for the generation of c-*fms* specific poly- and monoclonal antibodies (Taylor et al., 1989; Kacinski et al., 1991).

In several independent studies (Kacinski et al., 1990a; Baiocchi et al., 1991), CSF-1 and CSF-1R expression was clearly localized to malignant epithelial cells and stromal macrophages of high grade, advanced stage neoplasms and low, but detectable, levels of c-*fms* transcript expression, and high level CSF-1 expression has been demonstrated (both by Northern blotting and by reverse transcription and PCR amplification) in a variety of ovarian carcinoma-derived cultured cell lines. Another study suggested metastatic ovarian carcinomas were more likely to co-express both CSF-1 and CSF-1R than less aggressive lesions (Baiocchi et al., 1991). *In vitro*, ligand activation of epithelial carcinoma cell CSF-1Rs stimulates carcinoma cell invasion through basement membrane analogues and stimulates cell surface localization of endogenous urokinase (Filderman et al., 1992; Bruckner et al., 1992) while other studies have shown that the level of CSF-1R expression in epithelial carcinoma cells is increased by glucocorticoids and progestins (Taylor et al., 1990). Our own unpublished studies also suggest that the c-*fms* transcripts of several ovarian Ca-derived cultured cell lines differ from those expressed by macrophages, trophoblast, and breast carcinoma cells in significant ways which may alter the ligand affinity and basal tyrosine kinase activity of the proteins they encode.

Similarly, expression of IL-6 and IL-6R has been demonstrated in ovarian carcinomas and carcinoma-derived cell lines while the existance of a functional mitogenic IL-6/IL-6R mitogenic loop may have been demonstrated by the application of *antisense* techniques in several ovarian carcinoma-derived cultured cell lines.

As mentioned, both CSF-1 and IL-6 are abundantly synthesized by many ovarian carcinoma lines *in vitro*; and, *in vivo*; and patients with CSF-1 or IL-6 producing tumors have markedly elevated circulating (CSF-1) and ascitic fluid (CSF-1, IL-6) levels which rise and fall in parallel with changes in neoplastic disease activity. Such results suggest that determinations of the levels of CSF-1 and IL-6 in blood and/or ascites might be exploited as 'tumor markers' (complimentary to CA-125) of ovarian carcinoma disease progression, response to therapy, or recurrence (Kacinski et al., 1989b).

Autocrine or paracrine activation of CSF-1R by tumor or stromally produced CSF-1 may result in expression by tumor cells of activated macrophage-like phenotypes while paracrine stimulation of macrophages (in tumor stroma or ascites) may interfere with host immune anti-tumor responses as described above.

HER-2/NEU expression

HER-2/*neu* (c-*erbB2*) is a growth factor receptor *tyrosine kinase* structurally related to EGF-R which is activated by several distinct TGFα-like ligands (Samanta et al., 1994; Stern et al., 1986; Stern et al., 1988; Luttress et al., 1991; Dougall et al., 1993; Connelly and Stern, 1990; Williams et al., 1993; Peles et al., 1992; Peles et al., 1993; Ben-Levy et al., 1992). Binding of these ligands to the receptor leads to receptor autophosphorylation, receptor dimerization and internalization, and activation of downstream components of cytoplasmic signal transduction involving other tyrosine and serine/threonine kinases. Alternatively, at least in murine system, mutations in the transmembrane domain appear to facilitate receptor–receptor interactions and spontaneous aggregation to produce constitutive activation of the receptor's tyrosine kinase even in the absence of ligand (Peles et al., 1992; Ben-Levy et al., 1992). HER-2/*neu* receptor molecules have also been shown to interact and aggregate with EGF-R and other HER-family tyrosine kinases such that ligand activation of either HER-2/*neu* or of another HER-family receptor results in the *tyrosine-phosphorylation* and functional *activation* of the other (Samanta et al., 1994; Connelly and Stern, 1990; Williams et al., 1993).

HER-2/*neu* amplification and over-expression has been observed in a significant fraction ($\approx 1/3$) of ovarian carcinomas (Slamon et al., 1988) and in several ovarian carcinoma-derived cultured cell lines, one of which (SKOV3) expresses a novel high molecular weight HER-2/*neu* transcript (King et al., 1992). However, it is not clear that even high level expression of HER-2/*neu* correlates with the intraperitoneal tumorigenicity of ovarian Ca-derived cell lines in nude mice (King et al., 1992) nor is it obvious that any known ovarian Ca cell lines (even those which express very high levels of HER-2/*neu* receptors) respond to any of the putative HER-2 ligands (Peles et al., 1993). However, this apparent *in vitro* insensitivity to ligand may be more the consequence of constitutive co-expression of HER-regulin and other HER-family receptor ligands by ovarian carcinoma cell lines rather than due to any abnormalities in the structure of the over-expressed receptors or their response to ligand (Kacinski, B. - unpublished observations).

Several recent studies, suggest that HER-2/*neu* antigen over-expression by ovarian carcinomas may correlate with a poorer than average prognosis (Berchuck et al., 1990a). However, this finding contrasts with two other recent reports (Haldane et al., 1990; Rubin et al., 1993) of no significant adverse prognostic impact of HER-2/*neu* antigen over-expression for ovarian carcinoma patients and with a third study which suggests that HER-2/*neu* over-expression correlates more closely with specific histopathologic features (clear cell or endometroid histology) than with either relapse-free or overall survival (Kacinski et al., 1992). In addition, significant heterogeneity in HER-2/*neu* expression has been described in some primary ovarian tumors, metastatic lesions, and in different ovarian carcinoma-derived cell lines (King et al., 1992) — much as is the case for a variety of other ovarian carcinoma tumor antigens (Berchuck et al., 1990c). Of note, HER-2/*neu* expression by ovarian carcinoma cells can be downmodulated by Γ-interferon (Marth et al., 1990b), an observation which may account for at least some of the anti-proliferative effects of this cytokine on certain ovarian carcinomas and tumor-derived cultured cell lines.

In summary, it is clear that the proliferation and differentiation of benign and neoplastic ovarian epithelial cells are under the control of a host of paracrine, juxtacrine, and autocrine cytokine circuits which lead to the controlled events of the normal ovulatory cycle and which, when deranged, lead to the uncontrolled proliferation and dissemination of ovarian carcinomas. However, at the time of the preparation of this manuscript, it is clear that we understand only a small subset of these interactions. It will take much additional study of normal and abnormal ovarian epithelial and stromal cell physiology before we can hope to offer new rational therapeutic strategies which exploit the results of our research to block or, at least, interfere with the auto-, para-, and juxtacrine circuits which regulate ovarian carcinoma proliferation and dissemination, stromal proliferation and neovascularization and host immune anti-tumor responses.

References

Baiocchi, G., Kavavagh, J.J., Talpaz, M., Wharton, J.T., Gutterman, J.U. and Kurzrock, R. (1991) Expression of the macrophage colony-stimulating factor and its receptor in gynecologic malignancies. *Cancer* 67: 990–996.

Bauknecht, T., Birmelin, G. and Kommoss, F. (1990) Clinical significance of oncogenes and growth factors in ovarian carcinomas. *J. Steroid Biochem. Mol. Biol.* 37: 855–862.

Ben-Levy, R., Peles, E., Goldman-Michael, R. and Yarden, Y. (1992) An oncogenic point mutations confers high affinity ligand binding to the neu receptor: implications for the generation of site heterogeneity. *J. Biol. Chem.* 267: 17304–17313.

Berchuck, A., Kamel, A., Whitaker, R., Kearns, B., Olt, G., Kinney, R., Soper, J.T., Dodge, R., Clarke-Pearson, D.L. and Marks, P. (1990a) Overexpression of HER-2/neu is associated with poor survival in advanced epithelial ovarian cancer. *Cancer Res.* 50: 4087–4091.

Berchuck, A., Olt, G.J., Everitt, L., Soisson, A.P., Bast, R.C., Jr. and Boyer, C.M. (1990b) The role of peptide growth factors in epithelial ovarian cancer. *Obstet. Gynecol.* 75: 255–262.

Berchuck, A., Olt, G.J., Soisson, A.P., Kamel, A., Soper, J.T., Boyer, C.M., Clarke-Pearson, D.L., Lesie, D.S. and Bast, R.C., Jr. (1990c) Heterogeneity of antigen expression in advanced epithelial ovarian cancer. *Am. J. Obstet. Gynecol.* 162: 883–888.

Berchuck, A., Rodriguez, G.C., Kamel, A., Dodge, R.K., Soper, J.T., Clarke-Pearson, D.L. and Bast, R.C., Jr. (1991) Epidermal growth factor receptor expression in normal ovarian epithelium and ovarian cancer. I. Correlation of receptor expression with prognostic factors in patients with ovarian cancer. *Am. J. Obstet. Gynecol.* 164: 669–674.

Browning, P.J., Bunn, H.F., Cline, A., Shuman, M. and Nienhaus, A.W. (1986) Replacement of COOH-terminal truncation of v-fms sequence markedly reduces transformation potential. *PNAS* 83: 7800–7804.

Bruckner, A., Filderman, E., Kirchheimer, J.C., Binder, B.R. and Remold, H.G. (1992) Endogenous receptor-bound urokinase mediates tissue invasion of the human lung carcinoma cell lines A549 and Calu-1. *Cancer Res.* 52: 3043–3047.

Connelly, P.A. and Stern, D.F. (1990) The epidermal growth factor receptor and the product of the NEU protooncogene are members of a receptor tyrosine phosphorylation cascade. *PNAS* 87: 6054–6057.

Cuossens, L., VanBeveren, C.V., Smith, D., Chen, E., Mitchell, R.L., Isacke, C.M., Verma, I.M. and Ullrich, A. (1986) Structural alteration of viral homologue of the receptor protooncogene *fms* the carboxyl terminus. *Nature* 320: 277–280.

DiBlasio, A.M., Cremononesi, K. and Vigano, O. (1993) Basic FGF and its receptor mRNA are expressed in human ovarian neoplasms. *Am. J. Obstet. Gynecol.* 169: 1517–1523.

Dietl, J. and Marzusch, K. (1993) Ovarian surface epithelium and human ovarian cancer. *Gynecol. Obstet. Invest.* 35: 129–135.

Dougall, W.C., Qian, X. and Greene, M.I. (1993) Interaction of the p185/neu and EGF-R tyrosine kinases: implications for cellular transformation and tumor therapy. *J. Cell Biochem.* 53: 61–73.

Filderman, A.E., Bruckner, A. and Kacinski, B.M. (1992) Macrophage colony-stimulating factor (CSF-1) enhances invasiveness in CSF-1 receptor-positive carcinoma cell lines. *Cancer Res.* 52: 3661–3669.

Foekens, J.A., van Putten, W.L., Portengen, H., Rodenburg, C.J., Reubi, J.C., Berns, P.M., Henzen-Logmans, S.C., van der Burg, M.E., Alexleva-Figusch, J. and Kiljn, J.G. (1990) Prognostic value of pS2 protein and receptors for epidermal growth factor (EGF-R), insulin-like growth factor-1 (IGF-1-R) and somatostatin (SS-R) in patients with breast and ovarian cancer. *J. Steroid Biochem. Mol. Biol.* 37: 815–821.

Haldane, J.S., Hird, V., Hughes, C.M. and Gullick, W.J. (1990) c-*erbB-2* oncogene expression in ovarian cancer. *J. Pathol.* 162: 231–237.

Henriksen, R., Funa, K., Wilander, E., Backstrom, T., Ridderheim, M. and Oberg, K. (1993) Expression and prognostic significance of PDGF and its receptors in epithelial ovarian neoplasms. *Cancer Res.* 53: 4550–4554.

Kacinski, B.M., Carter, D., Kohorn, E.I., Mittal, K., Bloodgood, R.S., Donahue, J., Kramer, C.A., Fischer, D., Edwards, R., Chambers, S.K., Chambers, J.T. and Schwartz, P.E. (1989a) Oncogene expression in vivo by ovarian adenocarcinomas and mixed-mullerian tumors. *Yale J. Biol. Med.* 62: 379–392.

Kacinski, B.M., Stanley, E.R., Carter, D., Chambers, J.T., Chambers, S.K., Kohorn, E.I. and Schwartz, P.E. (1989b) Circulating levels of CSF-1 (M-CSF), a lymphohematopoietic cytokine, may be a useful marker of disease status in patients with malignant ovarian neoplasms. *Int. J. Rad. Onc. Biol. Phys.* 17: 159–164.

Kacinski, B.M., Carter, D., Mittal, K., Yee, L.D., Scala, K.A., Donofrio, K., Chambers, S.K., Wang, K.I., Yang-Feng, T., Rohrschneider, L.R. and Rothwell, V.M. (1990a) Ovarian adenocarcinomas express fms-complementary transcripts and fms antigen, often with coexpression of CSF-1. *Am. J. Pathol.* 137: 135–147.

Kacinski, B.M., Chambers, S.K., Carter, D., Filderman, A.E. and Stanley, E.R. (1990b) The macrophage colony stimulating factor CSF-1, an auto- and paracrine tumor cytokine, is also a circulating 'tumor marker' in patients with ovarian, endometrial and pulmonary neoplasms. *Prog. Leuk. Biol.* 10B: 393–400.

Kacinski, B.M., Scata, K.A., Carter, D., Yee, L.D., Sapi, E., King, B.L., Chambers, S.K., Jones, M.A., Pirro, M.H., Stanley, E.R. and Rohrschneider, L.R. (1991) *fms* (CSF-1 receptor) and CSF-1 transcripts and protein are expressed by human breast carcinomas *in vivo* and *in vitro*. *Oncogene* 6: 941–952.

Kacinski, B.M., Mayer, A.R., King, B.L., Canter, D.T. and Chambers, S.K. (1992) *neu* protein overexpression in benign, borderline and malignant ovarian neoplasms. *Gynecol. Oncol.* 44: 245–253.

King, B.L., Carter, D., Foellmer, H.G. and Kacinski, B.M. (1992) *neu* proto-oncogene amplification and expression in ovarian adenocarcinoma cell lines. *Am. J. Pathol.* 140(1): 23–31.

Kommoss, F., Wintzer, H.O., Von Kleist, S., Kohler, M., Walker, R., Langton, B., Van Tran, K., Pfleiderer, A. and Bauknecht, T. (1990) *In situ* distribution of transforming growth factor alpha in normal human tissues and in malignant tumours of the ovary. *J. Pathol.* 162: 223–230.

Kommoss, F., Wölfle, J., Bauknecht, T., Pfisterer, J., Kiechle-Schwarz, M., Pfleiderer, A., Sauerbrei, W., Kiehl, R. and Kacinski, B.M. (1994) Co-expression of M-CSF transcriptions and protein, *fms* (M-CSF receptor) transcripts and protein and steroid receptor content in adenocarcinomas of the ovary. *J. Pathol.* 174: 111–119.

Lupu, R., Colomer, R., Zugmaier, G., Sarup, J., Shepard, M., Slamon, D. and Lippmann, M.E. (1990) Direct interaction of a ligand for the erbB2 oncogene product with the EGF receptor and p185^{erbB2}. *Science* 249: 1552–1555.

Luttrell, D.K., Lee, A., Lansing, T.J., Crosby, R.M., Jung, K.D., Willard, D., Luther, M., Rodriguez, M., Berman, J. and Gilmer, T.M. (1991) Involvement of pp60^{c-src} with two major signaling pathways in human breast carcinoma. *PNAS* 91: 83–87.

Malik, S.T., Griffin, D.B., Naylor, M.S., Fiers, W., Oliff, A. and Balkwill, F.R. (1990) The complex effects of recombinant tumour necrosis factor-alpha (rhTNF-alpha) in human ovarian cancer xenograft models. *Prog. Clin. Biol. Res.* 49: 393–403.

Marth, C., Lang, T., Koza, A., Mayer, I. and Daxenbichler, G. (1990a) Transforming growth factor-beta and ovarian carcinoma cells: regulation of proliferation and surface antigen expression. *Cancer Lett.* 51: 221–225.

Marth, C., Muller-Holzner, E., Greiter, E., Crönauer, M.V., Zeimet, A., Doppler, W., Eibl, B., Hynes, N.E. and Daxenbichler, G. (1990b) Gamma-interferon reduces expression of the protooncogene c-erbB2 in human ovarian carcinoma cells Cancer Res. 50: 7037–7041.

Mills, G.B., May, C., Hill, M., Campbell, S., Shaw, P. and Marks, A. (1990) Ascitic fluid from human ovarian cancer patients contains growth factors necessary for intraperitoneal growth of human ovarian adenocarcinoma cells. J. Clin. Invest. 86: 851–855.

Mobus, V.J., Asphal, W., Knapstein, P.G. and Kreinberg, R. (1993) Effects of interferon gamma on the proliferation and modulation of cell-surface structure of human ovarian carcinoma cell lines. J. Cancer Res. Clin. Oncol. 120: 27–34.

Moradi, M.M., Carson, L.F., Weinberg, B., Haney, A.F., Twiggs, L.B. and Ramakrishnan, S. (1993) Serum and ascitic fluid levels of interleukin-1, interleukin-6 and tumor necrosis factor-α in patients with ovarian epithelial cancer. Cancer 72: 2433–2440.

Naylor, M.S., Malik, S.T., Stamp, G.W., Jobling, T. and Balkwill, F.R. (1990) In situ detection of tumour necrosis factor in human ovarian cancer specimens. Eur. J. Cancer 26: 1027–1030.

Nicosia, S.V., Narconis, R.J. and Saunders, B.O. (1989) Regulation and temporal sequence of surface epithelium morphogenesis in the postovulatory rabbit ovary. Prog. Clin. Biol. Res. 296: 111–119.

Peles, E., Lamprecht, R., Ben-Levy, R., Tzahar, E. and Yarden, Y. (1992) Regulated coupling of the neu receptor to PI-3'kinase and its release by oncogenic activation. J. Biol. Chem. 267: 12266–12274.

Peles, E., Ben-Levy, R., Tzahar, E., Liu, N., Wen, D. and Yarden, Y. (1993) Cell-type specific interaction of neu differentiation factor (NDF/heregulin) with neu/HER-2 suggests complex ligand-receptor relationships. EMBO J. 12: 961–971.

Price, F.V., Chambers, S.K., Chambers, J.T., Carcangiu, M.L., Schwartz, P.E., Kohorn, E.I., Stanley, E.R. and Kacinski, B.M. (1993) Colony-stimulating factor-1 in primary ascites of ovarian cancer is a significant predictor of survival. Am. J. Obstet. Gynecol. 168: 520–527.

Rodriguez, G.C., Berchuck, A., Whitaker, R.S., Schlossman, D., Clarke-Pearson, D.L. and Bast, R.C. (1991) Epidermal growth factor receptor expression in normal ovarian epithelium and ovarian cancer. II. Relationship between receptor expression and response to epidermal growth factor. Am. J. Obstet. Gynecol. 164: 745–750.

Rosen, E.M., Goldberg, I.D., Liu, D., Setter, E., Bhargava, M., Reiss, M. and Kacinski, B.M. (1991) Tumor necrosis factor stimulates epithelial tumor cell motility. Cancer Res. 51: 5315–5321.

Rubin, S.C., Finstad, C.L., Wong, G.Y., Almadrones, L., Plante, M. and Lloyd, K.O. (1993) Prognostic significance of HER-2/neu expression in advanced ovarian cancer. Am. J. Obstet. Gynecol. 168: 162–169.

Samanta, A., LeVea, C.M., Dougall, W.C., Qian, X. and Greene, M.I. (1994) Ligand and p185[c-neu] density govern receptor interactions and tyrosine kinase activation. PNAS 91: 1711–1715.

Sherr, C.J., Matsushima, H. and Roussel, M.F. (1992) Regulation of CYL/cyclin D genes by colony-stimulating factor-1. CIBA Foundation Symp. 170: 209–219.

Slamon, D.J., Godolphin, W., Jones, L.A., Holt, J.A., Wong, S.G., Keith, D.E., Levin, W.J., Stuart, S.G., Udove, J., Ullrich, A. and Press, M.F. (1988) Studies of the HER-2/neu proto-oncogene in human breast and ovarian cancer. Science 244: 707–712.

Stern, D.F., Heffernan, P.A. and Weinberg, R.A. (1986) p185, a product of the neu proto-oncogene, is a receptor-like protein associated with tyrosine kinase activity. Mol. Cell Biol. 6: 1729–1740.

Stern, D.F., Kamps, M.P. and Cao, H. (1988) Oncogenic activation of P185[neu] stimulates tyrosine phosphorylation in vivo. Mol. Cell Biol. 8: 3969–3973.

Taylor, G.R., Reedijk, M., Rothwell, V., Rohrschneider, L. and Pawson, T. (1989) The unique insert of cellular and viral fms protein tyrosine kinase is dispensable for enzymatic and transforming activity. EMBO J. 8: 2029–2037.

Torre-Amione, G., Beauchamp, R.D., Koeppen, H., Park, B.H., Schreiber, H., Moses, H.L. and Rowley, D.A. (1990) A highly immunogenic tumor transfected with a murine transforming growth factor type beta cDNA escapes immune surveillance. PNAS USA 87: 1486–1490.

Watson, J.M., Sensintaffar, J.L., Berek, J.S. and Martinez-Maza, O. (1990) Constitutive production of interleukin 6 by ovarian cancer cell lines and by primary ovarian tumor cultures. Cancer Res. 50: 6959–6965.

Williams, R., Sanghera J., Wu, F., Carbonaro-Hall, D., Campbell, D.L., Warburton, D., Pelech, S. and Hall, F. (1993) Identification of a human EGFR-associated protein kinase as a new member of the mitogen-activated protein kinase/extracellular signal-regulated protein kinase family. *J. Biol. Chem.* 268: 18213–18217.

Willman, C.L., Stewart, C.C., Miller, V., Yi, T.L. and Tomasi, T.B. (1989) Regulation of MHC class II gene expression in macrophages by hematopoietic colony-stimulating factors (CSF). *J. Exp. Med.* 170: 1559–1567.

Wimalasena, J., Meehan, D., Dostal, R., Foster, J.S., Cameron, M. and Smith, M. (1993) Growth factors interact with estradiol and gonadotrophins in the regulation of ovarian cancer cell growth and growth factor receptors. *Oncology Res.* 5: 325–327.

Wilson, A.P., Fox, H., Scott, I.V. and Lee, H. (1991) A comparison of the growth-promoting properties of ascitic fluids, cyst fluids and peritoneal fluids from patients with ovarian tumours. *Br. J. Cancer* 63: 102–108.

Wu, S., Boyer, C.M., Whitaker, R.S., Berchuck, A., Wiener, J.R., Weinberg, J.B. and Bast, R.C., Jr. (1993) Tumor necrosis factor alpha as an autocrine and paracrine growth factor for ovarian cancer: monokine induction of tumor cell proliferation and tumor necrosis alpha expression. *Cancer Res.* 53: 1939–1944.

Wu, S., Meeker, W.A., Wiener, J.R., Berchuck, A., Bast, R.C., Jr. and Boyer, C.M. (1994) Transfection of ovarian cancer cells with tumor necrosis factor-alpha (TNF-α) antisense mRNA abolishes the proliferative response to interleukin-1 (IL-1) but not TNF-α. *Gynecol. Oncol.* 53: 59–63.

Yarden, Y., Escobedo, J.A., Kuang, W.-J., Yang-Feng, T.L., Daniel, T.O., Tremble, P.M., Chen, E.Y., Ando, M.E., Harkins, R.N. and Francke, U. (1986) Structure of the receptor for PDGF helps define a family of closely-related receptors. *Nature* 323: 226–232.

Yarden, Y., Kuang, W.-J., Yang-Feng, T., Coussens, L., Munemitsu, S., Dull, T.J., Chen, E., Schlessinger, J., Francke, U. and Ullrich, A. (1987) Human proto-oncogene c-*kit*: A new cell surface receptor tyrosine kinase for an unidentified ligand. *EMBO J.* 6: 3341.

Epithelial–Mesenchymal Interactions in Cancer
ed. by I.D. Goldberg & E.M. Rosen
© 1995 Birkhäuser Verlag Basel/Switzerland

Keratinocyte growth factor as a cytokine that mediates mesenchymal–epithelial interaction

J.S. Rubin[1], D.P. Bottaro[1], M. Chedid[1], T. Miki[1], D. Ron[2],
G.R. Cunha[3] and P.W. Finch[4]

[1]*Laboratory of Cellular and Molecular Biology, National Cancer Institute, Building 37,
Room 1E24, Bethesda, MD 20892, USA*
[2]*Department of Biology, Technion-Israeli Institute of Technology, Technion City, Haifa-32000,
Israel*
[3]*Department of Anatomy and Reproductive Endocrinology Center, University of California,
San Francisco, CA 94143, USA*
[4]*Ruttenberg Cancer Center, Mount Sinai School of Medicine, New York, NY 10029, USA*

Summary. Keratinocyte growth factor (KGF) is a member of the heparin-binding fibroblast growth factor family (FGF-7) with a distinctive pattern of target-cell specificity. Studies performed in cell culture suggested that KGF was mitogenically active only on epithelial cells, though from a variety of tissues. In contrast, KGF was produced solely by cells of mesenchymal origin, leading to the hypothesis that it might function as a paracrine mediator of mesenchymal–epithelial communication. Biochemical analysis and molecular cloning established that the KGF receptor (KGFR) was a tyrosine kinase isoform encoded by the *fgfr-2* gene. Many detailed investigations of KGF and KGFR expression in whole tissue and cell lines largely substantiated the pattern initially perceived *in vitro* of mesenchymal and epithelial distribution, respectively. Moreover, functional assays in organ culture and *in vivo* and analysis of agents regulating KGF expression reinforced the idea that KGF acts predominantly on epithelial cells. While the data do not implicate a KGF autocrine loop in neoplasia, paracrine sources of factor or ligand-independent signaling by the KGFR might contribute to malignancy. Alternatively, because of its differentiation-promoting effects, KGF may retard processes that culminate in uncontrolled cell growth.

Introduction

Our search for epithelial growth factors was inspired by the knowledge that most human malignancies arise in epithelial cell populations (Wright and Allison, 1984), which continuously turn over while other cells in the same tissue are relatively quiescent. Given the association of growth factors and neoplasia, we reasoned such factors may play an important part in the process of epithelial cell transformation. Early experiments indicated that epithelial-specific mitogenic activity was present in medium conditioned by fibroblasts from a variety of embryonic and adult tissues. This was consistent with evidence that mesenchymal interactions presumably mediated by diffusible substances had a major impact on epithelial cell proliferation (Cunha et al., 1983; Sawyer and Fallows, 1983; Schor et al., 1987). By fractionating conditioned medium from M426 human embryonic lung fibroblasts, we succeeded in isolating two growth factors. The first, keratinocyte growth factor (KGF), was entirely novel (Rubin et al., 1989), while the second proved to be an isoform of hepatocyte growth factor/scatter factor (HGF/SF) (Rubin et al., 1991) which was purified in other laboratories while our work was in progress. KGF was the seventh member of the heparin-binding, fibroblast growth factor family to be identified (Finch et al., 1989), and was alternatively designated FGF-7 (Baird and Klagsbrun, 1991). In contrast to the other well-characterized members of the FGF family, its target-cell specificity *in vitro* was restricted to epithelial cell types (Rubin et al., 1989). In this article, after reviewing the salient properties of KGF and its high affinity receptor, we will discuss recent work which provides additional evidence that KGF functions primarily as a paracrine mediator of mesenchymal–epithelial communication.

KGF: Summary of physical and biological properties

KGF was purified from fibroblast culture as a monomeric polypeptide with an apparent molecular weight of 26–28 kDa (Rubin et al., 1989). Minor heterogeneity was likely due to differences in glycosylation. The KGF cDNA encodes a 194 amino acid protein, including a classical signal peptide for secretion and one potential N-linked glycosylation site (Finch et al., 1989). When expressed in bacteria, a biologically active protein was obtained with an apparent size of 21 kDa, providing indirect evidence that the factor synthesized in mammalian cells is glycosylated and that this post-translational modification is not required for potent activity (Ron et al., 1993a). Database analysis revealed that KGF was a member of the FGF family. The area of homology comprised approximately the carboxy-terminal two-thirds of the KGF coding sequence (Finch et al., 1989). In this region, the KGF sequence is 30–45% identical to the eight other proteins in the family (Baird and Klagsbrun, 1991; Tanaka et al., 1992; Miyamoto et al., 1993).

In addition to human embryonic lung fibroblasts, stromal cells from a variety of other sources express KGF in culture. This includes cells from human adult lung, skin, mammary gland, stomach, bladder and prostate (unpublished observations). No expression was detected in a diverse group of epithelial cell lines, endothelial cells or melanocytes (Finch, 1989 and unpublished observations), although one report claimed to detect KGF expression by polymerase chain reaction (PCR) analysis in melanocytes and melanoma cells (Albino et al., 1991). Consistent with prevailing observations, KGF transcript was detected in the dermis but not epidermis of skin removed from newborn mice (Finch et al., 1989). In contrast to the stromal pattern of expression, KGF's target-cell specificity was restricted to epithelial cell types. Initially, we demonstrated that KGF stimulated DNA synthesis in BALB/

MK mouse keratinocytes, B5/589 human mammary epithelial cells and CCL208 rhesus monkey bronchial epithelial cells (Rubin et al., 1989). Subsequently, additional responsive cells were identified such as human keratinocytes (Marchese et al., 1990), rat and human prostatic epithelial cells (Yan et al., 1992; Rubin et al., 1992), rat hepatocytes (Itoh et al., 1993; Suzuki et al., 1993; Strain et al., 1994), type II pneumocytes (Panos et al., 1993), corneal epithelial cells (Wilson et al., 1993) and bovine ovarian granulosa cells (Parrott et al., 1994). No activity was seen on fibroblasts, saphenous vein endothelial cells, melanocytes or rat myoblasts (Rubin et al., 1989; Halaban et al., 1991; Ron et al., 1993b), although KGF reportedly was active on corneal endothelial cells (Wilson et al., 1993). KGF not only stimulates DNA synthesis but also supports sustained cell growth. At least in the case of BALB/MK cells, proliferation required the presence of insulin or insulin-like growth factors (IGFs) (Rubin et al., 1989), as was the case for epidermal growth factor (EGF) and acidic and basic FGF (Falco et al., 1988). Such a synergy between insulin/IGFs and a variety of other growth factors has been reported in other systems (for instance, see Coppola et al., 1994 and references cited therein).

While KGF and other factors like EGF have a similar proliferative effect on certain epithelial cells, they appear to have distinct influences on selected phenotypic characteristics. For instance, human keratinocytes incubated in the presence of millimolar concentrations of calcium plus KGF expressed keratin 1 and filaggrin, markers of terminal differentiation, whereas cells exposed to either EGF or transforming growth factor alpha (TGFα) did not (Marchese et al., 1990). Colony morphology of prostatic epithelial cells grown *in vitro* varied depending on whether KGF or EGF was included in the medium (Hayward et al., in press; D. Peehl, personal communication). Recent work also suggests that KGF has specific effects on the synthesis of secretory products by prostatic and uterine epithelial cells (S. Glasser, personal communication, and unpublished observations) and pulmonary surfactant components by type II pneumocytes (J. Shannon, perconal communication). That KGF has other non-proliferative effects was documented in a study of human keratinocytes showing stimulation of cell migration and plasminogen activator activity (Tsuboi et al., 1993). Morphogenic effects observed in organ culture systems will be described below.

Biochemical analysis and molecular cloning of KGF receptors

Consistent with its target-cell specificity in mitogenesis bioassays, we detected saturable, specific high-affinity binding of ^{125}I-KGF to the surface of BALB/MK keratinocytes but not NIH/3T3 fibroblasts (Fig. 1, Bottaro et al., 1990). Radiolabeled KGF binding to BALB/MK was

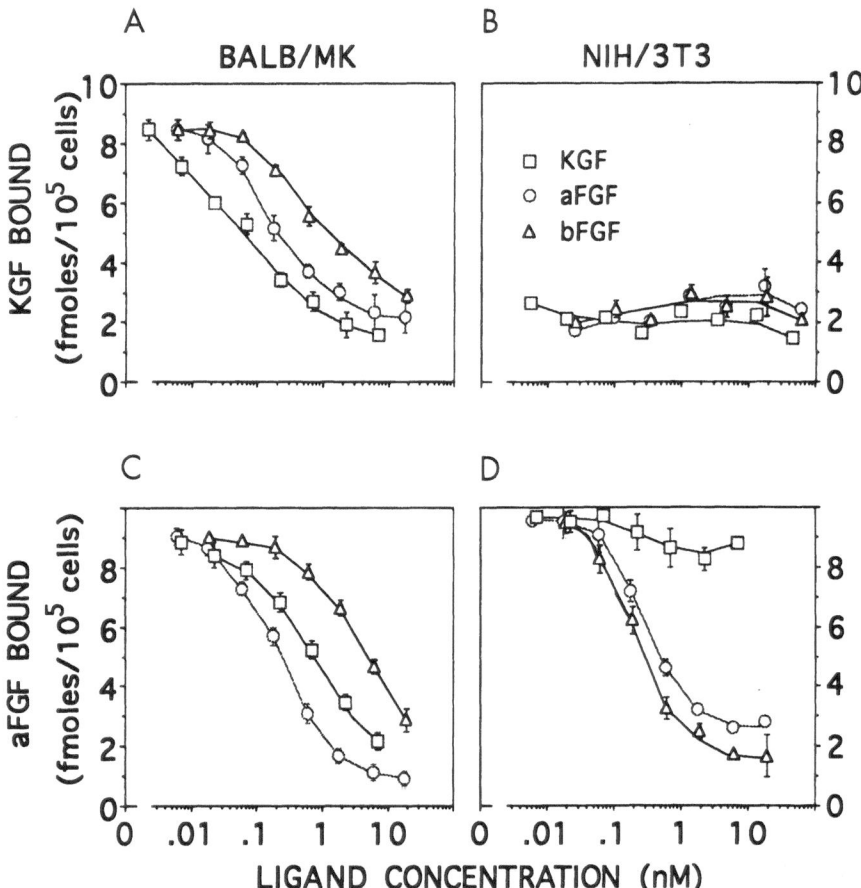

Figure 1. (A) Specific binding of ^{125}I-KGF (1 ng/ml) to BALB/MK cells, expressed as femtomoles bound per 10^5 cells, competed by increasing concentrations (nM) of unlabeled KGF (squares), aFGF (circles) or bFGF (triangles). Values shown are the mean of triplicate measurements $\pm$ standard deviation. Where no error bars are shown, the error is less than symbol size. Comparable results were obtained using either low concentrations of heparin (1–3 μg/ml) or brief salt extraction to block low-affinity ligand binding in all competition studies shown. (B) Specific ^{125}I-KGF binding on NIH/3T3 cells, displaced by unlabeled KGF, aFGF or bFGF. (C) Specific binding of ^{125}I-aFGF (1 ng/ml) to BALB/MK cells, displaced by unlabeled KGF, aFGF or bFGF. (D) Specific ^{125}I-aFGF binding on NIH/3T3 cells, displaced by unlabeled KGF, aFGF or bFGF. (Data were first reported in Bottaro et al. (1990)).

displaced efficiently by unlabeled KGF and aFGF, and less well by bFGF, in agreement with their relative potency in assays of DNA synthesis (Aaronson et al., 1991). In contrast, radiolabeled aFGF bound well to both keratinocytes and fibroblasts, but KGF only competed for binding to BALB/MK. On the other hand, bFGF was a significantly better competitor on the fibroblasts than the keratinocytes. These observations indicated that keratinocytes possess a high affinity receptor for KGF to which the FGFs also bind, while fibroblasts have a distinct FGF receptor(s) which does not interact with KGF.

The ability of fibroblasts such as NIH/3T3 cells to synthesize, but not bind KGF, offered a strategy for the cloning of the KGFR. Earlier studies, either with FGF family members such as hst/K-FGF/FGF-4 (Taira et al., 1987; Delli-Bovi and Basilico, 1987) or FGF-5 (Zhan et al., 1988) that possess signal peptides, or with bFGF constructs to which a signal peptide sequence had been added (Rogelj et al., 1988; Blam et al., 1988), demonstrated that transfection of NIH/3T3 cells with vectors encoding these secreted factors would result in transformation. The presumed mechanism for this effect was the creation of a functional transforming autocrine loop. By analogy, introduction of KGF receptors into NIH/3T3 cells via transfection might cause transformation by creation of a similar, KGF-dependent autocrine mechanism. Screening foci of transformed cells for evidence of high affinity KGF binding sites would identify those cells that had become transformed as a consequence of acquiring such an autocrine loop.

This approach was successfully executed, initially by expressing in NIH/3T3 cells a cDNA library prepared with mRNA from BALB/MK cells (Miki et al., 1991). Sequence analysis revealed that the KGF receptor (KGFR) was a transmembrane tyrosine kinase closely related to receptors for other FGF ligands. Like other FGF receptors, KGFR contains immunoglobulin-like (Ig) domains of approximately 50 amino acids in its extracellular portion and an intracellular tyrosine kinase domain that is divided by a 14 amino acid insert (Fig. 2). Although the first KGFR cloned contained two Ig domains, subsequent clones isolated from both the mouse and a human mammary epithelial cell library had three Ig loops (Miki et al., 1992). Isoforms having either two or three domains when expressed in mammalian cells possessed similar high affinity binding. The kinase domain of KGFR isoforms was identical to that present in the *bek* gene product (Kornbluth et al., 1988; Dionne et al., 1990), which also was identified as TK14 (Houssaint et al., 1990) and ultimately termed BEK (Champion-Arnaud et al., 1991) or FGFR-2 (Keegan et al., 1991). In fact, the KGFR is essentially identical to BEK/FGFR-2, except for a strikingly divergent stretch extending from the carboxy-terminal half of the Ig loop closest to the transmembrane region into the stem region (Fig. 2). In this area, KGFR and BEK/FGFR-2 sequences are only 47% identical. Another tyrosine kinase cloned from a human stomach tumor cell line and designated K-SAM shared the same sequence as KGFR in this region, but contained a carboxy-terminal deletion and substitution and a small deletion in the juxtamembrane domain (Hattori et al., 1990). Analysis of genomic DNA revealed that the divergent sequences present in KGFR and BEK/FGFR-2 resulted from alternative splicing of exons in the same gene (Champion-Arnaud et al., 1991; Johnson et al.,

KGFR   KGFR   FGFR-2
mouse   human   human

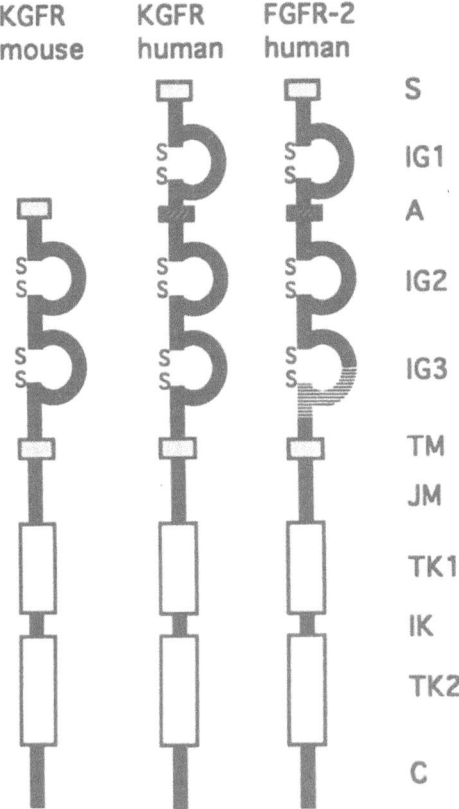

	S
	IG1
	A
	IG2
	IG3
	TM
	JM
	TK1
	IK
	TK2
	C

Figure 2. Schematic representation of KGFR and BEK/FGFR-2 isoforms. S, signal sequence; IG1, IG2 and IG3, immunoglobulin-like domains; A, acidic region; TM, transmembrane domain; JM, juxtamembrane domain; TK1 and TK2, tyrosine kinase domains; IK, interkinase domain; C, carboxy-terminus domain. Note that the two basic KGFR isoforms, containing either two or three IG domains as illustrated here, have been identified in both mouse and human epithelial cells. (Figure is from Miki et al. (1992)).

1991; Miki et al., 1992; Yayon et al., 1992; Dell and Williams, 1992).

Despite the overall structural similarity between KGFR and BEK/ FGFR-2, these two receptor isoforms have strikingly different ligand-binding properties. As noted above, KGFR binds both KGF and aFGF well but bFGF rather poorly. In contrast, BEK/FGFR-2 binds bFGF and aFGF well, but not KGF. Thus, alternative splicing was responsible for determining the ligand-binding specificity of these receptor molecules (Miki et al., 1992; Yayon et al., 1992; Dell and Williams, 1992). Both the *fgfr-1* and *fgfr-3* genes also contain alternative exons resembling the one (designated IIIb or K exon) in the *fgfr-2* gene that confers KGF binding (Johnson et al., 1991; Chellaiah et al., 1994). In the case of *fgfr-3*, the degree of homology is only moderate and the isoform containing this sequence does not bind KGF or bFGF while it does bind aFGF avidly (Chellaiah et al 1994). However, in the case of *fgfr-1 and fgfr-2*, the IIIb amino acid sequences are approximately 80% identical. Whether KGF can bind to the corresponding isoform of FGFR-1 remains to be determined.

While the results summarized above clearly indicated that ligand-binding specificity was dependent on the sequence present in the carboxy-terminal half of the third Ig loop and the adjacent stem region, the mechanism for this effect was not established. To address the possibility that all or a portion of the KGF binding site was located in this region, synthetic peptides corresponding to the part of this segment having the least homology to the alternative exon (IIIc or B exon) were generated and tested for their ability to block KGF mitogenic activity and binding to intact receptor. The receptor-derived peptides, which correspond to the carboxy-terminal portion of the third Ig loop, antagonized KGF mitogenic activity and competed with radiolabeled KGF for receptor binding in the micromolar concentration range (Bottaro et al., 1993). These data demonstrated that at least a portion of the binding site resides in this region of the receptor molecule.

Although the relatively low KGF binding affinity of these peptides compared to the intact KGFR (with a dissociation constant in sub-nanomolar range) might be due to the many alternative conformations available to a soluble peptide, it also could indicate that other structural elements were required for high affinity binding. The latter interpretation was advanced by Zimmer et al. (1993) who generated a chimeric molecule in which the sequence of the third Ig loop specified by FGFR-2 exon IIIb was substituted for the corresponding region of FGFR-1. The resulting protein, which in effect was FGFR-1 but with the carboxy-terminal half of the 'KGF-specific' Ig domain from the KGFR isoform introduced in a cognate location, bound KGF though with a 15-fold lower affinity than the KGF receptor. In contrast, when the second Ig loop from FGFR-2 and the carboxy-terminal portion of the third Ig loop were substituted into the corresponding region of FGFR-1, the chimeric molecule bound KGF as well as wild type KGFR. Thus, it was reasoned that the second Ig domain also contained sequence important for high affinity KGF binding. However, one could not exclude the possibility that the second Ig domain of FGFR-1 in some manner interfered with the binding of KGF. This matter was resolved by another series of experiments that involved the creation of chimeric molecules, consisting of either the individual second or third Ig loops, or both together, fused to the mouse Ig heavy chain Fc domain. Experiments clearly established that high affinity KGF binding was associated with the third Ig loop alone, although a modest increase in affinity was observed when both domains were present (Cheon et al., 1994). A similar conclusion was reached independently by Chellaiah et al. (1994) who substituted the third Ig loop from KGFR into the FGFR-3 background and obtained high affinity KGF binding.

A summary of KGF binding would not be complete without mention of proteoglycan. As with the other members of the FGF family, KGF is a heparin-binding molecule; approximately 0.6 M sodium chloride is required to elute it from heparin-Sepharose at neutral pH (Rubin et al., 1989). Low affinity, high capacity binding to heparin-like molecules was evident on fibroblasts as well as keratinocytes and, consequently did not correlate with mitogenic signaling (Bottaro et al., 1990). In contrast to aFGF (and to some extent, bFGF), heparin has an inhibitory rather than stimulatory effect on KGF mitogenic activity on BALB/MK cells (Ron et al., 1993a). Reduction in the amount of proteoglycan on the surface of BALB/MK cells or rat myoblast transfectants expressing KGFR resulted in decreased aFGF binding and mitogenicity, while the corresponding KGF activities were increased (Reich-Slotky et al., 1994). The KGF-specific inhibitory activity of heparin was attributed to overlapping binding sites for heparin and KGFR in the KGF molecule (Bottaro et al., 1993). Nonetheless, several reports suggest that heparan-type proteoglycan is required for high affinity binding and activity of aFGF and bFGF (Yayon et al., 1991; Rapraeger et al., 1991; Ornitz et al., 1992). The possibility remains that a component of proteoglycan is required for KGF activity. The validity of such a hypothesis is the subject of current investigation.

In situ hybridization analysis of KGF and KGFR expression

The analysis of KGF and KGFR expression in cell culture, as well as the pattern of KGF target-cell specificity established *in vitro*, was fully consistent with a proposed role for this signaling pathway in mesenchymal–epithelial interactions. In the past few years, several studies involving *in situ* hybridization analysis have provided a large amount of additional data that bear on this subject. Most of this work focused on expression during development, when molecules associated with cell proliferation, differentiation and migration would presumably be especially active. This also had the added benefit, particularly late in gestation, of facilitating a wide survey of well-defined, distinct structures in the terminal phases of organogenesis.

The first reports dealt with FGFR expression in the developing mouse and made use of probes that could not distinguish between different isoforms having unique ligand binding properties. Rather, their focus was to compare the patterns of FGFR-1 and FGFR-2 expression (Orr-Urtreger et al., 1991; Peters et al., 1992). Nonetheless, these articles are pertinent to the present discussion because they demonstrated a marked difference in the regulation of the two receptor genes. In particular, by the mid- and late-gestational periods, a striking dichotomy was documented: while FGFR-1 was seen primarily in mesenchyme, strong FGFR-2 expression was observed in the adjacent epithelia. The surface ectoderm, comprising skin and its appendages, and the lining of inner surfaces such as the respiratory and digestive tracts all had high levels of FGFR-2 expression. Earlier in gestation, FGFR-2 transcripts were detected in portions of the neural tube, body wall and developing limbs including the apical ectodermal ridge which is crucial for limb bud growth. FGFR-2 signal was evident in prebone and precartilage elements of the developing axial skeleton and the intramembranous bony structures of the head and face.

The analysis was taken a step further when probes that discriminated between FGFR-2 isoforms containing IIIb versus IIIc sequences were employed in a similar analysis (Orr-Urtreger et al., 1993). Both variants were expressed duting murine gastrulation, though the KGFR, IIIb-containing forms were more abundant. The spatial distribution of transcripts at this stage was diffuse and overlapping. However, as development progressed, a key distinction emerged: KGFR transcripts were prominent in the surface ectoderm and its derivative structures, including hair roots, enamel organ of teeth, cornea and mammary gland, and in the gut and its derivatives; BEK/FGFR-2 transcripts were primarily associated with osteogenesis, although they were also seen in the brain, adrenal gland and atrioventricular heart valve. Interestingly, KGFR was seen in the bones of the vertebral column, pelvis, limbs, skull, face and nasal cavity though at lower levels than BEK/FGFR-2. Once evident, these patterns persisted throughout organogenesis. Developmental studies in the chicken (Sato et al., 1992; Patstone et al., 1993) and amphibian (Shi et al., 1994) have documented similar differences in the distribution of FGFR-2 transcripts. Thus, expression analyses with whole tissue have reinforced the idea that alternatively spliced receptor isoforms not only have distinct ligand-binding properties but also a fundamental difference in their cellular and tissue distribution. Of special note, the KGFR isoforms were highly expressed in a wide array of epithelial tissues.

The pattern of KGF expression in the mouse during development was described in another study (Mason et al., 1994). Complementing the results just summarized for KGFR, these authors detected KGF transcript in the mesenchyme of many tissues adjacent to epithelia previously shown to contain KGFR. These included lung, salivary gland, esophagus, stomach, small intestine and dermis. In the case of the dermis, KGF transcripts were first seen between days 15.5 and 16.5 post coitus (p.c.) when the epidermis changes from a simple to a stratified epithelium, and thus, suggested a role for KGF in mediating this process. Our lab independently performed a comprehensive *in situ* hybridization analysis in which KGF and KGFR expression were

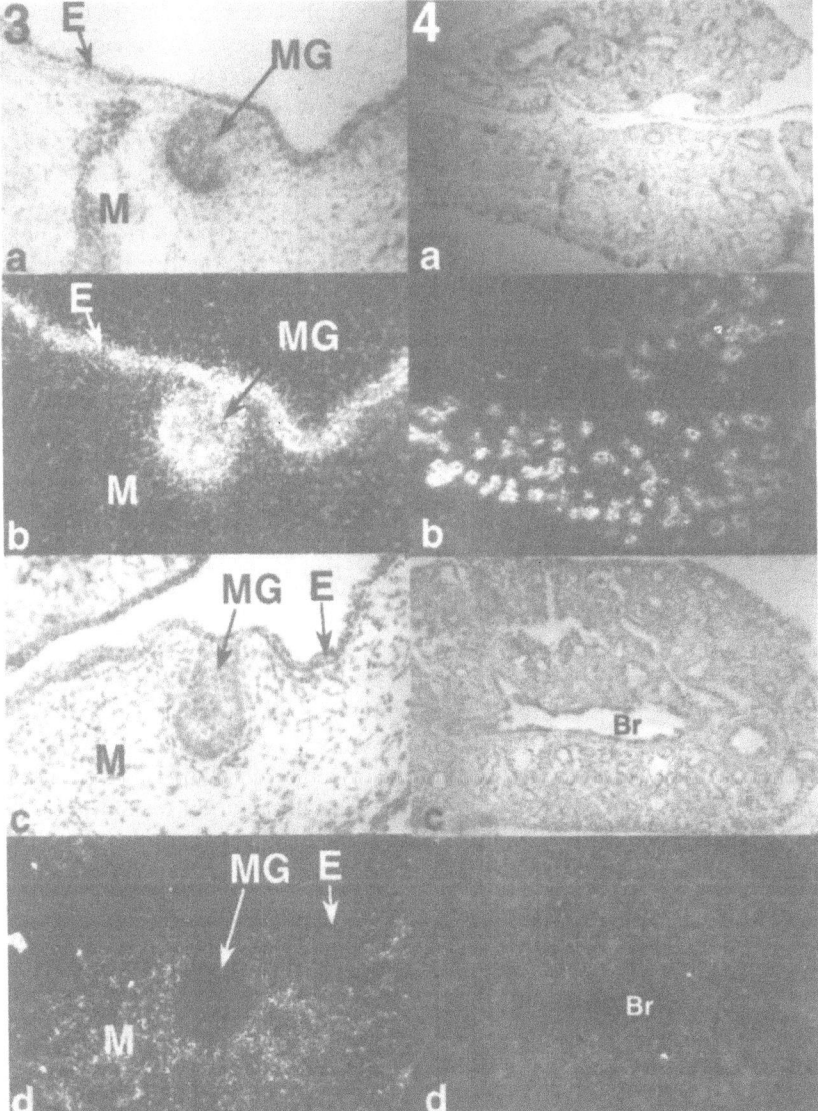

Figure 3. *In situ* hybridization analysis of KGFR (a) and (b) and KGF (c) and (d) transcripts in developing skin and mammary gland bud. Comparison of bright (a) and dark field (b) images indicates that KGFR is expressed in the epidermis (E) and mammary gland bud (MG), but not in the mesenchyme (M). A similar inspection of bright (c) and dark (b) field views demonstrates that KGF is expressed in the mesenchyme, but not in the epidermis or mammary gland bud; magnification × 125.

Figure 4. *In situ* hybridization analysis of KGFR (a) and (b) and KGF (c) and (d) transcripts in developing lung. Examination of bright (a) and (c) and dark field (c) and (d) images reveals that KGFR is expressed in branched epithelial ductal structures but not in surrounding mesenchyme, while KGF is expressed diffusely in the mesenchyme and is absent from bronchi (Br) and other airways; (a,b) magnification × 45; (c,d) magnification × 55.

examined in parallel (Finch et al., manuscript submitted). The findings largely confirm the results obtained by others who studied KGF or KGFR individually. For instance, see Figures 3 and 4 for an illustration of growth factor and receptor expression in developing skin, mammary gland and lung. Our study expanded the number of epithelial tissues exhibiting the juxtaposition of KGF and KGFR expression. Specifically, the mesenchymal–epithelial distribution of growth factor and receptor, respectively, were documented in the ureteric bud of the kidney rudiment, mammary gland, urogenital sinus, external genitalia, thyroid and thymus. Taken altogether, these analyses of expression in whole tissue provide strong evidence in support of the idea that KGF is a mesenchymally-derived mediator of epithelial growth and development.

However, both the work by Mason et al. (1994) and our own study indicated that KGF probably has other functions as well. Mason et al. state that the earliest detection of KGF transcript in the developing mouse is in the heart, beginning 8.5 day p.c. Signal is stronger in the primitive atrium than ventricle and persists until 11 day p.c. They point out that this may contribute to the development of the endocardial cushions, which ultimately become the fibrous part of the atrioventricular septum and cardiac valves, because FGFR-2 transcripts were detected in the endocardial cushions from 9.5 to 12 day p.c. (Peters et al., 1992). Transient regional expression was reported in the ventricular zone of the forebrain. KGF also may have a role in myogenesis, judging from its pattern of expression in somites and developing musculature. Both studies showed strong expression of KGF in skeletal muscle of the limbs, tongue and head; Orr-Urtreger et al. (1993) and we also observed KGFR in muscle. Similarly, we noted simultaneous expression of KGF and KGFR in the periosteal cells and mesenchyme surrounding developing intramembranous facial bone. Co-expression of the growth factor and its receptor was seen in portions of intestinal smooth muscle, while juxtaposition of their expression was noted in perichondrium and cartilage in various locations. Thus, it appears likely that KGF participates in other processes in addition to having a major role in mesenchymal–epithelial interactions.

Regulation of KGF expression: Further evidence for a role as paracrine mediator of cell–cell communication

A number of cytokines and growth factors have been tested for their effects in modulating KGF gene expression in fibroblasts. Of the molecules surveyed, interleukin 1 (IL-1) was the most potent inducer of KGF expression (Chedid et al., 1994; Brauchle et al., 1994). This stimulatory effect of IL-1 was observed in both lung and dermal fibroblasts, and was documented at both the RNA and protein levels. It was suggested to account, at least in part, for the dramatic induction of KGF transcript that had been observed in mouse skin following the creation of full-thickness wounds (Werner et al., 1992); presumably, the local release of IL-1 and serum factors increased production of KGF which could promote epithelial migration and proliferation as part of the healing process. In addition, Boismenu and Havran (1994) recently observed that $\gamma\delta$ T cells in skin produce KGF which may participate in the response to injury. We also have evidence that selected T cells express KGF (Finch, unpublished observations).

While infiltrating inflammatory cells would be one likely source of IL-1, keratinocytes themselves also produce this cytokine (Oppenheim et al., 1986). The potential significance of this latter source was highlighted in a study describing the consequences of coculturing keratinocytes and dermal mesenchymal cells (Smola et al., 1993). When human keratinocytes were cocultured with either irradiated human dermal microvascular endothelial cells or fibroblasts, proliferation of the keratinocytes was enhanced. As a possible explanation of this phenomenon, the authors observed that KGF expression was markedly elevated in the coculture preparations compared to cultures of the individual cell populations (as well as nonirradiated fibroblasts). They also detected IL-1 production by the keratinocytes and induction of the IL-1 receptor in the mesenchymal cells. This suggested a plausible mechanism whereby cells in the epithelial compartment of skin could instruct their mesenchymal neighbors to release factors that would have a stimulatory impact on proliferation and other activities of the epithelial population. Perhaps similar loops involving IL-1 or other agents known to promote KGF expression such as TGFα (or its EGF-like relatives) (Chedid et al., 1994) may be operative in other epithelial tissues.

Additional evidence implicating KGF as a mediator of mesenchymal–epithelial interactions was obtained in a series of studies involving sex steroid hormones. A large body of research has established that many effects of these hormones involve direct action of hormone on stroma or mesenchyme, with a subsequent response manifested by adjacent epithelia. This is true both during development and in the adult (Cunha et al., 1987; Cunha et al., 1992a,b). Yan et al. (1992) reported that testosterone stimulated the expression of KGF by rat prostatic stromal cells in culture, and that rat prostatic epithelial cells proliferated in response to recombinant KGF. They inferred that the androgen-dependent increase in epithelial cell division in stromal–epithelial cocultures was due to the release of KGF. We independently made similar observations with early passage human prostatic stromal cell preparations, and documented that KGF was responsible for a major portion of the epithelial growth-promoting activity released by

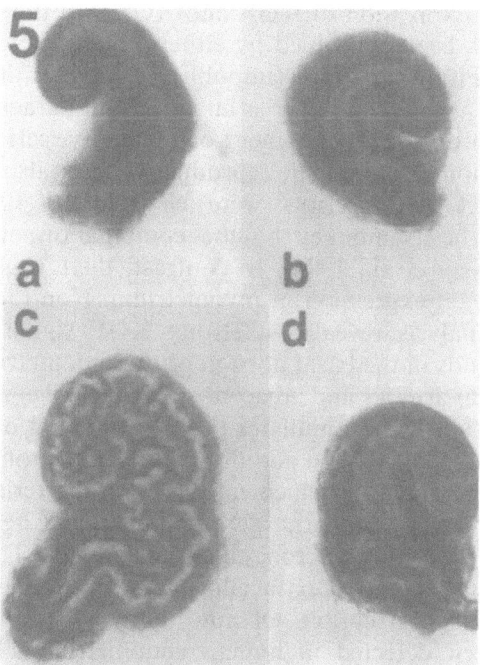

Figure 5. Effect of anti-KGF monoclonal antibody on mouse seminal vesicle morphology. Whole-mount photographs of neonatal mouse seminal vesicles at birth (a) or incubated for 4 days in the presence of basal medium supplemented with (b) control mouse monoclonal immunoglobulin (IgG, 6 µg/ml), (c) control IgG (6 µg/ml + testosterone (T, 10^{-7} M), or (d) anti-KGF (6 µg/ml) + T (10^{-7} M). The mouse seminal vesicle at birth is a simple, unbranched structure (a) which does not grow or develop in the absence of T (b). The dramatic stimulatory effect of T on epithelial growth and branching morphogenesis (c) is profoundly inhibited by inclusion in the medium of a neutralizing antibody to KGF (d). (See Alarid et al. (1994) for details; magnification ×20).

prostatic stromal cells (Rubin et al., 1992, and manuscript submitted).

Even more dramatic results were obtained with experiments in organ culture. Newborn mouse seminal vesicles (SV) placed in organ culture undergo androgen-dependent growth and differentiation (Shima et al., 1990). Addition of a KGF-specific, neutralizing monoclonal antibody to this system caused striking inhibition of both SV growth and branching morphogenesis (Alarid et al., 1994; Fig. 5). This inhibition was due to a decrease in epithelial proliferation and differentiation, as the mesenchymal layer was not affected by anti-KGF treatment. Similar observations have been made with neonatal rat ventral prostate in organ culture (Cunha et al., 1994). For mouse seminal vesicle, when KGF was substituted for testosterone in the culture medium, organ growth was partially restored while in the rat ventral prostate, both growth and differentiation were almost as pronounced as they were in the presence of testosterone. Expression of KGF and KGFR in these tissues during development has been confirmed by either RNase protection assays or *in situ* hybridization analysis (unpublished observations). All these results strongly suggest that KGF is an important paracrine mediator of androgen action during development of seminal vesicle and prostate.

KGF expression in the female reproductive tract also is regulated by sex steroids. KGF transcripts were elevated 70–100 fold in endometrium of rhesus monkey by the combination of estradiol plus progesterone (Koji et al., 1994). In contrast, there was relatively little hormonal effect on expression in myometrium. Non-radioactive *in situ* hybridization analysis revealed a strong KGF signal in the stroma surrounding glands in the basal portion of the endometrium, and also in the walls of spiral arteries and adjacent stroma (Koji et al., 1994). These results implied that KGF stimulates proliferation and/or differentiation of glandular epithelial cells. In addition, KGF may promote invasion of trophoblasts, epithelial cells which enter the spiral arteries as part of the process to establish the maternal-fetal circulation. Alternatively, KGF produced during the progesterone-dominant, luteal phase of the menstrual cycle may be stored in extracellular matrix and facilitate epithelial repair following menstruation. Of note, both KGF and KGFR transcripts have been detected in human endometrium (Pekonen et al., 1993).

Activity of recombinant KGF following *in vivo* administration

Most of the data summarized thus far were obtained with cells in culture or tissues during development. In this section, we will focus on the action of KGF in experimental systems involving adult animals. As in the other studies, this work reinforces the idea that KGF functions primarily to elicit responses in epithelial targets.

In view of the dramatic increase in KGF expression observed when full-thickness wounds were created in mouse skin (Werner et al., 1992), it was of interest to investigate the effects of exogenous KGF in the setting of skin injury. Using a porcine model, Staiano-Coico et al. (1993) observed that topically applied KGF increased the rate of reepithelialization of partial thickness wounds. While KGF did not speed the rate of wound closure in full-thickness wounds, its use was associated with epidermal thickening and a deep rete ridge patterning. There was an increase in the number of serrated basal cells, which had thicker bundles of tonofilaments and contributed to better developed hemidesmosomes. Increased deposition of collagen fibers in the superficial dermis was also noted, perhaps resulting from epidermal release of a secondary cytokine which could act on adjacent dermal elements. These findings indicated that KGF had significant effects on healing skin that were manifested predominantly in the newly-formed epidermis and might result in a better attachment of epidermis to dermis. In a model of deep partial thickness wounds in the rabbit ear, Pierce et al. (1994) documented not only epidermal thickening and an increased rate of reepithelialization, but also enhanced proliferation and differentiation of early progenitor cells within hair follicles and sebaceous glands. These results were consistent with *in situ* hybridization analyses which identified KGFR transcripts in these locations and KGF in corresponding regions of the dermis (Orr-Urtreger et al., 1993; Mason et al., 1994; Finch et al., manuscript submitted and other unpublished observations). Moreover, when expression of a dominant negative KGFR was targeted to keratinocytes in transgenic mice via the keratin 14 promoter, the skin exhibited epidermal atrophy, abnormal hair follicles and impaired reepithelialization following injury (Werner et al., 1994). Paradoxically, when KGF expression was similarly targeted to keratinocytes, hair follicle morphogenesis was markedly inhibited (Guo et al., 1993). However, this may have been a secondary effect due to the high density of keratinocytes in the skin of these animals, with consequent disruption of normal tissue architecture. Taken altogether, the findings suggest that both endogenous and exogenous KGF influence wound healing in skin, with a primary impact on the proliferation and differentiation of distinct epithelial components of this tissue.

Dramatic effects were observed in lung following intratracheal administration of KGF to rats (Ulich et al., 1994b). The alveolar epithelial lining is comprised of two cell types, Type I and Type II pneumocytes. The former are flat, teminally differentiated, non-mitotic cells that constitute most of the airway lining and mediate efficient gas exchange between air and the blood of the alveolar capillary bed. Type II pneumocytes can be viewed as stem cells because they retain the capacity to proliferate and differentiate into Type I pneumocytes. They also produce surfactant, a secreted material that functions to keep the airways open by reducing surface tension. In response to injury, Type II pneumocytes proliferate, subsequently line alveolar septae and differentiate into Type I pneumocytes as normal tissue architecture is restored. In accordance with a recent *in vitro* study of rat alveolar Type II cells (Panos et al., 1993), intratracheal instillation of KGF resulted in a profound stimulation of Type II pneumocyte proliferation. After 2 days, a single dose of 5 mg/kg KGF stimulated micropapillary epithelial hyperplasia in a multifocal pattern corresponding to the normal dispersed distribution of Type II cells in the airways. On day 3, the mitotically active cells had formed a cuboidal epithelium lining alveoli throughout much of the lung. Immunoreactive staining for surfactant protein B and ultrastructural analysis confirmed that these cells were Type II pneumocytes. On subsequent days, the histological pattern in the airways returned to normal, suggesting that Type II pneumocytes had differentiated into Type I cells. Proliferation was evident to a lesser extent in the epithelium of small bronchioles, though not in large airways, in agreement with the distribution of FGFR-2 transcripts in developing lung as determined by *in situ* hybridization analysis (Peters et al., 1992; no distinction was made between KGFR and BEK isoforms in the study by Peters et al.). In contrast to Type II pneumocytes and bronchial epithelium, KGF administration had no apparent effect on pulmonary stromal cells, goblet cells, connective tissue or vessels. RNase protection analysis confirmed that both KGF and KGFR transcripts were relatively abundant in lungs from adult rats. These findings imply that endogenous KGF may play a role in the homeostasis of lung epithelia and, especially, in the response to injury. Exogenous KGF, at sufficiently high doses and delivered directly into the lung elicited a major proliferative response in the Type II pneumocyte population with histological evidence of concomitant production of surfactant components. These effects suggest that the normal response to epithelial lung injury might be enhanced by local administration of KGF.

In this context, it is useful to cite a recent transgenic study which raised the possibility of a functional role for KGF in developing lung. After targeting expression of a dominant negative KGFR isoform to lung epithelium with the human surfactant protein C promoter, a severe defect was seen in branching morphogenesis of the airways which resulted in death during the perinatal period (Peters et al., 1994). This established an important role for FGFs in formation of the bronchial tree. Given our knowledge of KGF expression in lung and the ligand-binding specificity of the receptor, it is reasonable to propose that KGF is one, if not the crucial, factor involved in mediating this process.

After systemic administration of KGF to rats, either via intravenous, intraperitoneal or subcutaneous routes, many striking changes were observed in epithelial tissues. Among the most prominent consequences were those observed in the mammary glands of both female and male animals (Ulich et al., 1994a). In the nulliparous female, after three consecutive days of daily intraperitoneal injections (5 mg/kg) there was a substantial increase in the number and complexity of ducts. The ductal epithelium was hyperplastic, as manifested by stratification and a dramatic elevation in mitotic figures. Following KGF treatment for one week, mammary ducts were massively enlarged with such epithelial hyperplasia that lumina were often occluded. As in the normal nulliparous female, no acinar structures or lactational differentiation were noted. In male rats after one week of KGF treatment, the same dramatic hyperplastic and hypertrophic changes in mammary ducts were documented. However, in contrast to the nulliparous female, acinar growth with lactational differentiation were also evident. (This may reflect the fact that some acinar elements were seen in untreated males, while none were detected in the nulliparous females.) Pregnant rats treated for three consecutive days with KGF (intraperitoneal injections on days 17, 18 and 19 followed by sacrifice on day 20 of gestation) exhibited prominent ductal hyperplasia as described in the other rats, as well as acinar proliferation. Curiously, lactating animals similarly treated with KGF for three consecutive days during the first postpartum week did not show a response to KGF, even though RNase protection analysis confirmed the presence of KGFR transcripts at a level comparable to that present late in pregnancy. The basis for this lack of responsiveness to exogenous KGF in the lactating rat is unclear, but may be due to a maximal effect of endogenous growth factor because KGF expression was approximately seven-fold higher in lactating animals than pregnant controls. This elevated expression suggested that KGF might have a role in supporting lactation. In any case, the marked changes seen in non-lactating animals following KGF treatment demonstrate a remarkable sensitivity of the mammary gland to KGF. Combined with the evidence of both growth factor and receptor expression, these observations imply a likely role for KGF signaling in functioning of breast tissue.

Systemic treatment with KGF also produced profound changes in several organs of the gastrointestinal tract. Epithelial hyperplasia was seen in intralobular ducts of pancreas, adjacent to and within pancreatic islets (Yi et al., 1994). One day after a single intraperitoneal injection, proliferation of epithelial cells was detected from the foregut to the colon and in liver (Housley et al., 1994). Following a series of daily doses, there was a selective induction of mucin-producing cells throughout the GI tract. Besides a stimulatory effect on hepatocyte mitogenesis, numerous serum parameters of liver function such as albumin concentration were elevated. Again, RNA analysis documented that KGF and KGFR were expressed throughout the GI tract of adult rats, indicating a likely role for endogenous factor in the maintenance and function of epithelial populations. Interestingly, KGF transcripts in liver were low in comparison to KGFR, while in spleen the reverse was the case. This may indicate that another ligand such as aFGF normally is operative in liver; alternatively, KGF from spleen may travel via the portal circulation to act in liver. This was proposed earlier in a report describing the isolation of KGF protein from bovine spleen (Suzuki et al., 1993). As with the effects on mammary tissue described above, all the effects observed in the GI tract were entirely reversible by discontinuing KGF administration (Ulich et al., 1994a,b; Yi et al., 1994; Housley et al., 1994).

KGF and KGFR involvement in cancer

As mentioned at the outset, a primary motivation behind the effort to purify an epithelial-specific mitogen was the possibility that it might

Table 1. KGF and KGFR expression in human tumor cell lines[1]

	KGF RNA (# positive/# tested)	KGFR (FGFR-2) RNA (# positive/# tested)
Bladder	0/8	3/8
Breast	0/16	8/16
Colon	0/6	4/6
Glioblastoma	0/4	1/4
Kidney	0/9	5/9
Liver	0/2	1/2
Lung	0/6	1/6
Ovary	0/4	2/4
Stomach	0/13	12/13
Total	0/68	37/68

[1]Transcripts were detected by Northern blot analysis of 20 μg total RNA from each cell line, using a probe for either KGF or KGFR. In the latter case, the probe corresponded to the extracellular domain of the molecule. Thus, cross-hybridization to BEK/FGFR-2 transcripts was not definitively excluded. Positive and negative controls included in each of the blots validated the interpretation of hybridization results.

participate in carcinogenesis. However, we did not detect any examples of KGF expression after analyzing RNA from more than 60 carcinoma lines originating in several different organs (Tab. 1). This indicated that it would be unlikely for KGF to act as an autocrine factor in epithelial tumors. On the other hand, about 50% of the tumor lines contained transcripts that hybridized to a probe corresponding to the extracellular domain of KGFR (Tab. 1). Moreover, others have used reverse tran scription-PCR (RT-PCR) specifically to document the presence of KGFR isoforms (having FGFR-2 exon IIIb sequence) in a high percentage of carcinomas, especially from esophagus and stomach, though not in any sarcomas tested (Ishii et al., 1994; Iida et al., 1994). KGF stimulated DNA synthesis by one such line *in vitro*, suggesting that carcinoma cells *in vivo* might respond to paracrine sources of KGF (Iishi et al., 1994). In support of this hypothesis, RT-PCR analysis demonstrated KGF transcripts in tissue from 14/15 breast tumors (Koos et al., 1993). However, normal breast tissue also exhibited KGF expression, making it difficult to attribute malignant changes to KGF activity.

A K-SAM isoform (subsequently designated K-SAM type II) was amplified ∼ 50 fold in two poorly differentiated stomach carcinoma lines (Hattori et al., 1990; Katoh et al., 1992). This protein differs from KGFR primarily in the carboxy-terminal portion of the molecule, a distinction resulting from gene rearrangement. Because the corresponding portion of other tyrosine kinases has a negative regulatory function (Ullrich and Schlessinger, 1990), such a structural change could enhance basal tyrosine kinase activity. Over-expression itself might stimulate proliferation through either ligand-dependent or -independent mechanisms. It is interesting to note that the intraductal hyperplasia seen in mammary tissue following systemic KGF treatment was reminiscent of changes associated with early malignancy (Ulich et al., 1994a). Whereas this abnormality resolved after withdrawal of exogenous KGF, one could imagine that persistent signaling through the KGFR might contribute to neoplasia.

Alternatively, there is reason to believe that KGF signaling might retard processes leading to malignancy, especially virulent forms of neoplasia. Using a rat Dunning tumor model of tumor progression, Yan et al. (1993) observed that a shift from a well-differentiated, slow-growing phenotype to a poorly differentiated, highly aggressive pattern correlated with a switch from expression of an FGFR-2 isoform containing exon IIIb sequence (KGFR) to one with exon IIIc. This transition also corresponded to a loss of responsiveness to stroma, which might be attributable to the absence of KGFR (Yan et al., 1993), although a clear causal relationship was not established. The highly malignant properties of the cells lacking KGFR could be due to additional changes in gene expression; in particular, *de novo* expression by the tumor cells of FGFR-1 and ligands capable of binding to it were thought to play a role in the dramatically altered phenotype (Yan et al., 1993).

Conclusions

The hypothesis that KGF is a paracrine mediator of mesenchymal–epithelial interaction was originally based largely on cell culture data. Subsequently, a wealth of information from *in situ* hybridization analyses of KGF and KGFR, expression studies showing KGF regulation especially by sex steroid hormones and many examples of KGF activity in organ cultures and whole animals provided strong additional support for this idea. The apparent exceptions to this paradigm are noteworthy and should be further explored. Future investigations of clinical specimens should determine whether a departure from the normal pattern of KGF and KGFR expression correlates with any disorders. In this regard, development of reagents for immunohistochemical analysis of KGF and its receptor would be valuable to verify and extend results obtained by *in situ* hybridization analysis. We believe that KGF functions primarily to stimulate development of epithelial tissues, and in the adult to maintain tissue homeostasis. The latter would encompass normal epithelial differentiation and cell turnover, as well as heightened responses to injury and toxic exposures. Part of the challenge of future studies will be to learn if recombinant KGF might have a therapeutic benefit by augmenting the actions of endogenous factor. Alternatively, if conditions are identified in which KGF signaling is detrimental, then use of specific antagonists would be a strategic option.

Acknowledgements

We thank Glenn Pierce for providing us with a manuscript prior to publication (Housley et al., 1994); Elaine Alarid for permission to use a figure adapted from Alarid et al., 1994; Joel Brody for assistance in the preparation of Figures 3–5; and Stuart A. Aaronson for his support and encouragement of KGF research. Portions of this article are from a review in Cell Biology International (Rubin et al., 1995). Supported in part by NIH grants DK32157, CA49996, CA59831, DK47517, CA64872 and DK45861.

References

Aaronson, S.A., Bottaro, D.P., Miki, T., Ron, D., Finch, P.W., Fleming, T.P., Ahn, J., Taylor, W.G. and Rubin, J.S. (1991) Keratinocyte growth factor. A fibroblast growth factor family member with unusual target cell specificity. *Ann. NY Acad. Sci.* 638: 62–77.

Alarid, E.T., Rubin, J.S., Young, P., Chedid, M., Ron, D., Aaronson, S.A. and Cunha, G.R. (1994) Keratinocyte growth factor functions in epithelial induction during seminal vesicle development. *Proc. Natl. Acad. Sci. USA* 91: 1074–1078.

Albino, A.P., Davis, B.M. and Nanus, D.M. (1991) Induction of growth factor RNA expression in human malignant melanoma: markers of transformation. *Cancer Res.* 51: 4815–4820.

Baird, A. and Klagsbrun, M. (1991) The fibroblast growth factor family; Nomenclature meeting report and recommendations, January 17, 1991. *Ann NY Acad. Sci.* 638: xiii–xvi.

Blam, S.B., Mitchell, R., Tischer, E., Rubin, J.S., Silva, M., Silver, S., Fiddes, J.C., Abraham, J.A. and Aaronson, S.A. (1988) Addition of growth hormone secretion signal to basic fibroblast growth factor results in cell transformation and secretion of aberrant forms of the protein. *Oncogene* 3: 129–136.

Boismenu, R. and Havran, W.L. (1994) Modulation of epithelial cell growth by intraepithelial γδ T cells. *Science* 266: 1253–1255.

Bottaro, D.P., Rubin, J.S., Ron, D., Finch, P.W., Florio, C. and Aaronson, S.A. (1990) Characterization of the receptor for keratinocyte growth factor. Evidence for multiple fibroblast growth factor receptors. *J. Biol. Chem.* 265: 12767–12770.

Bottaro, D.P., Fortney, E., Rubin, J.S. and Aaronson, S.A. (1993) A keratinocyte growth factor receptor-derived peptide antagonist identifies part of the ligand binding site. *J. Biol. Chem.* 268: 9180–9183.

Brauchle, M., Angermeyer, K., Hubner, G. and Werner, S. (1994) Large induction of keratinocyte growth factor expression by serum growth factors and pro-inflammatory cytokines in cultured fibroblasts. *Oncogene* 9: 3199–3204.

Champion-Arnaud, P., Ronsin, C., Gilbert, E., Gesnel, M.C., Houssaint, E. and Breathnach, R. (1991) Multiple mRNAs code for proteins related to the BEK fibroblast growth factor receptor. *Oncogene* 6: 979–987.

Chedid, M., Rubin, J.S., Csaky, K.G. and Aaronson, S.A. (1994) Regulation of keratinocyte growth factor gene expression by interleukin 1. *J. Biol. Chem.* 269: 10753–10757.

Chellaiah, A.T., McEwen, D.G., Werner, S., Xu, J. and Ornitz, D.M. (1994) Fibroblast growth factor receptor (FGFR) 3. Alternative splicing in immunoglobulin-like domain III creates a receptor highly specific for acidic FGF/FGF-1. *J. Biol. Chem.* 269: 11620–11627.

Cheon, H.G., LaRochelle, W.J., Bottaro, D.P., Burgess, W.H. and Aaronson, S.A. (1994) High-affinity binding sites for related fibroblast growth factor ligands reside within different receptor immunoglobulin-like domains. *Proc. Natl. Acad. Sci. USA* 91: 989–993.

Coppola, D., Ferber, A., Miura, M., Sell, C., D'Ambrosio, C., Rubin, R. and Baserga, R. (1994) A functional insulin-like growth factor I receptor is required for the mitogenic and transforming activities of the epidermal growth factor receptor. *Mol. Cell Biol.* 14: 4588–4595.

Cunha, G.R., Chung, L.W.K., Shannon, J.M., Aguchi, O. and Fuji, H. (1983) Hormone-induced morphogenesis and growth: Role of mesenchymal–epithelial interactions. *Recent Prog. Horm. Res.* 39: 559–598.

Cunha, G.R., Donjacour, A.A., Cooke, P.S., Mee, S., Bigsby, R.M., Higgins, S.J. and Sugimura, Y. (1987) The endocrinology and developmental biology of the prostate. *Endocrine Rev.* 8: 338–362.

Cunha, G.R. and Young, P. (1992a) Role of stroma in oestrogen-induced epithelial proliferation. *Epithelial Cell Biol.* 1: 18–31.

Cunha, G.R., Alarid, E.T., Turner, T., Donjacour, A.A., Boutin, E.L. and Foster, B.A. (1992b) Normal and abnormal development of the male urogenital tract: Role of androgens, mesenchymal–epithelial interactions and growth factors. *J. Andrology* 13: 465–475.

Cunha, G.R., Sugimura, Y., Foster, B.A., Rubin, J.S. and Aaronson, S.A. (1994) The role of growth factors in the development and growth of the prostate and seminal vesicle. *Biomed. and Pharmacother. (Suppl. 1)*, 48; *in press*.

Dell, K.R. and Williams, L.T. (1992) A novel form of fibroblast growth factor receptor 2. Alternative splicing of the third immunoglobulin-like domain confers ligand binding specificity. *J. Biol. Chem.* 267: 21225–21229.

Delli Bovi, P. and Basilico, C. (1987) Isolation of a rearranged human transforming gene following transfection of Kaposi sarcoma DNA. *Proc. Natl. Acad. Sci. USA* 84: 5660–5664.

Dionne, C.A., Crumley, G., Bellot, F., Kaplow, J.M., Searfoss, G., Ruta, M., Burgess, W.H., Jaye, M. and Schlessinger, J. (1990) Cloning and expression of two distinct high-affinity receptors cross-reacting with acidic and basic fibroblast growth factors. *EMBO J.* 9: 2685–2692.

Falco, J.P., Taylor, W.G., Di Fiore, P.P., Weissman, B.E. and Aaronson, S.A. (1988) Interactions of growth factors and retroviral oncogenes with mitogenic signal transduction pathways of Balb/MK keratinocytes. *Oncogene* 2: 573–578.

Finch, P.W., Rubin, J.S., Miki, T., Ron, D. and Aaronson, S.A. (1989) Human KGF is FGF-related with properties of a paracrine effector of epithelial cell growth. *Science* 245: 752–755.

Guo, L., Yu, Q.C. and Fuchs, E. (1993) Targeting expression of keratinocyte growth factor to keratinocytes elicits striking changes in epithelial differentiation in transgenic mice. *EMBO J.* 12: 973–986.

Halaban, R., Funasaka, Y., Lee, P., Rubin, J., Ron, D. and Birnbaum, D. (1991) Fibroblast growth factors in normal and malignant melanocytes. *Ann. NY Acad. Sci.* 638: 232–243.

Hattori, Y., Odagiri, H., Nakatani, H., Miyagawa, K., Naito, K., Sakamoto, H., Katoh, O., Yoshida, T., Sugimura, T. and Terada, M. (1990) K-*sam*, an amplified gene in stomach cancer, is a member of the heparin-binding growth factor receptor genes. *Proc. Natl. Acad. Sci. USA* 87: 5983–5987.

Hayward, S.W., Dahiya, R., Cunha, G.R., Bartek, J., Despande, N. and Narayan, P. (1995) Establishment and characterization of an immortalized but non-tumorigenic human prostate epithelial cell line: BPH-1. *Cell. Dev. Biol. Anim.* 31: 14–24.

Housley, R.M., Morris, C.F., Boyle, W., Ring, B., Biltz, R., Tarpley, J.E., Aukerman, S.L., Devine, P.L., Whitehead, R.H. and Pierce, G.F. (1994) Keratinocyte growth factor induces proliferation of hepatocytes and epithelial cells throughout the rat gastrointestinal tract. *J. Clin. Invest.* 94: 1764–1777.

Houssaint, E., Blanquet, P., Champion-Arnaud, P., Gesnel, M.C., Torriglia, A., Courtois, Y. and Breathnach, R. (1990) Related fibroblast growth factor receptor genes exist in the human genome. *Proc. Natl. Acad. Sci. USA* 87: 8180–8184.

Iida, S., Katoh, O., Tokunaga, A. and Terada, M. (1994) Expression of fibroblast growth factor gene family and its receptor gene family in the human upper gastrointestinal tract. *Biochem. Biophys. Res. Commun.* 199: 1113–1119.

Ishii, H., Hattori, Y., Itoh, H., Kishi, T., Yoshida, T., Sakamoto, H., Oh, H., Yoshida, S., Sugimura, T. and Terada, M. (1994) Preferential expression of the third immunoglobulin-like domain of K-*sam* product provides keratinocyte growth factor-dependent growth in carcinoma cell lines. *Cancer Res.* 54: 518–522.

Itoh, T., Suzuki, M. and Mitsui, Y. (1993) Keratinocyte growth factor as a mitogen for primary culture of rat hepatocytes. *Biochem. Biophys. Res. Commun.* 192: 1011–1015.

Johnson, D.E., Lu, J., Chen, H., Werner, S. and Williams, L.T. (1991) The human fibroblast growth factor receptor genes: a common structural arrangement underlies the mechanisms for generating receptor forms that differ in their third immunoglobulin domain. *Mol. Cell Biol.* 11: 4627–4634.

Katoh, M., Hattori, Y., Sasaki, H., Tanaka, M., Sugano, K., Yazaki, Y., Sugimura, T. and Terada, M. (1992) K-*sam* gene encodes secreted as well as transmembrane receptor tyrosine kinase. *Proc. Natl. Acad. Sci. USA* 89: 2960–2964.

Keegan, K., Johnson, D.E., Williams, L.T. and Hayman, M.J. (1991) Isolation of an additional member of the fibroblast growth factor receptor family, FGFR-3. *Proc. Natl. Acad. Sci. USA* 88: 1095–1099.

Koji, T., Chedid, M., Rubin, J.S., Slayden, O.D., Csaky, K.G., Aaronson, S.A and Brenner, R.M. (1994) Progesterone-dependent expression of keratinocyte growth factor mRNA in stromal cells of the primate endometrium: keratinocyte growth factor as a progestomedin. *J. Cell Biol.* 125: 393–401.

Koos, R.D., Banks, P.K., Inkster, S.E., Yue, W. and Brodie, A.M. (1993) Detection of aromatase and keratinocyte growth factor expression in breast tumors using reverse transcription–polymerase chain reaction. *J. Steroid Biochem. Mol. Biol.* 45: 217–225.

Kornbluth, S., Paulson, K.E. and Hanafusa, H. (1988) Novel tyrosine kinase identified by phosphotyrosine antibody screening of cDNA libraries. *Mol. Cell Biol.* 8: 5541–5544.

Marchese, C., Rubin, J., Ron, D., Faggioni, A., Torrisi, M.R., Messina, A., Frati, L. and Aaronson, S.A. (1990) Human keratinocyte growth factor activity on proliferation and differentiation of human keratinocytes: Differentiation response distinguishes KGF from EGF family. *J. Cell Physiol.* 144: 326–332.

Mason, I.J., Fuller-Pace, F., Smith, R. and Dickson, C. (1994) FGF-7 (keratinocyte growth factor) expression during mouse development suggests roles in myogenesis, forebrain regionalisation and epithelial–mesenchymal interactions. *Mech. Dev.* 45: 15–30.

Miki, T., Fleming, T.P., Bottaro, D.P., Rubin, J.S., Ron, D. and Aaronson, S.A. (1991) Expression cDNA cloning of the KGF receptor by creation of a transforming autocrine loop. *Science* 251: 72–75.

Miki, T., Bottaro, D.P., Fleming, T.P., Smith, C.L., Burgess, W.H., Chan, A.M. and Aaronson, S.A. (1992) Determination of ligand-binding specificity by alternative splicing: two distinct growth factor receptors encoded by a single gene. *Proc. Natl. Acad. Sci. USA* 89: 246–250.

Miyamoto, M., Naruo, K.-I., Seko, C., Matsumoto, S., Kondo, T. and Kurokawa, T. (1993) Molecular cloning of a novel cytokine cDNA encoding the ninth member of the fibroblast growth factor family, which has a unique secretion property. *Mol. Cell Biol.* 13: 4251–4259.

Oppenheim, J.J., Kovacs, E.J., Matsushima, K. and Duru, S.K. (1986) There is more than one interleukin 1. *Immunol. Today* 7: 45–56.

Ornitz, D.M., Yayon, A., Flanagan, J.G., Svahn, C.M., Levi, E. and Leder, P. (1992) Heparin is required for cell-free binding of basic fibroblast growth factor to a soluble receptor and for mitogenesis in whole cells. *Mol. Cell Biol.* 12: 240–247.

Orr-Urtreger, A., Givol, D., Yayon, A., Yarden, Y. and Lonai, P. (1991) Developmental expression of two murine fibroblast growth factor receptors, *flg*, and *bek*. *Development* 113: 1419–1434.

Orr-Urtreger, A., Bedford, M.T., Burakova, T., Arman, E., Zimmer, Y., Yayon, A., Givol, D. and Lonai, P. (1993) Developmental localization of the splicing alternatives of fibroblast growth factor receptor-2 (FGFR2). *Dev. Biol.* 158: 475–486.

Panos, R.J., Rubin, J.S., Csaky, K.G., Aaronson, S.A. and Mason, R.J. (1993) Keratinocyte growth factor and hepatocyte growth factor/scatter factor are heparin-binding growth factors for alveolar type II cells in fibroblast-conditioned medium [published erratum appears in *J. Clin. Invest.* 1994 Mar, 93(3): 1347]. *J. Clin. Invest.* 92: 969–977.

Parrott, J.A., Vigne, J.L., Chu, B.Z. and Skinner, M.K. (1994) Mesenchymal–epithelial interactions in the ovarian follicle involve keratinocyte and hepatocyte growth factor production by thecal cells and their action on granulosa cells. *Endocrinology* 135: 569–575.

Patstone, G., Pasquale, E.B. and Maher, P.A. (1993) Different members of the fibroblast growth factor receptor family are specific to distinct cell types in the developing chicken embryo. *Dev. Biol.* 155: 107–123.

Pekonen, F., Nyman, T. and Rutanen, E.M. (1993) Differential expression of keratinocyte growth factor and its receptor in the human uterus. *Mol. Cell Endocrinol.* 95: 43–49.

Peters, K.G., Werner, S., Chen, G. and Williams, L.T. (1992) Two FGF receptor genes are differentially expressed in epithelial and mesenchymal tissues during limb formation and organogenesis in the mouse. *Development* 114: 233–243.

Peters, K., Werner, S., Liao, X., Wert, S., Whitsett, J. and Williams, L. (1994) Targeted expression of a dominant negative FGF receptor blocks branching morphogenesis and epithelial differentiation of the mouse lung. *EMBO J.* 12: 973–986.

Pierce, G.F., Yanagihara, D., Klopchin, K., Danilenko, D.M., Hsu, E., Kenney, W.C. and Morris, C.F. (1994) Stimulation of all epithelial elements during skin regeneration by keratinocyte growth factor. *J. Exp. Med.* 179: 831–840.

Rapraeger, A.C., Krufka, A. and Olwin, B.B. (1991) Requirement of heparan sulfate for bFGF-mediated fibroblast growth and myoblast differentiation. *Science* 252: 1705–1708.

Reich-Slotky, R., Bonneh-Barkay, D., Shaoul, E., Berman, B. and Ron, D. (1994) Differential effect of cell associated heparan sulfates on the binding of aFGF and KGF to the KGF receptor. *J. Biol. Chem.* 269: 32279–32285.

Rogelj, S., Weinberg, R.A., Fanning, P. and Klagsbrun, M. (1988) Basic fibroblast growth factor fused to a signal peptide transforms cells. *Nature* 331: 173–175.

Ron, D., Bottaro, D.P., Finch, P.W., Morris, D., Rubin, J.S. and Aaronson, S.A. (1993a) Expression of biologically active recombinant keratinocyte growth factor. Structure/function analysis of amino-terminal truncation mutants. *J. Biol. Chem.* 268: 2984–2988.

Ron, D., Reich, R., Chedid, M., Lengel, C., Cohen, O.E., Chen, A.M.-L., Neufeld, G., Miki, T. and Tronick, S.R. (1993b) Fibroblast growth factor receptor 4 is a high affinity receptor for both acidic and basic fibroblast growth factor but not for keratinocyte growth factor. *J. Biol. Chem.* 268: 5388–5394.

Rubin, J.S., Osada, H., Finch, P.W., Taylor, W.G., Rudikoff, S. and Aaronson, S.A. (1989) Purification and characterization of a newly identified growth factor specific for epithelial cells. *Proc. Natl. Acad. Sci USA* 86: 802–806.

Rubin, J.S., Chan, A.M., Bottaro, D.P., Burgess, W.H., Taylor, W.G., Cech, A.C., Hirschfield, D.W., Wong, J., Miki, T., Finch, P.W. and Aaronson, S.A. (1991) A broad-spectrum human lung fibroblast-derived mitogen is a variant of hepatocyte growth factor. *Proc. Natl. Acad. Sci. USA* 88: 415–419.

Rubin, J.S., Peehl, D.M., Chedid, M., Alarid, E.T., Cunha, G.R., Ron, D. and Aaronson, S.A. (1992) KGF is a paracrine mediator of epithelial cell growth and development in the male reproductive tract. *Second International Symposium on the Biology of Prostate Growth NIH Publication: 14 (Abst.).*

Rubin, J.S., Bottaro, D.P., Chedid, M., Miki, T., Ron, D., Cheon, H.-G., Taylor, W.G., Fortney, E., Sakata, H., Finch, P.W. and LaRochelle, W.J. (1995) Keratinocyte growth factor. *Cell Biol. Int.* 19: 399–411.

Sato, M., Kitazawa, Y., Katsumata, A., Mukamoto, M., Okada, T. and Takeya, T. (1992) Tissue-specific expression of two isoforms of chicken fibroblast receptor, *bek* and *cek3*. *Cell Growth & Diff.* 4: 355–361.

Sawyer, R.H. and Fallows, J.F. (1983) *Epithelial–Mesenchymal Interactions During Development.* Praeger, New York, N.Y.

Schor, S.L., Schor, A.M., Howell, A. and Crowther, D. (1987) Hypothesis: persistent expression of fetal phenotypic characteristics by fibroblasts is associated with an increased susceptibility to neoplastic disease. *Exp. Cell Biol.* 55: 11 17.

Shi, D.L., Launay, C., Fromentoux, V., Feige, J.J. and Boucaut, J.C. (1994) Expression of fibroblast growth factor receptor-2 splice variants is developmentally and tissue-specifically regulated in the amphibian embryo. *Dev. Biol.* 164: 173–182.

Shima, H., Tsuji, M., Young, P.F. and Cunha, G.R. (1990) Postnatal growth of mouse seminal vesicle is dependent on 5α-dihydrotestosterone. *Endocrinology* 127: 3222–3233.

Smola, H., Thiekotter, G. and Fusenig, N.E. (1993) Mutual induction of growth factor gene expression by epidermal-dermal cell interaction. *J. Cell. Biol.* 122: 417–419.

Staiano-Coico, L., Krueger, J.G., Rubin, J.S., D'limi, S., Vallat, V.P., Valentino, L., Fahey, T., Hawes, A., Kingston, G., Madden, M.R., Mathwich, M., Gottlieb, A.B. and Aaronson, S.A. (1993) Human keratinocyte growth factor effects in a porcine model of epidermal wound healing. *J. Exp. Med.* 178: 865–878.

Strain, A.J., McGuinness, G., Rubin, J.S. and Aaronson, S.A. (1994) Keratinocyte growth factor and fibroblast growth factor action on DNA synthesis in rat and human hepatocytes: modulation by heparin. *Exp. Cell. Res.* 210: 253–259.

Suzuki, M., Itoh, T., Osada, H., Rubin, J.S., Aaronson, S.A., Suzuki, T., Koga, N., Saito, T. and Mitsui, Y. (1993) Spleen-derived growth factor, SDGF-3, is identified as keratinocyte growth factor (KGF). *FEBS Lett.* 328: 17–20.

Taira, M., Yoshida, T., Miyagawa, K., Sakamoto, H., Terada, M. and Sugimura, T. (1987) cDNA sequence of human transforming gene *hst* and identification of the coding sequence required for transforming activity. *Proc. Natl. Acad. Sci. USA* 84: 2980–2984.

Tanaka, A., Miyamoto, N., Minamino, M., Takeda, M., Sato, H., Matsuo, H. and Matsumoto, K. (1992) Cloning and characterization of an androgen-induced growth factor essential for the androgen-dependent growth of mouse mammary carcinoma cells. *Proc. Natl. Acad. Sci. USA* 89: 8928–8932.

Tsuboi, R., Sato, C., Kurita, Y., Ron, D., Rubin, J.S. and Ogawa, H. (1993) Keratinocyte growth factor (FGF-7) stimulates migration and plasminogen activator activity of normal human keratinocytes. *J. Invest. Dermatol.* 101: 49–53.

Ulich, T.R., Yi, E.S., Cardiff, R., Yin, S., Bikhazi, N., Biltz, R., Morris, C.F. and Pierce, G.F. (1994a) Keratinocyte growth factor is a growth factor for mammary epithelium in vivo. The mammary epithelium of lactating rats is resistant to the proliferative action of keratinocyte growth factor. *Am. J. Pathol.* 144: 862–868.

Ulich, T.R., Yi, E.S., Longmuir, K., Yin, S., Biltz, R., Morris, C.F., Housley, R.M. and Pierce, G.F. (1994b) Keratinocyte growth factor is a growth factor for type II pneumocytes in vivo. *J. Clin. Invest.* 93: 1298–1306.

Ullrich, A. and Schlessinger, J. (1990) Signal transduction by receptors with tyrosine kinase activity. *Cell* 61: 203–212.

Werner, S., Peters, K.G., Longaker, M.T., Fuller-Pace, F., Banda, M.J. and Williams, L.T. (1992) Large induction of keratinocyte growth factor expression in the dermis during wound healing. *Proc. Natl. Acad. Sci. USA* 89: 6896–6900.

Werner, S., Smola, H., Liao, X., Longaker, M.T., Krieg, T., Hofschneider, P.H. and Williams, L.T. (1994) The function of KGF in morphogenesis of epithelium and reepithelialization of wounds. *Science* 266: 819–822.

Wilson, S.E., Walker, J.W., Chwang, E.L. and He, Y.G. (1993) Hepatocyte growth factor, keratinocyte growth factor, their receptors, fibroblast growth factor receptor-2, and the cells of the cornea. *Invest. Ophthalmol. Vis. Sci.* 34: 2544–2561.

Wright, N. and Allison, M. (1984) *The Biology of Epithelial Cell Population.* Oxford University Press, New York, NY.

Yan, G., Fukabori, Y., Nikolaropoulos, S., Wang, F. and McKeehan, W.L. (1992) Heparin-binding keratinocyte growth factor is a candidate stromal-to-epithelial-cell andromedin. *Mol. Endocrinol.* 6: 2123–2128.

Yan, G., Fukabori, Y., McBride, G., Nikolaropolous, S. and McKeehan, W.L. (1993) Exon switching and activation of stromal and embryonic fibroblast growth factor (FGF)-FGF receptor genes in prostate epithelial cells accompany stromal independence and malignancy. *Mol. Cell Biol.* 13: 4513–4522.

Yayon, A., Klagsbrun, M., Esko, J.D., Leder, P. and Ornitz, D. M. (1991) Cell surface, heparin-like molecules are required for binding of basic fibroblast growth factor to its high affinity receptor. *Cell* 64: 841–848.

Yayon, A., Zimmer, Y., Shen, G.H., Avivi, A., Yarden, Y. and Givol, D. (1992) A confined variable region confers ligand specificity on fibroblast growth factor receptors: implications for the origin of the immunoglobulin fold. *EMBO J.* 11: 1885–1890.

Yi, E.S., Yin, S., Harclerode, D.L., Bedoya, A., Bikhazi, N.B., Housley, R.M., Aukerman, S.L., Morris, C.F., Pierce, G.F. and Ulich, T.R. (1994) Keratinocyte growth factor induced pancreatic ductal epithelial proliferation. *Am. J. Pathol.* 145: 80–85.

Zhan, X., Bates, B., Hu, X.G. and Goldfarb, M. (1988) The human FGF-5 oncogene encodes a novel protein related to fibroblast growth factors. *Mol. Cell Biol.* 8: 3487–3495.

Zimmer, Y., Givol, D. and Yayon, A. (1993) Multiple structural elements determine ligand binding of fibroblast growth factor receptors. Evidence that both Ig domain 2 and 3 define receptor specificity. *J. Biol. Chem.* 268: 7899–7903.

Epithelial–Mesenchymal Interactions in Cancer
ed. by I.D. Goldberg & E.M. Rosen
© 1995 Birkhäuser Verlag Basel/Switzerland

Stromal–epithelial interaction in type IV collagenase expression and activation: The role in cancer metastasis

Y.E. Shi[1] and Y. Liu[2]

[1]Department of Pediatrics and Pathology,
[2]Department of Pediatrics, Long Island Jewish Medical Center, Albert Einstein College of Medicine, New Hyde Park, N.Y. 11042, USA

Introduction

It is the process of metastasis in cancer which impairs the ability of surgery to cure all but its earliest stages. In histologic observations of carcinomas, it is generally noted that as epithelial growth becomes dysregulated, the stromal–epithelial tissue compartmentalization also can become disturbed. The dissolution of the epithelial basement membrane signals local invasion of the disease; blood vessel infiltration (angiogenesis) is thought to allow the tumor cells to spread via lymphatic and hematogenous means to nearby and to distant organs. Metastasis is thought to depend upon alterations in cell adhesion, motility, proteolysis, and the ability to survive in another organ's environments.

Degradation of the extracellular matrix and basement membrane is an essential component of several steps in tumor invasion and metastasis. (Liotta and Stetler-Stevenson, 1991); this event also occurs during some physiologic processes, such as embryo morphogenesis (Alexander et al., 1991), blastocyst implantation (Lola and Graham, 1990), angiogenesis and tissue remodeling (Banda et al., 1988; Alexander et al., 1991). The malignant cancer cells are generally characterized by the capacity to invade and degrade tissue barriers, such as basement membrane, by enhanced proteolysis. The proteinases are a large family, grouped into 4 main classes: serine–proteinases such as plasminogen activators, plasmin and elastase; cysteine-proteinases (principally cathepsin B and L); metalloproteases, which include interstitial or type-I collagenases, type-IV collagenases or gelatinases, and stromelysins; and aspartylprotenases (eg., cathepsin D). Although enzymes from all four classes of proteinases have been implicated in the process of cancer invasion and metastasis, specific emphasis has been placed on matrix metalloproteinases (MMPs), particularly type-IV collagenases,

which degrades type IV collagen, the major structural component
of basement membrane (Duffy, 1992; Goldberg and Eisen, 1990).
MMPs are thought to be required for degradation of the basement
membrane and stromal matrix as cancer cells invade locally, enter
nearby vasculature (intravasation), exit vasculature in meta-static sites
(extravasation), and expand through the sites of metastasis (Dickson, Shi and Johnson, 1994). In addition, the process of
angiogenesis itself probably requires somewhat similar proteolytic
events.

The ability of metastatic tumor cells to induce MMPs has been
implicated in the process of tumor progression. Although extensive
studies have been conducted to try to associate tumor production of
a particular MMP such as 72 kDa type IV collagenase (MMP-2) and
92 kDa type IV collagenase (MMP-9) with the metastatic phenotype
(Brown et al., 1993; Kawashima et al., 1994; Naylor et al., 1994; Shi
et al., 1993a; Bernhard et al., 1990, 1994; Ballin et al., 1988, 1991),
there is a growing body of evidence which suggests that production
of MMPs by the host may play a critical role in human tumor progression. This new view is based on the three main lines of evidence:
(1) predominant stromal (especially the stroma surrounding invading
malignant epithelium) expression of type IV collagenases (Pyke et al.,
1993; Naylor et al., 1994; Newell et al., 1994) and stromelysin-3 (Basset et al., 1990; Basset et al., 1994) demonstrated by in situ hybridization and immunohistochemical studies; (2) stimulation of stromal-
derived MMP expression by direct cell-cell contact with metastatic
cancer cells (Kataoka et al., 1993; Himelstein et al., 1994) and (3)
association of metastatic phenotype of malignant tumor cells with
MMP activation potential but not MMP production (Azzam et al.,
1993; Brown et al., 1993; Sato et al., 1994b). Therefore, tumor cell-
host stroma mediated MMP production and activation may be a
more appropriate model in terms of the relevance of MMPs in cancer
invasion and metastasis.

Based on limited in vitro studies using established cancer cell lines
and in vivo studies using human malignant cancer samples and experimental animal models, we have attempted to summarize the literature
on potential mechanisms which might modulate the host stroma–tumor epithelium interaction in expression and activation of MMPs,
particularly type IV collagenases, during cancer metastasis. The pur-
pose of this chapter is to amplify these points in an attempt to emphasize the host effect on MMP production and cancer metastasis.
It is emphasized, however, that much more research needs to be
carried out on the various mechanisms which might modulate
MMP expression, secretion, presentation and activation during cancer
metastasis.

Metalloproteinase in cancer metastasis

Cancer invasion and metastasis

In breast cancer and most other malignant human tumors, the most deadly aspect of the oncogenic process is the local invasion and subsequent metastasis of the primary tumor. The process of metastasis is thought to be multi-factorial, involving both the acquisition of positive effectors (oncogenes) and the loss of factors (metastasis suppressor genes), such that only a very small proportion of the original tumor cells may achieve full metastatic potential (Liotta et al., 1991). It is now proposed that development of metastatic cancer proceeds through early phases of a transformed phenotype with uncontrolled growth leading to invasion. The loss of controlled cell growth does not in itself produce invasion and metastasis. The acquisition of the metastatic phenotype requires such additional genetic changes as high binding affinity for extracellular matrix, increased proteolytic activity and high cellular motility.

Tumor cell interaction with the extracellular matrix, and in particular with the basement membrane at multiple stages, has emerged as a specific critical aspect of metastatic cascade. The basement membrane is an organized, continuous extracellular matrix structure that separates epithelia from their stroma. Its major constituents are laminin, collagen type IV, and heparan sulfate proteoglycans. The basement membrane is considered as a barrier that must be crossed during tumor invasion. Benign tumors, in some proliferative disorders, are always characterized by a continuous basement membrane which surrounds the tumorigenic cells of the epithelial compartment and separates them from the stromal tissue. However, during the complex multistep process that leads to the formation of metastasis, malignant cells encounter and penetrate numerous basement membranes as they exit the primary tumor and intravasate into the preferred sites. The interaction with basement membrane is particularly important at two stages of metastasis: at the primary tumor site during initial invasion; then following the completion of the intervening stages, attachment to and penetration of the endothelial basememt membrane of capillaries and lymphatics in distant organs (Poste et al., 1980; Liotta et al., 1986; Hunt et al., 1989; Liotta et al., 1991). Critical determinants in basement membrane traversion have been identified. Tumor cells first attach to the basement membrane glycoprotein laminin through cell-surface laminin-binding proteins (Hunt, 1989; Castronovo et al., 1989; Shi et al., 1993b; Shi et al., 1994a) which, in turn, has been shown to induce a metalloproteinase cascade resulting in activity of type IV collagenase (Kanemoto et al., 1990). This activity degrades the type IV collagen structural network of the basement membrane, allowing subsequent traversion by the tumor cell

(Liotta et al., 1991). Metastasis is proposed to depend on five major activities: angiogenesis, cellular attachment (to basement membrane and subsequently stromal matrix), proteolysis (of basement membrane and subsequently stromal matrix), migration through the barrier into the secondary sites, and, of course, colonization and proliferation in the distant organs (Liotta et al., 1991; Kohn, 1991).

Matrix metalloproteinase in cancer invasion and metastasis

Tumor cell invasion is thought to depend critically upon *proteolytic events* (Liotta, 1986; Gottesman, 1990). Proteinases appear to be required for the degradation of the basement membrane and stromal matrix as cancer cells invade locally, enter nearby vasculature, exit the vasculature in metastatic sites and expand through the site of metastasis. Enzymes from the four general classes of proteinase are probably involved in cancer metastasis: cathepsin B and L represent the cysteine proteinase (Kane et al., 1990; Sloane, 1990); cathepsin D is a member of the aspartyl proteinase class (Rochefort et al., 1990); collagenases, most specifically the type IV collagenases (MMP-2, MMP-9), interstitial collagenase (MMP-1) and stromelysins (MMP-3, MMP-11) represent the principal metal-dependent proteinases (Liotta et al., 1980; Stetler-Stevenson et al., 1991). Finally, urokinase and plasminogen represent the serine proteinase family (Duffy, 1992). As with the proteolytic cascade of blood coagulation, multiple enzymes appear to be involved in the metastatic process. This is evidenced by the fact that inhibitors for cysteinyl, metalloproteinases, and serine proteinases are all capable of blocking tumor cell invasion of native or reconstituted connective tissue barriers *in vitro* and metastasis *in vivo* (Liotta et al., 1991; Docherty et al., 1994). Although each of these enzymes is over-expressed or inappropriately expressed in metastatic cancers or surrounding stromal tissues (Goldberg et al., 1990), the most extensive work has been on type IV collagenases which degrade type IV collagen, the major structural component of basement membrane and, presumably, one of the first barriers to metastases.

The proteolytic process of metastases are thought to initially depend upon type IV collagenase, a metalloproteinase activity encoded by two distinct genes. 72 kDa (MMP-2) and 92 kDa (MMP-9) enzymes have been described in a variety of cancers (Monteahudo et al., 1990; D'Errico et al., 1991; Kawashima et al., 1994; Naylor et al., 1994). Each is secreted in an inactive form and requires an activational cleavage yielding 68 kDa and 86 kDa active enzymes, respectively. The cleavage is either autocatalytic or dependent upon other proteinases in the tumor environment (Springman et al., 1990; Azzam et al., 1993; Sato et al., 1994b). Three natural inhibitors of these enzymes have been characterized: TIMP-1, TIMP-2 and TIMP-3 (Stetler-Stevenson et al., 1989; Unemori et al., 1990; Apte et al., 1994). The inactive, unclipped collagenases are generally secreted as a complex with a TIMP. TIMP-1 is secreted with MMP-9 and TIMP-2 is associated with MMP-2. Enzyme activity seems to depend upon other local determinants in the tumor or an imbalance of collagenase/TIMP secreted ratios.

Correlative evidence for the involvement of type IV collagenase in the invasive phenotype has been demonstrated in a variety of different human cancer systems particularly in human breast and colon cancers (Tryggvason, 1993). Nonetheless, measurements of the MMP-2 itself in breast and colon cancer are developing as useful prognostic indicators (Tryggvason et al., 1993; Pyke et al., 1993). Davies et al. (1993) observed an association between expression MMP-9 progelatinase and activated MMP-2 with malignant progression of breast cancer to stage III. Immunohistochemistry demonstrated that ductal epithelial cells stain positive for MMP-2 in carcinoma *in situ*, as do the majority of invasive breast carcinomas and metastatic lymph node deposits (Monteahudo et al., 1990; D'Errico et al., 1991). Immunchemical staining has also suggested a prognostic role of MMP-2 for local recurrence of the breast cancer (Tryggvason, 1993). *In situ* hybridization revealed a strong MMP-2 mRNA expression in numerous fibroblasts in the stroma surrounding the invasive human colon adenocarcinomas; either no expression or only very weak expression of MMP-2 in normal-appearing colonic mucosa (Pyke et al., 1993). A major potential application of MMP studies is in the area of detection and monitoring by serum or plasma assay tumor onset, response to therapy and relapse. Indeed, accumulation of procollagen I carboxyterminal propeptide fragments has been reported to be a useful, blood-borne marker of osteoblastic bone metastasis (Francini et al., 1993). Detection of MMPs themselves may be of even more use (Zucker et al., 1992; Moutsiakis et al., 1992). It is of interest that elevated levels of MMP-9, MMP-2, and the MMP inhibitor, TIMP-2, have been detected in the plasma of patients with breast, lung, bladder or colon cancer (Zucker, 1993; Garbisa et al., 1992; Marguilies et al., 1992).

Another group of enzymes of the matrix metalloproteinase gene family which may be involved in cancer metastases particularly in breast cancer metastasis are the stromelysins, which include stromelysin 1 and 2, matrilysin, and most recently, stromelysin 3. Data on the recently described stromelysin 3 suggests that its expression is associated with human breast cancer progression (Basset et al., 1990; Basset et al., 1994). In these studies, stromelysin 3 gene expression was localized to the stromal cells surrounding invasive, but not *in situ*, human breast carcinoma. The substrate specificity of stromelysin 3 has yet to be defined. Regulation of its activity by other proteinases and particularly by steroid hormones will be an extremely interesting future study.

Much recent attention has focused on cell membrane associated matrix-degrading proteinases and their role in the activation of pro-MMPs (Strongin et al., 1993; Azzam et al., 1993; Monsky et al., 1993). It is thought that matrix metalloproteinases may function in invasion after binding to and being activated by cell surface extensions termed 'invadopodia' on invasive tumor cells (Strongin et al., 1993; Monsky et al., 1993); and such 'receptors' are sensitive to metalloproteinase inhibitors (Strongin et al., 1993; Azzam et al., 1993). It is of interest that MMP-2 activation is restricted to highly invasive estrogen receptor-negative, vimentin-positive human breast cancer cell lines, is independent to MMP-2 production, and is associated with metastatic potential (Azzam et al., 1993). Recently, a new matrix metalloproteinase with a potential transmembrane domain has been cloned (Sato et al., 1994a,b). Expression of the enzyme on the cell surface induces specific activation of pro-MMP2 *in vitro* and enhances cellular invasion of the reconstituted basement membrane.

Tumor angiogenesis and metalloproteinase

Tumor angiogenesis refers to the growth of new vessels toward and within the tumor. A recent clinical study has emphasized the potential importance of angiogenesis to metastatic spread of some human cancers and particularly in breast cancer (Hayes et al., 1994; Weidner et al., 1991). In that study, the metastatic spread of breast tumors was quantitatively proportional to their degree of angiogenesis. A number of other model systems, including the recently developed spontaneous metastatic MCF-7 (Mcleskey et al., 1993; Kuebayashi et al., 1993) and T47D (Shi et al., 1994b) human breast cancer cell models, have led investigators to similar conclusions (Weidner et al., 1993; Folkman, 1994; Hayes et al., 1994). In addition to the requirement of angiogenesis for initial tumor dissemination, for a metastasis to proliferate, the same vascular requirement are present. Extravasated tumor cells proliferate as colonies, but require a new vascular supply to grow larger than $0.5 \, mm^2$ (Kohn, 1991). Therefore angiogenesis is necessary at the beginning and at the end of the metastasis, and antiangiogenesis is becoming an appealing modality for the treatment of breast cancer (Folkman, 1994; Horak et al., 1993).

The growth of new blood vessels involves adhesion, degradation of extracellular matrix, migration, proliferation and differentiation of vascular endothelial cells, a process similar to cancer metastasis. Like cancer metastasis, matrix degradation metalloproteinases are also required for angiogenesis (Fisher et al., 1994).

Stromal–epithelial interaction in regulation of MMP expression

Stromal expression of MMPs

Malignant epithelial cells are characterized by their ability to invade the adjacent fibroconnective tissue. A large body of experimental evidence supports the hypothesis that the expression of matrix degrading MMP correlates with tumor invasion, and that metastasis arise from neoplastic cell subpopulations with an enhanced ability to degrade tissue barriers (Liotta et al., 1991). Highly invasive tumor cell lines secrete both MMP-2 (Mori et al., 1994; Abe et al., 1994), and MMP-9 (Kawashima et al., 1994), and their inhibitors TIMP-1 and TIMP-2, and transfected cell lines displaying augmented MMP-2 or MMP-9 (Ballin et al., 1988; Bernhard et al., 1990; Bernhard et al., 1994) activity are associated with a metastatic phenotype. Because of this tumor cell-carried invasive ability, the view that tumor cell-derived matrix degrading activity makes a significant contribution to cancer invasion and metastasis has dominated thinking in this area. Much emphasis has been placed on the neoplastic cells being the principal source of MMPs to degrade the barriers to tissue invasion thus permitting spreading and distant metastasis. The cellular source of MMPs is of primary importance in the understanding of neoplastic cell behavior; recent evidence has indicated that the desmoplastic stroma may play an important and complementary role to the neoplastic cell in tissue remodeling in carcinomas. The original view of tumor-derived-MMP has been challenged with the discovery of stromal-derived stromelysin-3 from metastatic breast cancer in which stromelysin-3 expression is limited to the stromal cells surrounding invasive, but not *in situ*, human breast carcinoma (Basset et al., 1990; Basset et al., 1994). Most carcinoma cell lines express MMP-2 and MMP-9 type IV collagenases and other MMPs. It was anticipated that the highest expression of MMP would localize to the epithelial component; unexpectedly however, recent studies using *in situ* hybridization and/or immunohistochemical stainings localize MMP-2 and MMP-9 to the cells of tumor stroma and occur mainly in fibroblasts and vascular endothelial cells. This stromal expression pattern has been demonstrated in a variety of different human malignant diseases including lung cancer (Soini et al., 1993), colon cancer (Pyke et al., 1993; Poulsom et al., 1992), skin cancer (Karelina et al., 1993), ovarian cancer (Autio-Harmainen et al., 1993) and breast cancer (Polette et al., 1993). Either no expression or very weak signals of MMP-2 or MMP-9 were detected within the tumor epithelium. The occurrence of the matrix-degrading MMPs such as MMP-2, MMP-9 and stromelysin-3 in stromal fibroblasts and endothelial cells indicates that the stromal cells of tumors have a more pronounced impact on the spread of the neoplastic disease than previously thought. These observations may help to explain why the progression of malignant epithelial tumors in breast depends at least in part on their interaction with the stromal component (Salomon and Kidwell, 1988) and may be particularly relevant in tumors characterized by a florid desmoplastic reaction, such as breast and colon carcinomas (Pyke et al., 1993). The results further show that in their ability to synthesize these proteinases the stromal cells of tumors resemble those of developing embryonic tissues, this resemblance is probably connected with the constant remodeling of extracellular matrix in response to the proliferative activity of carcinoma cells.

However, there are arguments that could be proposed against the hypothesis that stromal cells make any significant contribution to basement membrane remodeling in carcinomas. The arguments are mainly based on immunohistochemical localization of MMP-2 protein to the epithelium of colon (Poulsom et al., 1992) and breast (Monteahudo et al., 1990; Polette et al., 1994) cancers. It is of interest that, although no MMP-2 protein stains were observed in stromal cells, expression of MMP-2 mRNA was detected in stromal cells (Poulsom et al., 1992; Polette et al., 1994). It seems unlikely that the cells labeled by *in situ* hybridization would maintain increased steady-state levels of mRNA without significant translational activity. With the recent emphasis of tumor cell-derived, plasma membrane-mediated activation of proMMP-2 (Azzam et al., 1993; Sato et al., 1994; Strongin et al., 1993), type IV collagenases may function in invasion only after binding to and being activated by plasma membrane invadopodia on invasive tumor cells (Monsky et al., 1993); it is conceivable to propose that some of the immunoreactivity seen in type IV collagenases in neoplastic cells is a result of the uptake of tumor cell processed activated enzymes initially derived from peritumoral stroma.

Stimulation of stromal-derived MMP production by cancer cells

Although tumor cells themselves are capable of producing some MMPs, the results from *in situ* hybridization studies clearly demonstrate that the majority of the expression of mRNA corresponding to MMP-2, MMP-9 and stromelysin-3 within a variety of malignant human cancers, is in fibroblasts at the tumor-stroma interface rather than in the tumor cell themselves. The observation that the strongest signal of type IV collagenases mRNA is in peritumoral stromal fibroblasts suggests a mechanistic tumor cell–stromal interaction in stimulation of MMP expression. If the major source of MMP for invasive carcinoma cells is not from malignant epithelial cells but rather from surrounding stromal cells, one could expect that the peritumoral fibroblasts are induced to produce higher levels of MMPs, by tumor cell-derived factors. This interaction of tumor cells with surrounding fibroblasts may amplify MMP production in the region of the tumor-stroma interface to the point where the tissue inhibitors of MMPs no longer balance their activity.

The role of tumor cell–fibroblast interactions in the regulation of collagenase production in neoplasms has been demonstrated in a variety of systems (Kataoka et al., 1993; Himelstein et al., 1994; Ellis et al., 1989; Golsen et al., 1985; Biswas, 1984). So far, the best studied one is the tumor cell-derived collagenase-stimulating factor (TCSF). TCSF is 58 kDa membrane bound glycoprotein, initially purified from human lung carcinoma cell line, LX-1 [Ellis et al., 1989). TCSF can increase the expression of MMP-1, MMP-2 and MMP-3 in a variety of human fibroblast lines (Kataoka et al., 1993). Most interestingly, TCSF also leads to increased activation of latent MMP-2, in addition to an increase in total amount of the enzyme. Although stimulation of MMPs in fibroblasts has been demonstrated by a variety of reagents such as phorbol esters (Frisch et al., 1987), cytochalasin B (Frisch et al., 1987), interleukin 1 (Conca et al., 1989), heat shock (Vance et al., 1989), and some growth factors (Ovarall et al., 1990; Matrisian et al., 1986), most reagents stimulate only one or two MMPs and some reagents such as TGF-β (Brown et al., 1990) stimulate one MMP (MMP-2) and inhibit another MMP (MMP-1). However, TCSF is the first reported single biological factor that stimulates MMP-1, MMP-2 and MMP-3 production at both the mRNA and protein levels. Tumor cell interaction with fibroblasts via TCSF could lead to increased degradation of interstitial or basement membrane matrix components and thus to enhanced tumor cell invasion.

Zucker et al. (1994) recently investigated the expression of TCSF in malignant human breast tissue. Immunohistochemical studies show intense staining of TCSF in invasive breast cancer cells; in contrast fibroblasts surrounding the cancer and nonmalignant ducts were weakly stained. However, the intraductal in situ breast cancer tissue was also intensely stained with anti-TCSF. Use of TCSF as a marker for invasive cancer needs to be further investigated.

The mechanism(s) whereby cancer cells stimulate expression of stromal-derived MMPs is unknown but is of considerable interest. Several studies suggest that stimulation of expression of stromal-derived MMPs requires direct contact with metastatic tumor cells. Himelstein et al. (1994) characterized a group of transformed rat embryo tumor cell lines whose metastatic phenotype correlates with their ability to stimulate stromal-derived MMP-9. MMP-9 activity was detectable in tumors of a metastatic v-myc- and H-ras-transformed cell line, 2.8, but was not found when these cells were grown alone in tissue culture. RA3, an H-ras and E1A transfectant, was not metastatic and did not produce MMP-9 in culture or in tumors. Speculating that the metastatic capacity of 2.8 tumor cells was related to their ability to induce expression of

MMP-9 by host cells, an *in vitro* primary rat embryo fibroblast (REF) and tumor cell co-culture model was established to examine potential tumor–host interactions. MMP-9 expression was induced by co-culture of metastatic 2.8 with REF; but co-culture of the non-metastatic line RA3 with REF did not result in induction of the MMP-9. The MMP-9 enzyme in these co-cultures was released by the fibroblasts, because the methanol-fixed 2.8 cells induced MMP-9 expression in REF, but fixed REF cells did not induce enzyme expression in 2.8 cells. In addition, *in situ* hybridization of nude mouse tumors derived from metastatic 2.8 line revealed MMP-9 expression mainly in the stromal cells. This induction of MMP-9 in fibroblasts seems to be dependent on cell–cell contact with metastatic tumor line 2.8, because (1) soluble factors were not sufficient for MMP-9 induction in fibroblasts by tumor cells (tumor cell-derived conditioned media did not induce expression of the MMP-9 by REF); (2) co-culture of tumor cells and fibroblasts separated by solute-permissive membranes abolished by the MMP-9 induction in REF. These results together support the hypothesis that metastatic 2.8 cells, but not non-metastatic RA3 cells, can induce normal fibroblasts to express the MMP-9 to facilitate tumor metastatic potentials, and that cell–cell contact may be necessary for this induction.

Evidence of high expression of MMP-2 and MMP-9 in fibroblasts at the tumor-stroma interface rather than in the tumor cells themselves, and the low levels of MMP-1 and MMP-2 in normal fibroblasts (Gray et al., 1992; Pyke et al., 1992), strongly suggests that peritumoral fibroblasts are induced to produce higher levels of MMPs, presumably by tumor cell-derived inducers such as TCSF. Much more basic biological and molecular research on tumors is needed to address fundamental questions, such as the specificity of the tumor–host stroma interaction and the molecular mechanism(s) underlying the cell–cell contact mediated the induction of MMPs.

Activation of metalloproteinase

Secreted extracellular matrix metalloproteinases initiate turnover of extracellular matrix macromolecules. The extracellular activity of these enzymes is controlled at the level of gene expression, proenzyme activation, and interaction with specific tissue inhibitors of metalloproteinase, TIMP. All members of the matrix metalloproteinase family are secreted as inactive zymogens and require an extracellular activation. All the members in MMP family have three definable structural domains of related primary sequence, including a propeptide domain, a catalytic NH_2-terminal domain and a hemopexin-like COOH-terminal domain. Both 72 kDa and 92 kDa type IV collagenases have a further domain inserted into the catalytic domain with sequence similarities to the gelatin binding region of fibronectin (O'Connell et al., 1994). The question of the biochemical and biological function of the matrix metalloproteinase domains has been the subject of a number of recently described studies. The propeptide is a determinant of proenzyme latency, functioning by means of a highly conserved sequence with an unpaired cysteine residue which is thought to interact with the active site zinc atom (reviewed by Birkedal-Hansen et al., 1993). The catalytic domain contains identifiable sequences for zinc and calcium binding and is involved in substrate cleavage, and autolytic activation cleavages (Okada et al., 1986; Marcy et al., 1991; Murphy et al., 1992a). The COOH-terminal domain is not involved in autolytic or cellular activation but affects proenzyme-TIMP complex formation (O'Connell et al., 1994).

Although mechanisms of enzyme activation in solution have been intensively investigated, a physiologically relevant pathway involved in the activation and compartmentalization of the enzymes in the tumor environment is poorly understood.

Activation of MMP in a cell-free system

The control of MMP activity has been attributed to a cysteine-switch mechanism, in which a cysteine residue contained in the propeptide of all members of the MMP family forms a complex with the zinc atom located in the active site. The interaction between a sulfhydryl side chain of the cysteine residue and the active zinc atom results in a catalytically inert active center (Birkedal-Hansen et al., 1993). The sulfhydryl group in this interaction is donated by the cysteine-73 within a highly conserved sequence PRCGVPDV located immediately adjacent to the proenzyme cleavage site (1, 30, 31). Disruption of this interaction by proteolytic or autocatalytic removal of the Cys-73 containing sequence results in conformational rearrangement and rapid attainment of proteinase activity (1). *In vitro* activation of MMP-2 and MMP-9 type IV collagenase can be achieved artificially by detergents, chaotropic agents, certain proteinases (Springman et al., 1990) such as trypsin, stromelysin-1 (O'Connell et al., 1994) and plasmin (Reith and Rucklidge, 1992), and by autocatalytic reaction induced by organomercurial compound p-aminophenylmercuric acetate (APMA) (Morodomi et al., 1992). These activational cleavages yield 68 kDa and 73 kDa active enzymes, respectively. Using MMP-9 purified from neutrophils, Triebel et al. (1992) demonstrated that autoprocessing of MMP-9 in the presence of AMPA occurred with a four-step truncation of the NH_2-terminal propeptide to Met^{75} and a three-step loss of COOH-terminal fragments to Ala^{506}, generating a 73-kDa active enzyme. Stromelysin treatment of proMMP-9 generated an 82 kDa active form, NH_2 terminus Phe[88] (Ogata et al., 1992). In contrast, Okada et al. (1992) obtained active species of 68 kDa with APMA and 64 kDa with stromelysin treatment of proMMP-9. Although the *in vitro* proenzyme activation is well characterized there is no evidence that *in vitro* autocatalytic or proteolytic cleavage is a physiologically relevant pathway of MMP activation in the tumor environment.

Cell-mediated (plasma membrane-dependent) activation of MMP-2

The degradation of extracellular matrix (ECM) is often confined to localized sites at the cell surface that are in contact with the ECM. However, members of the MMP family, such as MMP-2, are secreted in zymogen form and are bound to their natural inhibitors, TIMPs. Physiological activation and centralization of the active enzymes are required to achieve matrix degradation; this might occur after binding of proenzymes or proenzyme-TIMP complexes to the plasma membrane. Several morphological and biochemical observations have suggested that activation of MMP-2 takes place after binding of proenzymes or their complexes with TIMPs to the plasma-membrane (Monsky et al., 1993; Azzam et al., 1993; Strongin et al., 1993); and this membrane-mediated activation of MMPs is inducible by factors such as concanavalin A in fibroblasts (Murphy et al., 1992a; Overall, 1990), TPA and TGF-β in certain sarcoma cells (Brown et al., 1990), or collagen I in carcinomas (Azzam et al., 1992; Azzam et al., 1993). These studies also demonstrated that MMP-2 is activated in the invasive tumor tissues through the proteolytic process mediated by the tumor cell membrane fraction that is sensitive to metalloproteinase inhibitors (Azzam et al., 1992; Strongin et al., 1993). Thus, a hypothetical activator of pro-MMP-2 is a membrane-bound matrix metalloproteinase.

This plasma membrane-dependent activation of MMP-2 was recently confirmed by the identification of a new membrane type–matrix metalloproteinase (MT-MMO) (Sato et al., 1994b). The MT-MMP gene was cloned from human placenta with PCR generated probes corresponding to the highly conserved amino acid residues among MMP family. Expression of MT-MMP induced specific activation of pro-MMP-2 *in vitro*, and resulted in the enhanced cellular invasion of reconstituted basement membrane Matrigel; no MMP-9 activation was observed. To further evaluate the biological relevance of MT-MMP in cancer metastasis, experiments were conducted to examine the possible co-existence of activated form of MMP-2 and MT-MMP expression in lung carcinoma tissues. As expected, a strong correlation between the existence of the activated form of MMP-2 and the expression of MT-MMP was observed. Thus, it is very likely that previously demonstrated a hypothetical MMP-2 activator corresponds to MT-MMP which triggers tumor invasion by activating pro-MMP-2 on the cell surface.

Unlike most matrix degrading metalloproteinases, the MT-MMP is the first metalloproteinase having the transmembrane domain and localized on plasma membrane. The membrane localization of MT-MMP was first demonstrated by immunofluoresence staining and biochemical fractionation (Sato et al., 1994a; Sato et al., 1994b). Recently, the membrane localization was further confirmed by chimeric protein expression (Cao et al., 1994). In that study, a deletion of transmembrane domain of MT-MMP abolished not only its membrane localization but also its ability to activate pro-MMP-2. Fusion of the IL-2 receptor transmembrane domain to the truncated MT-MMP results in recovery of the membrane localization and pro-MMP-2 activation property. The membrane localization and the ability to activate pro-MMP-2 represent two unique features of the MT-MMP. Compared to the secreted MMPs, the membrane-associated matrix degrading metalloproteinases seem to have more correlation with tumor invasion (Monsky et al., 1993). The importance of MT-MMP and other unidentified membrane-associated MMP activation in cancer invasion should be emphasized, because they may play a critical role in centralization of activated MMPs at the tumor cell surface that are in contact with the extracellular matrix.

Association of MMP activation with metastatic progression

Expression of MMP-2 by cancer cells has been implicated in metastasis via cancer cell invasion of basement membranes mediated by degradation of matrix proteins. However, the abundance of latent MMP-2 proenzyme in normal tissues and fluids suggests that MMP-2 proenzyme utilization is limited by its physiological activation rather than by expression alone.

Using a panel of human breast cancer cell lines, Azzam et al. (1993) have examined the association of MMP-2 production and activation with breast cancer metastatic potential. The panel of cell lines constitutes a model for metastatic progression in human breast cancer (Thompson et al., 1992). There are strong correlations between invasive activity *in vitro*, ability to metastasize in nude mice, absence of estrogen receptors, and expression of the intermediate filament protein vimentin. Very little activation of endogenous MMP-2 progelatinase was observed when highly invasive estrogen receptor-negative and vimentin-positive human breast cancer cell lines such as Hs578T and BT549 were cultured on plastic. However, when the cells were plated on collagen I gels, active mature MMP-2 bands of 62-kDa and/or 59-kDa gelatinase were apparent. This type I collagen-induced activation of MMP-2 was limited to the invasive subset of the human breast cancer cell lines. No MMP-2 activation was observed in poorly invasive, estrogen receptor-positive, vimentin-negative lines such as MCF-7, T47D and ZR75-1 cells cultured on type I collagen gels. The putative MMP-2 activator appears to be localized on the cell surface, because the plasma membranes isolated from invasive Hs578T and BT549 cells cultured on type I collagen gels can activate the serum-derived MMP-2; but conditioned medium from the same cells does not induce activation of MMP-2.

The observation that MMP-2 does not induce MMP-2 activation by type I collagen or concanavalin A is restricted to invasive breast cancer cell lines and lack of MMP-2 production by some invasive breast cancer cell lines (Azzam et al., 1993) suggest that MMP production by invading cells may be less important than expression of MMP activator(s). Tumor cells may use exogenous MMPs derived from surrounding stromal tissue, inflammatory cells, or endothelial cells, which is abundant in healthy individuals (Vartio and Baumann, 1989). Acquisition of MMP activating capacity may be a rate-limiting step in the metastatic progression of some malignant diseases such as breast cancer.

Conclusions

A general aspect of neoplasia is the capacity to degrade and penetrate through tissue barriers, such as basement membrane, by enhanced proteolysis. Accumulated evidence suggests that proteinases important in the metastatic process include metalloproteinases as well as serine, cysteine and aspartyl proteinases (Duffy, 1992). Of these, most of the evidence to date in early stages of cancer metastasis implicates type IV collagenases MMP-2 and MMP-9 metalloproteinases. Although the association of MMP-2, MMP-9 and other MMPs with cancer invasion and metastasis has been well established, recent biological and molecular research has emphasized two fundamental questions regarding matrix-degrading MMPs and their role in cancer metastasis: (1) where is the major source for MMPs during cancer invasion and metastasis?; and (2) how is the proMMP activated?.

With regard the origin of MMPs, the original view that tumor cell-derived matrix degrading proteinases play an essential role in mediating cancer invasion and metastasis has been questioned. At present, there is circumstantial evidence suggesting that peritumoral fibroblast-derived MMPs play a critical role in cancer metastasis. If the peritumoral stroma-derived MMPs, but not the tumor-derived MMPs, are the driving force for degradation of basement membrane and stromal tissue, one may expect that such MMPs are under control by stroma-tumor cell interactions. In fact, direct stromal cell–tumor cell contact in inducing MMP-2 expression from stromal cells has been observed (Kataoka et al., 1993; Himelstein et al., 1994).

On the other hand, the original proposed physiological pathway of collagenase activation emphasizes a urokinase-dependent proteolytic cascade in which urokinase plasminogen activator converts plasminogen into plasmin. Plasmin is capable of activating purified procollagenase and prostromelysin. Plasmin-dependent activation of procollagenase generates enzyme species identical to those generated by limited proteolysis with trypsin or treatment with organo-mercurial compounds (Strongin et al., 1993). Less is known about the activation of the type IV procollagenases compared with that of collagenase and stromelysin. However, the recent biological and molecular studies that proenzymes or proenzyme-TIMP complexes may function in degradation of basement membrane only after binding to and being activated by the plasma membrane invadopodia of invasive cancer cells. The existence of tumor cell membrane-mediated activation of proMMPs is based on three lines of evidence: (1) discovery of cell surface-associated metalloproteinase MT-MMP and its ability to activate proMMP-2 (Sato et al., 1994); (2) unidentified plasma membrane-dependent and collagen-induced activation of proMMP-2 in normal and malignant human fibroblastoid cells (Azzam and Thompson, 1992; Strongin et al., 1993) and in malignant human breast cancer cells (Azzam et al., 1993); and (3) association of MMP-2 activation potential with metastatic progression in human breast cancer cells (Azzam et al., 1993). The possibility of tumor cell–plasma membrane-mediated activation of MMP-2 deserves further study as it has considerable effect not only on the activation of pro-MMPs but also on the localization of activated MMPs at the tumor cell surface that are in contact with extracellular matrix.

In summary, the points discussed here indicate a potential role of tumor epithelium–host stroma interaction in regulation of MMP expression and activation. The following three-point model is proposed: (1) peritumoral stroma provides the major source of MMPs for cancer invasion and metastasis; (2) stroma-derived MMPs are induced by invasive tumor cell-derived factor(s); and (3) these inducible, stroma-derived MMPs bind to and are activated by tumor cell plasma membrane proteins. Although more basic biological and molecular research on tumors is needed to understand the complex mechanism of extracellular matrix degradation during tumor metastasis, studies of tumor-derived MMP inducers and tumor membrane-associated MMP activators may provide promising diagnostic and therapeutic tools in cancer metastasis.

References

Abe, T., Mori, T., Kohno, K., Seiki, M., Hayakawa, T., Welgus, H.G., Hori, S. and Kuwano, M. (1994) Expression of 72 kDa type IV collagenase and invasion activity of human glioma cells. *Clin. Exp. Metastasis* 12: 296–304.

Alexander, C.M. and Werb, Z. (1991) Extracellular matrix degradation. *In:* E.D. Hay (ed.): *Cell Biology of Extracellular Matrix*, Second Edition. Plenum Press, New York, pp 255–302.

Apte, S.S., Mattei, M.G. and Olsen, B.R. (1994) Cloning of the cDNA encoding human tissue inhibitor of metalloproteinases-3 (TIMP-3) and mapping of the TIMP3 gene to chromosome 22. *Genomics* 19: 86–90.

Autio-Harmainen, H., Karttunen, T., Hurskainen, T., Hoyhtya, M. and Kauppila, A. (1993) Expression of 72 kDa type IV collagenase (gelatinase A) in benign and malignant ovarian tumors. *Lab. Invest.* 69: 312–321.

Azzam, H.S. and Thompson, E.W. (1992) Collagen-induced activation of the Mr 72,000 type IV collagenase in normal and malignant human fibroblastoid cells. *Cancer Res.* 52: 4540–4544.

Azzam, H.S., Arand, G., Lippman, M.E. and Thompson, E.W. (1993) Association of MMP-2 activation potential with metastatic progression in human breast cancer cell lines independent of MMP-2 production. *J. Nat. Cancer Inst.* 85: 1758–1764.

Ballin, M., Gonez, D.E., Sinha, C.C. and Thorgeirsson, U.P. (1988) Ras oncogene mediated introduction of a 92 kDa metalloproteinase; strong correlation with the malignant phenotype. *Biochem. Biophys. Res. Commun.* 154: 832–838.

Ballin, M., Mackay, A.R., Hartzler, J.L., Nason, A., Pelina, M.D. and Thorgeirsson, U.P. (1991) *Ras* level and metalloproteinase activity in normal versus neoplastic rat mammary tissue. *Clin. Exp. Metastasis* 9: 179–189.

Banda, M.J., Herron, G.S., Murphy, G., Werb, Z. and Dwyer, K.S. (1988) Proteinase induction by endothelial cells during wound repair. *Prog. Clin. Biol. Res.* 266: 117–130.

Basset, P., Bellocq, J.P., Wolf, C., Stoll, I., Hutin, P., Limacher, J.M., Podhajcer, O.L., Chenard, M.P., Rio, M.C. and Chambon, P. (1990) A novel metalloproteinase gene specifically expressed in stromal cells of breast carcinomas. *Nature* 348: 699–704.

Basset, P., Wolf, C., Rouyer, N., Bellocq, J.P., Rio, M.C. and Chombon, P. (1994) Stromelysin-3 in stromal tissue as a control factor in breast cancer behavior. *Cancer* 74: 1045–49.

Bernhard, E.J., Muschel, R. and Hughes, E.N. (1990) Mr 92,000 gelatinase release correlates with the metastatic phenotype in transformed rat embryo cells. *Cancer Res.* 50: 3872–3877.

Bernhard, E.J., Gruber, S.B. and Muschel, R.J. (1994) Direct evidence linking MMP-9 expression to the metastatic phenotype in transformed rat embryo cells. *Proc. Nat. Acad. Sci. USA* 91: 4293–4297.

Birkedal-Hansen, H., Moore, W.G.I., Bodden, M.K., Windsor, L.J., Birkedal-Hansen, B., DeCarlo, A. and Engler, J.A. (1993) Matrix metalloproteinase: a review. *Crit. Rev. Oral. Biol. Med.* 4: 197–250.

Biswas, C. (1984) Collagenase stimulation in cocultures of human fibroblast and human tumor cells. *Cancer Lett.* 24: 201–207.

Brown, P.D., Levy, A.T. and Margulies, I.M. (1990) Independent expression and cellular processing of Mr 72,000 type IV collagenase and interstitial collagenase in human tumorigenic cell lines. *Cancer Res.* 50: 6184–6191.

Brown, P.D., Bloxidge, R.E., Anderson, F. and Howell, A. (1993) Expression of activated gelatinase in human invasive breast carcinoma. *Clin. Exp. Metastasis* 11: 183–189.

Cao, J., Sato, H., Takino, T. and Seiki, M. (1994) The transmembrane domain of MT-MMP is essential for its localization to plasma membrane and for pro-gelatinase A activation function. *Clin. Exp. Metastasis* 12: 12.

Castronovo, V., Taraboletti, G., Liotta, L.A. and Sobel, M.E. (1989) Modulation of laminin receptor expression by estrogen and progestins in human breast cancer cell lines. *J. Natl. Cancer Inst.* 81: 781–786.

Conca, W., Kaplan, P.B. and Krane, S.M. (1989) Increases in levels of procollagenase messenger RNA in cultured fibroblast induced by human recombinant interleukin 1β or serum follow c-*jun* expression and are dependent on new protein synthesis. *J. Clin. Invest.* 83: 1753–1757.

Davies, B., Miles, D.W., Happerfield, L.C., Naylor, M.S., Bobrow, L.G., Rubens, R.D. and Balkwill, F.R. (1993) Activity of type IV collagenases in benign and malignant breast disease. *Br. J. Cancer* 67: 1126–1131.

D'Errico, A., Garbisa, S., Liotta, L.A., Castronova, V., Stetler-Stevenson, W.G. and Grigioni, W.F. (1991) *Mod. Pathol.* 4: 239–246.

Dickson, R.B., Shi, Y.E. and Johnson, M.D. (1994) A novel matrix-degrading protease in hormone dependent breast cancer. *Biochem. Soc. Transact.* 22(1): 49–52.

Duffy, M.I. (1992) The role of proteolytic enzymes in cancer invasion and metastasis. *Clin. Exp. Metastasis* 10: 145–155.

Ellis, S.M., Nabeshima, K. and Biswas, C. (1989) Monoclonal antibody preparation and purification of a tumor cell collagenase-stimulatory factor. *Cancer Res.* 49: 3385–3391.

Fisher, C., Gilbertson-Beadling, S., Powers, E.A., Petzold, G., Poorman, R. and Mitchell, M.A. (1994) Interstitial collagenase is required for angiogenesis *in vitro*. *Developmental Biol.* 162(2): 499–510.

Folkman, J. (1994) Angiogenesis and breast cancer. *J. Clin. Oncol.* 12(3): 441–443.

Francini, G., Gonnelli, S., Petrioli, R. and Camporeale, A. (1993) Procollagen type I carboxy-terminal propeptide as a marker of osteoblastic bone metastasis. *Cancer, Epidemiology, Biomarkers and Prevention* 2: 125–129.

Frisch, S.M., Clark, E.J. and Werb, Z. (1987) Coordinate regulation of stromelysin and collagenase genes determined with cDNA probes. *Proc. Natl. Acad. Sci. USA* 84: 2600–2604.

Garbisa, A., Scagliotti, G., Masiero, L., Di Francesco, C., Stetler-Stevenson, W.G. and Liotta, L.A. (1992) Correlation of serum metalloproteinase levels with lung cancer metastasis and response to therapy. *Cancer Res.* 52: 4548–4549.

Goldberg, G.I. and Eisen, A.Z. (1990) Extracellular matrix metalloproteinase in tumor invasion and metastasis, *In*: M.E. Lippman and R.B. Dickson (eds): *Regulatory Mechanisms in Breast Cancer*. Kluwer Academic Publishers, Boston, pp 421–440.

Goslen, J.B., Eisen, A.Z. and Bauer, E.A. (1985) Stimulation of skin fibroblast collagenase production by a cytokine derived from basal cell carcinoma. *J. Invest. Dermatol.* 85: 161–164.

Gottesman, M. (1990) The role of proteases in cancer. *Semin. Cancer Biol.* 1: 97–100.

Gray, S.T., Wikins, R. and Yun, K. (1992) Interstitial collagenase gene expression in oral squamous cell carcinoma. *Am. J. Pathol.* 141: 301–309.

Hayes, D.F. (1994) Angiogenesis and breast cancer. *Hematol. Oncol Clin. North. Am.* 8(1): 51–71.

Himelstein, B.P., Canete-Soler, R., Bernhard, E.J. and Muschel, R.J. (1994) Induction of fibroblast 92 kDa gelatinase/type IV collagenase expression by direct contact with metastatic tumor cells. *J. Cell Science* 107: 477–486.

Horak, E.R., Harris, A.L., Stuart, N. and Bicknell, R. (1993) Angiogenesis in breast cancer. Regulation, prognostic aspects, and implications for novel treatment strategies. *Ann. NY Acad. Sci.* 698: 71–84.

Hunt, G. (1989) The role of laminin in cancer invasion and metastasis. *Exp. Cell. Biol.* 57(3): 165–176.

Kane, S.E. and Gottesman, M.M. (1990) The role of cathepsin L in malignant transformation. *Sem. Cancer Biol.* 1: 127–136.

Kanemoto, T., Reich, R., Royce, L., Greatorex, D., Adler, S.H., Shiraishi, N., Martin, G.R., Yamada, Y. and Kleinman, H.K. (1990) Identification of an amino acid sequence from the laminin A chain that stimulates metastasis and collagenase IV production. *Proc. Natl. Acad. Sci. USA* 87: 2279–2283.

Karelina, T.V., Hruza, G.J., Goldberg, G.I. and Eisen, A.Z. (1993) Localization of 92-kDa type IV collagenase in human skin tumors: comparison with normal human fetal and adult skin. *J. Invest. Dermatol.* 100: 159–165.

Kataoka, H., DeCastro, R., Zucker, S. and Biswas, C. (1993) Tumor cell-derived collagenase-stimulatory factor increases expression of interstitial collagenase, stromelysin, and 72-kDa gelatinase. *Cancer Res.* 53: 3154–3158.

Kawashima, A., Nakanishi, I., Roessner, A., Obata, K. and Okada, Y. (1994) Expression of matrix metalloproteinase 9 (92-kDa gelatinase/type IV collagenase) induced by tumor necrosis factor alpha correlates with metastatic ability in a human osteosarcoma cell line. *Virchows Arch.* 424: 547–52.

Kohn, E.C. (1991) Invasion and metastasis: Biology and clinical potential. *Pharmac. Ther.* 52: 235–244.

Kuebayashi, J., Mcleskey, S.W., Johnson, M.D., Lippman, M.E., Dickson, R.B. and Fern, F.G. (1993) Quantitative demonstration of spontaneous metastasis by MCF-7 human breast cancer cells co-transfected with fibroblast growth factor-4 and *lacZ*. *Cancer Res*. 53: 2178–2187.

Liotta, L.A., Tryggvason, K., Garbisa, S., Hart, I., Foltz, C.M. and Schaffie, S. (1980) Metastasis potential correlates with enzymatic degradation of basement membrane collagen. *Nature* 284: 67–68.

Liotta, L.A. (1986) Tumor invasion and metastasis-role of the extracellular matrix: Rhodes Memorial Award Lecture. *Cancer Res*. 47: 1–7.

Liotta, L.A., Steeg, P.S. and Stetler-Stevenson, W.G. (1991) Cancer metastasis and angiogenesis: an imbalance of positive and negative regulation. *Cell* 64: 327–336.

Lola, P.K. and Graham, C.H. (1990) Mechanisms of trophoblast invasiveness and their control: the role of proteases and protease inhibitors. *Cancer Metastasis Rev*. 9: 369–380.

Marcy, A.I., Eiberger, L.L., Harrison, R., Chan, H.K., Hutchinson, N.I., Hagmann, W.K., Cameron, P.M., Boulton, D.A. and Hermer, J.D. (1991) Human fibroblast stromelysin catalytic domain: expression, purification, and characterization of a C-terminally truncated form. *Biochemistry* 30: 6476–6483.

Marguilies, I.M.K., Hoyhtya, M., Evans, C., Stracke, M.L., Liotta, L.A., Stetler-Stevenson, W.G. (1992) Urinary type IV collagenase: Elevated levels are associated with bladder transitional cell carcinoma. *Cancer Epidemiology, Biomarkers and Prevention* 1: 467–474.

Matrisian, L.M., Leroy, P., Tuhlmann, C., Gesnel, M.C. and Breathnach, R. (1986) Isolation of the oncogene and epidermal growth factor-induced transin gene: complex control in rat fibroblasts. *Mol. Cell. Biol*. 6: 1679–1686.

Mcleskey, S.W., Kurebayashi, J., Honing, S.F., Zweibel, J.A., Lippman, M.E., Dickson, R.B. and Kern, F.G. (1993) Development of an estrogen-independent, antiestrogen-resistant and metastatic breast carcinoma. *Cancer Res*. 53: 2168–2177.

Monsky, W.L., Kelly, T., Lin, C.Y., Yeh, Y., Stetler-Stevenson, W.G., Mueller, S.C. and Chen, W.T. (1993) Binding and localization of Mr 72,000 matrix metalloproteinase at cell surface invadopodia. *Cancer Res*. 53: 3159–3164.

Monteahudo, C., Merino, M.J., San-Juan, J., Liotta, L.A. and Stetler-Stevenson, W.G. (1990) Immunohistochemical distribution of type IV collagenase in normal, benign, and malignant breast tissue. *Am. J. Pathol*. 136: 585–592.

Mori, Y., Mizuuchi, H., Sato, K., Okamura, N. and Kudo, R. (1994) The factors involved in invasive ability of endometrial carcinoma cells. *Nippon Sanka Fujinka Gakkai Zasshi* 46: 509–516.

Morodome, T., Ogata, Y., Sasaguri, Y., Morimatsu, M. and Nagase, H. (1992) Purification and characterization of matrix metalloproteinase 9 from U937 monocytic leukaemia and HT1080 fibrosarcoma cells. *Biochem. J*. 285: 603–611.

Moutsiakis, D., Mancuso, P., Stetler-Stevenson, W. and Zucker, S. (1992) Characterization of metalloproteinase and tissue inhibitors of metalloproteinase in human plasma. *Connec. Tissue Res*. 28: 213–230.

Murphy, G., Willenbrock, F., Ward, R.V., Cockett, M.I., Eaton, D. and Docherty, A.J. (1992a) The C-terminal domain of the 72 kDa gelatinase A is not required for catalysis but is essential for membrane activation and modulates interactions with tissue inhibitors of metalloproteinase. *Biochem. J*. 283: 637–641.

Murphy, G., Allan, J.A., Willenbrock, F., Cockett, M.I., Eaton, D. and Docherty, A.J.P. (1992b) The role of the C-terminal domain in collagenase and stromelysin specificity. *J. Biol. Chem*. 267: 9612–9618.

Naylor, M.S., Stamp, G.W., Davies, B.D. and Balkwill, F.R. (1994) Expression and activity of MMPs and their regulators in ovarian cancer. *Int. J. Cancer* 58: 50–56.

Newell, K.J., Witty, J.P., Roders, W.H. and Matrisian, L.M. (1994) Expression and localization of matrix-degrading metalloproteinase during colorectal tumorigenesis. *Mol. Carcinog*. 10: 199–206.

O'Connell, J.P., Willenbrock, F., Docherty, A.J., Eaton, D. and Murphy, G. (1994) Analysis of the role of the COOH-terminal domain in the activation, proteolytic activity, and tissue inhibitor of metalloproteinase interactions of gelatinase. *B. J. Biol. Chem*. 269: 14967–14973.

Okada, Y., Nagase, H. and Harris, E.D. (1986) *J. Biol. Chem*. 261: 14245–14255.

Okada, Y., Enghild, J.J. and Nagase, H. (1992) Matrix metalloproteinase 3 (stromelysin) activates the precursor for the human matrix metalloproteinase-9. *J. Biol. Chem.* 267: 3581–3584.

Overall, C.M. (1990) Concanavalin A produces a matrix-degenerative phenotype in human fibroblasts. Induction and endogeneous activation of collagenase, 72 kDa gelatinase, and Pump-1 is accompanied by the tissue inhibitor of matrix metalloproteinases. *J. Biol. Chem.* 265: 21141–21151.

Polette, M., Clavel, C., Cockett, M., Murphy, G. and Birembaut, P. (1993) Detection and localization of mRNAs encoding matrix metalloproteinases and their tissue inhibitor in human breast pathology. *Invasion Metastasis* 13: 31–37.

Polette, M., Gilbert, N., Stas, I., Nawrocki, B., Noel, A., Stetler-Stevenson, W.G., Birembaut, P. and Foidart, M. (1994) Gelatinase A expression and localization in human breast cancers. An *in situ* hybridization study and immunohistochemical detection using conformal microscopy. *Virchows Arch.* 424: 641–645.

Poste, G. and Fidler, I. (1980) The pathogenesis of cancer metastasis. *Nature* 283: 139–146.

Poulsom, R., Pignatelli, M., Stetler-Stevenson, W.G., Liotta, L.A., Rogers, L. and Stamp, G.W.H. (1992) Stromal expression of 72 kDa type IV collagenase (MMP-2) and TIMP-2 mRNA in colorectal neoplasia. *Am. J. Pathol.* 141: 389–396.

Pyke, C., Dano, K. and Tryggvason, K. (1992) Localization of messenger RNA for 72,000 and 92,000 type IV collagenase in human skin cancers by *in situ* hybridization. *Cancer Res.* 52: 1336–1341.

Pyke, C., Ralfkiaer, E., Tryggvason, K. and Dano, K. (1993) Messenger RNA for two type IV collagenase is located in stromal cells in human colon cancer. *Am. J. Pathol.* 142: 359–365.

Reith, A. and Rucklidge, G.J. (1992) Invasion of brain tissue by primary glioma: evidence for the involvement of urokinase-type plasminogen activator as an activator of type IV collagenase. *Biochem. Biophys. Res. Commun.* 186: 348–354.

Rochefort, H., Capony, F. and Garcia, M., Cathepsin, D. (1990) A protease involved in breast cancer metastasis. *Cancer Metast. Rev.* 9: 321–331.

Salomon, D.S., Kidwell, W.R. (1988) *In:* M.E. Lippman and R.B. Dickson (eds): *Breast Cancer. Cellular and Molecular Biology.* Kluwer Academic Publishers, Boston, pp 363–389.

Sato, H., Takino, T., Cao, J., Okada, Y., Shinagawa, A. and Seiki, M. (1994a) Activation of gelatinase A by membrane-type matrix metalloproteinase which is specific for tumor tissue. *Clin. Exp. Metastasis* 12: 68.

Sato, H., Takino, T., Okada, Y., Cao, J., Shinagawa, A., Yamamoto, E. and Seiki, M. (1994b) A matrix metalloproteinase expressed on the surface of invasive tumor cells. *Nature* 370: 61–65.

Shi, Y.E., Tori, H., Yieh, L., Wellstein, A., Lippman, M.E. and Dickson, R.B. (1993a) Isolation and characterization of a novel matrix-degrading protease from hormone dependent human breast cancer cells. *Cancer Res.* 53: 1409–1415.

Shi, Y.E., Lippman, M.E., Torri, J., Yieh, L., Sobel, M.E., Yamada, Y., Dickson, R.B. and Thompson, E.W. (1993b) Expression of 67 kDa laminin receptor in human breast cancer cells: Regulation by progestins. *J. Exp. Clinical Metastasis* 11(3): 251–261.

Shi, Y.E., Henderson, D., Lippman, M.E. and Dickson, R.B. (1994a) Progestin and antiprogestin, in apoptosis, cell-matrix adhesion and metastasis. *In:* R.B. Dickson and M.E. Lippman (eds): *Drug and Hormonal Resistance in Breast Cancer: Cellular and Molecular Mechanisms*, p. 195.

Shi, Y.E., Liu, Y., Lippman, M.E and Dickson, R.B. (1994b) Progestins and antiprogestins, in mammary tumor growth and metastasis. *Human Reproduction* 9: 162–173.

Sloane, B.F. (1990) Cathepsin B and cystatins: Evidence for a role in cancer progression. *Sem. Cancer Biol.* 1: 137–152.

Soini, Y., Paakko, P. and Autio-Harmainen, H. (1993) Genes of lamnin B1 chain, alpha 1 (IV) chain of type IV collagen, and 72 kDa type IV collagenase are mainly expressed by the stromal cells of lung carcinomas. *Am. J. Pathol.* 142: 1622–1630.

Springman, E.B., Angleton, E.L. and Birkedal-Hansen, H. (1990) Multiple modes of activation of latent human fibroblast collagenase: Evidence for the role of a Cys-73 active-site zinc complex in latency and a 'cysteine switch' mechanism for activation. *Proc. Natl. Acad. Sci. USA* 87: 364–368.

Stetler-Stevenson, W.G., Krutzsch, H.C. and Liotta, L.A. (1989) Tissue inhibitor of metalloproteinase (TIMP-2). A new member of the metalloproteinase inhibitor family. *J. Biol. Chem.* 264: 17374–17378.

Stetler-Stevenson, W.G., Liotta, L.A. and Brown, P.D. (1991) *In:* R.B. Dickson and M.E. Lippman (eds): *Genes, Oncogenes and Hormones: Advances in Cellular and Molecular Biology of Breast Cancer.* Kluwer Academic Publishers, Boston, pp 21–41.

Strongin, A.Y., Marmer, B.L., Grant, G.A. and Goldverg, G.I. (1993) Plasma membrane-dependent activation of the 72 kDa type IV collagenase is prevented by complex formation with TIMP-2. *J. Biol. Chem.* 268: 14033–14039.

Thompson, E.W., Paik, S. and Brunner, N. (1992) Association of increased basement membrane invasiveness with absence of estrogen receptor and expression of vimentin in human breast cell lines. *J. Cell Physiol.* 150: 534–544.

Triebel, S., Blaser, J., Reinke, H., Knauper, V. and Tschesche, H. (1992) Mercurial activation of human PMN leucocyte type IV procollagenase (gelatinase). *FEBS Lett.* 298: 280–284.

Tryggvason, K. (1993) Type IV collagenase in invasive tumors. *Breast Cancer Res. Treat.* 24: 209–218.

Unemori, E.N., Bouhana, K.S. and Werb, Z. (1990) Vectorial secretion of extracellular matrix proteins, matrix-degrading proteinases, and tissue inhibitor of metalloproteinases by endothelial cells. *J. Biol. Chem.* 265: 445–451.

Vance, B.A., Kowalski, C.G. and Brinckerhoff, C.E. (1989) Heat shock of rabbit synovial fibroblasts increase expression of mRNAs for two metalloproteinases, collagenase and stromelysin. *J. Cell Biol.* 108: 2037–2043.

Vartio, T. and Baumann, M. (1989) Human gelatinase/type IV procollagenase is a regular plasma component. *FEBS Lett.* 255: 285–289.

Weidner, N., Semple, J.P., Welch, W.R. and Folkman, J. (1991) Tumor angiogenesis and metastasis: Correlation in invasive breast carcinoma. *J. New Engl. Med.* 324: 1–8.

Weidner, N., Carroll, P.R., Flax, J., Blumenfeld, W. and Folkman, J. (1993) Tumor angiogenesis correlates with metastasis in invasive prostate carcinoma. *Am. J. Pathol.* 143(2): 401–409.

Zucker, S., Lysik, R.M., Stetler-Stevenson, W.G., Liotta, L.A., Birkedal-Hansen, H. and Mann, W. (1992) Immunoassay of type IV collagenase/gelatinase (MMP-2) in human plasma. *J. Immunol. Meth.* 148: 189–198.

Zucker, S. (1993) Mr 92,000 type IV collagenase is increased in plasma of patients with colon cancer and breast cancer. *Cancer Res.* 53: 140–146.

Zucker, S., Elias, S., Lyubsky, A., Heimann and Biswas, C. (1994) Tumor collagenase stimulating factor (TCSF): The missing link connecting fibroblast of matrix metalloproteinases (MMP) *in vivo* and cancer cell production of MMP in vitro. *Clin. Exp. Metastasis* 12: 60.

Epithelial–Mesenchymal Interactions in Cancer
ed. by I.D. Goldberg & E.M. Rosen
© 1995 Birkhäuser Verlag Basel/Switzerland

Angiogenesis as a component of epithelial–mesenchymal interactions

D.S. Grant[1], R. Wesley Rose[1], J.K. Kinsella[3] and M.C. Kibbey[2]

[1]*Cardeza Foundation for Hematological Research, Thomas Jefferson University, College of Medicine, 1015 Walnut St., Curtis 703, Philadelphia, PA 19107*
[2]*Gerontology Research Center, National Institute on Aging, NIH, Baltimore, MD 21224*
[3]*Laboratory of Development Biology, National Institute of Dental Research, NIH, Bethesda, MD 20892, USA*

Summary. Here we review the role of angiogenesis as it pertains to the interactions between the epithelium and the mesenchyme, especially during tumor growth and metastasis. We illustrate and discuss several models of angiogenesis including endothelial tube formation on Matrigel. Finally, we examine angiogenic factors using the Matrigel model and investigate several other matrix molecules for their importance in angiogenesis and epithelial/stromal interactions.

Introduction

Maintenance of normal epithelium and mesenchymal structures *in vivo* is dependent upon many factors, for example, the extracellular matrix. In the skin the proliferating cells of the epidermis are supported by a specialized matrix designated, the basement membrane. Below this basement membrane is a complex network of stabilizing structures composed of collagens (primarily types I and III), elastins, proteoglycans, and several glycoproteins. Within this meshwork are numerous cell types such as fibroblasts, smooth muscle cells, pericytes, fat, and nerve cells, all of which produce different forms of extracellular matrix components. This extracellular matrix provides a scaffold for cell migration and for the storage and regulation of cytokines (such as fibroblast growth factor, FGF). The stability of structures such as blood vessels is normally self-maintained, but can be shifted in response to a wound, inflammation, and tumor growth. We have been interested in the role of the matrix in the regulation of vessel stability and growth. In this review we will present several findings that illustrate the role which the basement membrane plays in regulating vessel growth.

Structure of the epithelium

In order to understand the role of angiogenesis as a model of epithelial–mesenchymal interactions, it is necessary to have a basic understanding of the structural relationship of the epithelium to the mesenchyme. All epithelial cells are maintained by, and interact with the underlying mesenchymal layer. The integument is a good example of the interplay between epithelial and mesenchymal cells. Three distinct layers are present in the skin: epidermis, dermis, and hypodermis. The epidermis is the outermost layer, providing protection against friction and the sun's rays, as well as acting as a waterproof barrier. It consists of 4–5 layers of different cell types including the basal cell layer, which generates new epidermis cells. Since the epidermis is not vascularized, the cells undergo programmed cell death (apoptosis) as they migrate towards the upper layers of the epidermis. The dermis, located deep to the epidermis, helps to regulate body temperature. Since the layer of the skin consists of various collagenous, elastic and reticular fibers, it also conveys tone to the skin. The innermost layer of the integument, the hypodermis, connects the skin to the organs below and, like the dermis, is vascularized.

The maintenance of the epithelium is dependent on the basement membrane, which is formed both by basal epithelial cells and some mesenchymal cells and helps to form the boundary between the epidermis and the dermis. It has been shown that the basement membrane not only acts as a support mechanism structurally, but also plays an important role in the induction of cellular behavior (Paulsson, 1992). Transmembrane receptors, such as the integrins, are important in the cellular response to the proteins of the basement membrane (Paulsson, 1992), and it is through transmembrane signalling events that epithelial cells are induced to proliferate. This basement membrane layer also provides a regulatory barrier for blood vessel growth and is also important in the maintenance of vessel structure. For example, in psoriasis, the overabundant growth of keratinocytes is due in part of increased cytokine production by fibroblasts, which in turn are induced by leaky and proliferative blood vessels (Malhotra et al., 1989). The pivotal point is that this disruption in the normal functioning of the dermis may be caused and initiated by a localized breakdown of the vascular basement membrane.

Angiogenesis

Most blood vessels consist of a continuous layer of endothelial cells surrounded by smooth muscle cells which are usually stable and do not proliferate under normal conditions (Grant et al., 1992). Angiogenesis refers to the complex process of events that disrupts vessel stability and culminate in the formation of new blood vessels from existing parental vessels; it occurs in response to a variety of extracellular stimuli or tissue injury. The stability of a vessel is dependent upon the presence of the basement membrane (Grant et al., 1992). Several steps are involved in angiogenesis, including (1) breakdown of the underlying basement membrane which is influenced strongly by proteases derived from the endothelial layer and the basement membrane and perhaps from nearby mesenchymal cells, (2) initiation of endothelial migration and proliferation through the extracellular matrix, which leads to (3) organization of the cells and formation of tube-like structures with their own basement membrane, the new vessels (Bauer et al., 1992).

Several factors have been experimentally identified which have a substantial impact on the control of angiogenesis. For example, growth factors such as fibroblast growth factor (FGF), tumor necrosis factors (TNF) (Sato et al., 1990), transforming growth factor (TGF alpha and beta) (Pepper et al., 1990), platelet derived growth factor (PDGF) (Heldin et al., 1991), vascular endothelial cell growth factor (VEGF) (Kim et al., 1992), and hepatocyte growth factor (HGF) (Bussolino et al., 1992; Grant et al., 1993) have been shown to stimulate angiogenesis *in vivo* and affect endothelial cells either directly via the serum, via basement membrane, or through additional factors derived from the extravascular stroma. It has also been shown recently that some of these substances produced by the endothelial and muscle cells themselves can regulate their differentiation and function in an autocrine manner. The constituents of the basement membrane itself have been shown to play an important role in the initiation and support of the process (Kleinman et al., 1993). Basement membrane components include type IV collagen, heparan sulfate proteoglycan, and several glycoproteins, including laminin (Grant et al., 1992). All three constituents are biologically active and have the ability to bind to each other. Collagen IV, the most abundant component, makes up the structural scaffold of the basement membrane. Heparan sulfate proteoglycan plays an integral role in the maintenance of the ionic charge of the basement membrane, as well as in the binding of growth factors (Grant et al., 1992). Laminin (which consists of 3 chains, α, β and γ chains) is important in the attachement of human umbilical vein endothelial cells (HUVEC) to several substrates, as well as in the growth and differentiation of certain cell types (Grant et al., 1992).

We have a better understanding of the role of laminin in stimulating differentiation due to the discovery of a basement membrane-producing tumor, the Engelbreth-Holm-Swarm (EHS) tumor. A basement membrane extract, Matrigel, enriched in laminin, can be isolated from this tumor. Matrigel is a liquid at 4°C and a solid 37°C; therefore, it can be used to coat tissue culture plates while cold, polymerized in an incubator, then cells may be seeded upon the basement membrane substrate. When epithelial cells such as mammary or Sertoli cells are placed on this matrix they rapidly differentiate into three-dimensional structures re-

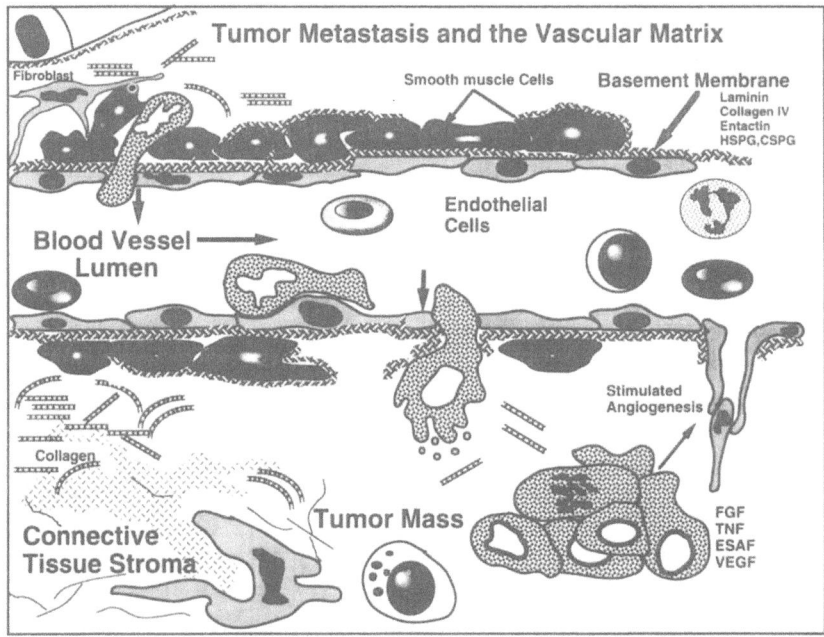

Figure 1. A diagrammatic representation of tumor metastasis in the mesenchyme and the vascular matrix.

sembling organs *in vivo*. When isolated HUVEC are seeded onto Matrigel they attach, migrate, and reorganize to form structures that look like a network of capillaries. This occurs within 18 hours and can be used as an *in vitro* assay to test the effect of several substances and cytokines on angiogenesis. This model has proven to be quite useful in the study of potentially angiongenic factors and their effect on tube formation by endothelial cells.

The role of angiogenesis in tumor growth and metastasis

Access to a blood supply is necessary for proliferating cells. This is especially true for the growth and spread of epithelial tumors. Tumor metastasis occurs when the basement membrane is breached by the growing tumor mass, causing a disruption in the integrity of basement membrane (Fig. 1). This allows the proliferating tumor cells access to newly growing blood vessels which have been stimulated by local angiogenic factors produced by mesenchyme and by the tumor itself. Tumor growth and angiogenesis are interdependent. In order to increase in size, a blood supply is necessary to support the proliferating tumor cells. Certain avascular tumors cannot increase their size beyond few cubic millimeters, but once neovascularization is induced, the tumor mass can increase its volume 1000-fold (Folkman, 1992). Therefore, each increase in tumor size must be accompanied by an increase in neovascularization (Folkman, 1992). Tumors can be classified as either prevascular or vascular, and transformation occurs when the tumor cells switch to an angiogenic phenotype and begin to produce factors which stimulate neovascularization, such as TNF, VEGF, TFGβ, and bFGF (Folkman, 1992). These factors are also found in the matrix itself, and can be mobilized by the tumor cells (Folkman, 1992). As a result, a proliferating tumor can increase its available blood supply and metastasize via several different pathways (Fig. 1). It has been the goal of our laboratory to study angiogenesis in relation to tumor growth. We have used several models of angiogenesis to define the role of matrix components.

Methods

Several models have been developed to test angiogenic substances and to examine the mechanism of blood vessel formation (Auerbach et al., 1994). Some of these models involve the stimulation of existing endothelial cells *in situ* whereas other models have been developed which mimic some or all of the events occurring *in vivo* and *in vitro* assay systems. The mechanisms involved in both endothelial cell attachment and vessel formation are difficult to examine *in vitro* due to the complex relationship of endothelial cells to both soluble and insoluble factors *in situ*. Endothelial cells from various sources have been isolated and maintained in culture (Jaffe et al., 1973). If the cells are permitted to become superconfluent without a change of the culture medium, structures form above the monolayer which resemble capillary-like vessels (Madri et al., 1991; Montesano et al., 1992). This differentiation can be accelerated by the additon of a matrix such a gelatin, fibrin or collagen to the surface of the plastic (Ingber, 1991; Liu et al., 1990; Madri et al., 1991). The induction of capillary-like tube formation *in vitro* can also be initiated by the addition of phorbol esters or by incubating the cells within a collagen I gel (Kinsella et al., 1992; Montesano and Orci, 1985). These are good models of angiogenesis but they usually require three or more days for vessels to form, and in many cases only half of the cells participate in the differentiation process. Others have shown that if cultured endothelial cells are incubated in a fibrin clot or with collagen IV, vessel formation will occur within a week (Madri and Williams, 1983). With these systems, capillary-like structures are observed but the response is generally slow, requiring several days to several weeks. In some cases, the vessels are inside out, secreting basement membrane material and interstitial collagen into the lumen.

The model of angiogenesis we have been using in the past five years involves Matrigel. Although this model does not demonstrate all aspects of angiogenesis (proliferation and invasion of the cells are not seen), it demonstrates tube formation and the steps which include migration and differentiation. Matrigel is a mixture of basement membrane components and growth factors (Vukicevic et al., 1992), extracted from the Englebreth-Holm-Swarm (EHS) tumor and has been found to induce and/or maintain the differentiation of a wide variety of cells (Kleinman et al., 1987). The components are extracted with 2.0 M urea and then dialyzed into a physiological buffer. At 40°C, the components remain in solution but polymerize when warmed to 24–37°C. The gelled extract has the appearance of authentic basement membrane in the electron microscope.

Results and Discussion

Tube formation with Matrigel

Previous work in our laboratory has shown that endothelial cells attach, migrate, and assemble on this laminin-rich reconstituted matrix, Matrigel, to form tube-like structures resembling capillaries within 18 hours (Fig. 2) (Grant et al., 1992; Grant et al., 1990; Lawley and Kubota, 1989). Work in our lab has shown that there is significant remodeling of the Matrigel during the first 4 hours of migration and organization of the cells. Within a period of 8–10 hours, capillary-like structures are already apparent (Fig. 2). When these structures are cut in cross-section and viewed in the electron microscope, they are cylindrical and a lumen is visible (Grant et al., 1990; Grant et al., 1989). We obtain similar tube formation if the endothelial cells are cultured within a rat tail collagen gel (containing normal growth factors) over a period of two days (Fig. 2). Both methods result in tubes of similar dimensions and with an organized cytoskeleton. This illustrates that the matrix, whether it be from basement membrane or interstitial collagen, has the ability to induce tube formation in cultured endothelial cells.

This Matrigel *in vitro* model system has also been used to define the cellular interactions with the basement membrane and to identify several intracellular events occurring during capillary vessel formation. The endothelial cells use multiple receptors to interact with basement membrane components. Protein synthesis, an intact cytoskeleton, and the activities of protein kinase C and collagenase IV are all required for tube formation (Grant et al., 1990). Using this model system, inhibitors and stimulators of angiogenesis can be screened prior to *in vivo* testing.

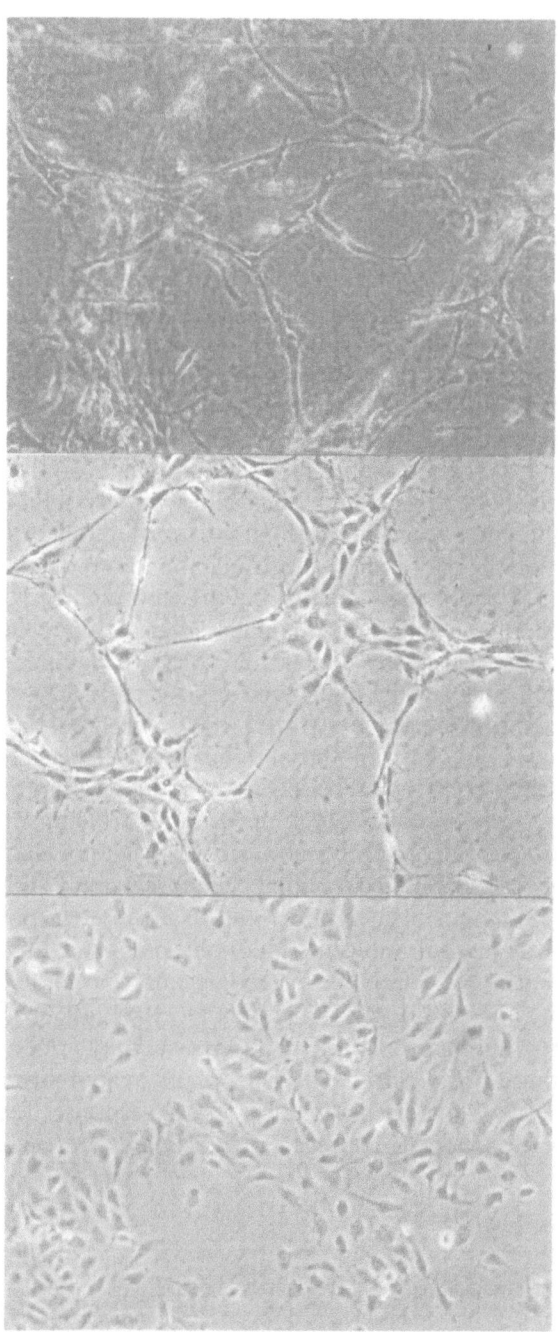

PLASTIC, 5 h   MATRIGEL, 5 h   COLLAGEN I GEL, 2 d

Figure 2. Matrix-induced vessel formation. HUVEC form a cobblestone layer on plastic, whereas tube formation is clearly visible in the cells plated on Matrigel and collagen I. Note that the vessel size and dimensions are similar.

To date we have examined several stimulators of angiogenesis such as scatter factor (hepatocyte growth factor), haptoglobin, neuropeptide Y and several synthetic peptides derived from cell binding sites in laminin (RGD, YIGSR, SIKVAV).

Proteins in Matrigel that induce tube formation

Matrigel is a complex mixture of many different biologically active proteins and thus provides many signals that induce endothelial cell differentiation. In order to understand which matrix proteins are involved in tube formation on Matrigel, we first seeded HUVEC after preincubating the solidified Matrigel with antibodies to laminin, collagen IV, vitronectin, entactin, and fibronectin (Fig. 3). As previously illustrated (Grant et al., 1990) antibodies to laminin could block tube formation by almost 100% at a serum dilution of 1:10. When anti-collagen IV was added, flattening of the cells and tube inhibition was also demonstrable. Tube formation could also be inhibited by antibodies to vitronectin, entactin, and fibronectin; but each antibody inhibited the tube formation differently resulting in different morphologies (Fig. 3).

Laminin is one of the most important and abundant substances in basement membranes and it has a direct role in cell attachment, migration, and induction of the differentiated phenotype of many cells. We have examined and at least partly defined the role(s) of laminin and its specific cell-binding sites at the biochemical level using an *in vitro* angiogenic model (Grant et al., 1990, 1992). We observed that RGD and YIGSR sites are involved in the stabilization of the vessel and epithelium; however, the SIKVAV site disrupts this equilibrium and induces branching. Furthermore, based on *in vitro* attachment, enzymatic and morphological data (Grant et al., 1993), it is postulated that presentation of the SIKVAV site to stable cells *in vivo* may induce the cells to penetrate the basement membrane and to invade the underlying substratum. Since the laminin molecule is usually polymerized within a meshwork of collagen IV, entactin, proteoglycans and other glycoproteins (Heathcote and Grant, 1981; Timpl et al., 1981, 1983; Tsilibary and Charonis, 1986; Yurchenco et al., 1986), this site may not be readily available for binding to the endothelial cells, or its availability may be reduced to levels which do not elicit the sprouting responses seen with the SIKVAV peptide. There is a precedent for latent sites in laminin

Figure 3. The effects of antibodies to several matrix components on tube formation by HUVEC. Differing degrees of tube formation are observed when antibodies to each matrix component are introduced.

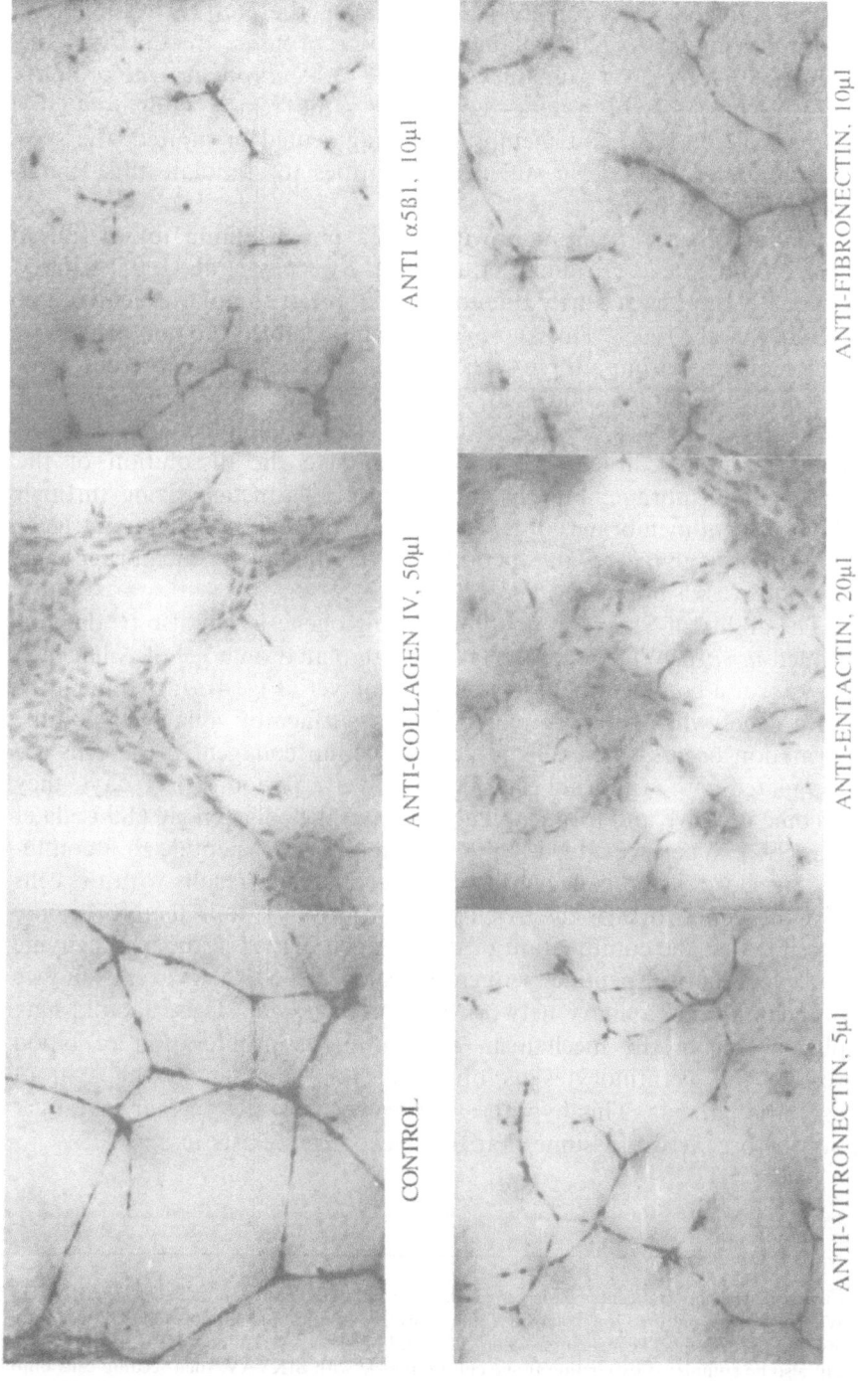

being 'activated' by proteases. For example, the RGD site on the A chain is thought to be available only after protease treatment of the laminin P1 fragment (Aumailley et al., 1990). A cryptic neurite promoting site has also been reported in the same region (Edgar, 1985). Perhaps if the basement membrane is fragmented or injured, the sites may become available in sufficient quantities to stimulate this migratory-invasive activity.

SIKVAV has also been shown to be a potent stimulator of tumor growth and spread (Fridman et al., 1991, Sweeney et al., 1991; Kibbey et al., 1993), at least partly due to stimulation of tumor vascularization (Kibbey et al., 1993). This angiogenic ability enabled the tumor mass to double in size within a three-week period. Since laminin is present in all basement membranes and is associated with interstitial structures such as vessels and nerves, we suspect that this site in laminin can be exposed under pathological conditions which involve the dissolution of the basement membrane. For instance, tumor cells metastasizing through the basement membranes of a vessel release proteases which may trigger the latent activity of the peptide in the basement membrane and promote vessel branching.

The ability of SIKVAV to stimulate angiogenesis can also be demonstrated *in vitro*. When endothelial cells are plated on a gel of collagen I, they form a confluent cobble-stone monolayer (Fig. 4). It is difficult to form tubes when cells are plated on the surface of collagen, but tube formation occurs when cells are plated within collagen. If the cells are stimulated with a phorbol ester (PMA) over a period of 1–2 days, they become invasive and form capillaries within the collagen gel (Kinsella et al., 1992). When we added 200 μg of the SIKVAV peptide to a confluent layer of cells on collagen I, we obtained similar results with the cells forming a network of tubes within 12 h (Fig. 4). It is therefore quite possible that the combination of the collagen matrix in the mesenchyme and stimulatory peptide sequences, such as SIKVAV, can induce branching and capillary network formation *in situ*. These factors may play a role in the mechanism of continuous proliferation of blood vessels and keratinocytes as observed, for example, in the skin of psoriatic patients. This hypothesis, however, requires further investigation before we can assume that this mechanism exists *in vitro*.

Figure 4. The top and center panels are light micrographs of cells on collagen I ± SIKVAV. When cells are cultured on collagen I they form a cobblestone monolayer. The cells are induced to form tubelike structures when 200 µg/ml of SIKVAV is added. Network formation can also be stimulated by coating tissue culture plastic with SIKVAV, then seeding cells onto it. The bottom panel is a scanning electron micrograph of the formation of a SIKVAV-stimulated HUVEC.

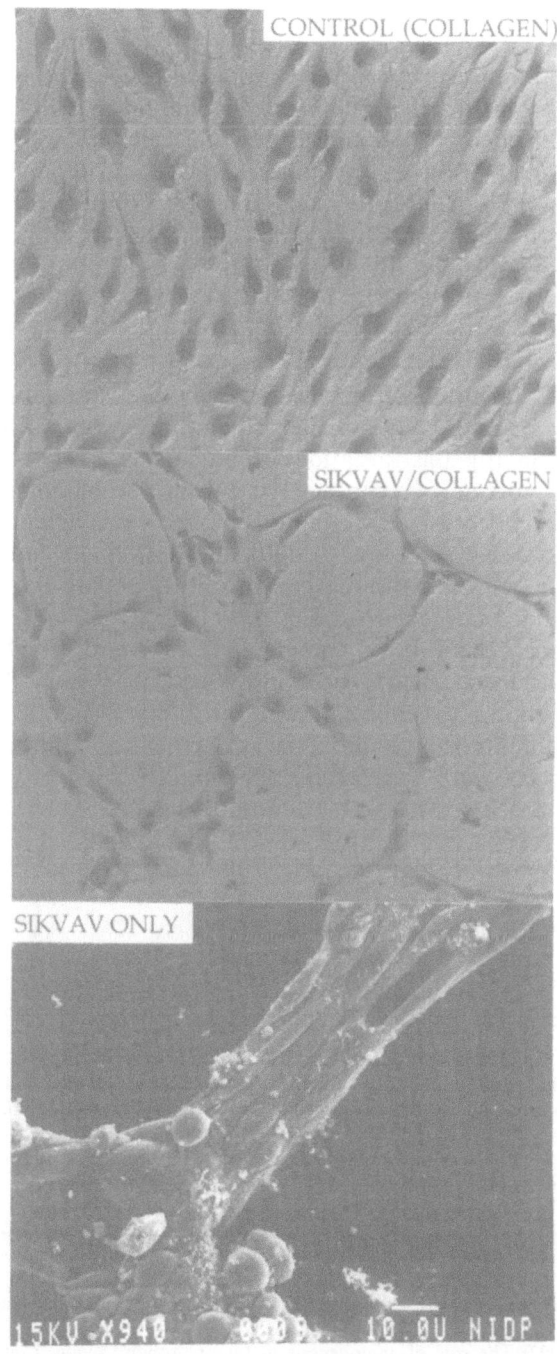

Conclusions

It is clear that angiogenesis plays an integral role in the stability of the epithelium and the interstitial matrix. The importance of the extracellular matrix is evident in the role it plays in cellular migration and the storage and regulation of cytokines, as well as in angiogenesis. Its support allows the maintenance and constant turnover of the overlying epithelium. In the epithelium the basement membrane forms a boundary between the dividing keratinocytes and underlying mesenchyme. A discontinuity in the epithelial basement membrane can result in a shift in the established equilibrium, as is seen in tumor cell proliferation and mestastasis. Invasive and proliferative tumor cells can elicit numerous responses from the mesenchymal vasculature. This shift in angiogenic activity is due to the production of several factors which are responsible for the initiation of angiogenesis. These potentially angiogenic factors can be characterized using the basement membrane extract Matrigel as a model. Using this model we have identified several factors that stimulate endothelial differentiation *in vitro* and angiogenesis *in vivo*. In addition, we have demonstrated that the extracellular matrix molecules interact to direct blood vessel formation. With this Matrigel model, the role of laminin, one of the most abundant and important basement membrane constituents, has been further elucidated. In addition to collagen I and IV, entactin, fibronectin, and vitronectin, laminin is integral to the attachment, migration, and induction of biological activity in many cell types.

Acknowledgements
We would like to thank Barbara Piasecki for her help as a part-time student in the lab.

References

Auerbach, R., Auerbach, W. and Polakowski, I. (1994) *In vivo assays for angiogenesis.* Plenum Press, New York.

Aumailley, M., Timpl, R. and Sonnenberg, A. (1990) Antibody to integrin alpha 6 subunit specifically inhibits cell-binding to laminin fragment 8. *Exp. Cell Res.* 188: 55–60.

Bauer, J., Margolis, M., Schreiner, C., Edgell, C.J., Azizkhan, J., Lazarowski, E. and Juliano, R.L. (1992) *In vitro* model of angiogenesis using a human endothelium-derived permanent cell line: contributions of induced gene expression, G-proteins, and integrins. *J. Cell. Physiol.* 153: 437–449.

Bussolino, F., Di, R.M., Ziche, M., Bocchietto, E., Olivero, M., Naldini, L., Gaudino, G., Tamagnone, L., Coffer, A. and Comoglio, P.M. (1992) Hepatocyte growth factor is a potent angiogenic factor which stimulates endothelial cell motility and growth. *J. Cell. Biol.* 119: 629–641.

Edgar, D. (1985) Nerve growth factors and molecules of the extracellular matrix in neuronal development. *J. Cell Sci.* 3: 107–113.

Folkman, J. (1992) The role of angiogenesis in tumor growth. *Semin. Cancer Biol.* 3: 65–71.

Fridman, R., Kibbey, M.C., Royce, L.S., Zain, M., Sweeney, T.M., Jicha, D.L., Yannelli, J.R., Martin, G.R. and Kleinman, H.K. (1991) Enhanced tumor growth of both primary and established human and murine tumor cells in athymic mice after coinjection with Matrigel. *J. Natl. Cancer Inst.* 83: 769–774.

Grant, D.S., Tashiro, K.I., Segui-Real, B., Yamada, Y., Martin, G.R. and Kleinman, H.K. (1989) Two different laminin domains mediate the differentiation of human endothelial cells into capillary-like structures *in vitro*. *Cell* 58: 933–943.

Grant, D.S., Kleinman, H.K. and Martin, G.R. (1990) The role of basement membranes in vascular development. *Ann. NY Acad. Sci.* 588: 61–72.

Grant, D.S., Kinsella, J.L., Fridman, R., Auerbach, R., Piasecki, B.A., Yamada, Y., Zain, M. and Kleinman, H.K. (1992) Interaction of endothelial cells with a laminin A chain peptide (SIKVAV) *in vitro* and induction of angiogenic behavior *in vivo*. *J. Cell. Physiol.* 153: 614–625.

Grant, D.S., Kleinman, H.K., Goldber, I.D., Bhargava, M.M., Nickoloff, B.J., Kinsella, J.L., Polverini, P. and Rosen, E.M. (1993) Scatter factor induces blood vessel formation *in vivo*. *Proc. Natl. Acad. Sci. USA* 90: 1937–41.

Heathcote, J.G. and Grant, M.E. (1981) The molecular organization of basement membrane. *Int. Rev. Conn. Tissue Res.* 9: 191–264.

Heldin, C.H., Usuki, K. and Miyazono, K. (1991) Platelet-derived endothelial cell growth factor. *J. Cell Biochem.* 47: 208–210.

Ingber, D.E. (1991) Control of capillary growth and differentiation by extracellular matrix. Use of a tensegrity (tensional integrity) mechanism for signal processing. *Chest.* 99: 34S–40S.

Jaffe, E.A., Nachman, R.L., Becker, C.G. and Minick, C.R. (1973) Culture of human endothelial cells derived from umbilical veins – identification by morphological and immunological criteria. *J. Clin. Invest.* 52: 2745–2756.

Kibbey, M.C., Grant, D.S. and Kleinman, H.K. (1992) Role of the SIKVAV site of laminin in promotion of angiogenesis and tumor growth: an *in vivo* Matrigel model. *J. Natl. Cancer Inst.* 84: 1633–1638.

Kim, K.J., Li, B., Houck, K., Winer, J. and Ferrara, N. (1992) The vascular endothelial growth factor proteins: identification of biological relevant regions by neutralizing monoclonal antibodies. *Growth Factors* 7: 53–64.

Kinsella, J.L., Grant, D.S., Weeks, B.S. and Kleinman, H.K. (1992) Protein kinase C regulates endothelial cell tube formation on basement membrane matrix, Matrigel. *Exp. Cell Res.* 199: 56–62.

Kleinman, H.K., Graf, J., Iwamoto, Y., Kitten, G.T., Ogle, R.C., Sasaki, M., Yamada, Y., Martin, G.R. and Luckenbill-Edds, L. (1987) Role of basement membranes in cell differentiation. *Ann. NY Acad. Sci.* 513: 134–145.

Kleinman, H.K., Weeks, B.S., Schnaper, H.W., Kibbey, M.C., Yamamura, K. and Grant, D.S. (1993) The laminins: a family of basement membrane glycoproteins important in cell differentiation and tumor metastases. *Vitam. Horm.* 47: 161–186.

Lawley, T.J. and Kubota, Y. (1989) Induction of morphologic differentiation of endothelial cells in culture. *J. Invest. Dermatol.* 93S: 59S–61S.

Liu, H.M., Wang, D.L. and Liu, C.Y. (1990) Interactions between fibrin, collagen and endcothelial cells in angiogenesis. *Adv. Exp. Med. Biol.* 281: 319–331.

Madri, J.A. and Williams, S.K. (1983) Capillary Endothelial Cell Cultures: Phenotypic Modulation by Matrix Components. *J. Cell. Biol.* 97: 153–165.

Madri, J.A., Bell, L., Marx, M., Merwin, J.R., Basson, C. and Prinz, C. (1991) Effects of soluble factors and extracellular matrix components on vascular cell behavior *in vitro* and *in vivo*: models of de-endothelialization and repair. *J. Cell Biochem.* 45: 123–130.

Malhotra, R., Stenn, K.S., Fernandez, L.A. and Braverman, I.M. (1989) Angiogenic properties of normal and psoriatic skin associate with epidermis, not dermis. *Lab Inv.* 61: 162–165.

Montesano, R. and Orci, L. (1985) Tumor-promoting phorbol esters induce angiogenesis *in vitro*. *Cell* 42: 469–477.

Montesano, R., Pepper, M.S., Vassalli, J.-D. and Orci, L. (1992) Modulation of antiogenesis *in vitro*. In: R. Steiner, P.B. Weisz and R. Langer (eds): *Antiogenesis: Key Principles – Science, Technology, Medicine*, Birkhäuser Verlag, Basel, Switzerland, pp 129–136.

Paulsson, M. (1992) Basement membrane proteins: structure, assembly, and cellular interactions. *Crit. Rev. Biochem. Mol. Biol.* 27: 93–127.

Pepper, M.S., Belin, D., Montesano, R., Orci, L. and Vassalli, J.D. (1990) Transforming growth factor-beta 1 modulates basic fibroblast growth factor-induced proteolytic and angiogenic properties of endothelial cells *in vitro*. *J. Cell Biol.* 111: 743–755.

Sato, N., Nariuchi, H., Tsuruoka, N., Nishihara, T., Beitz, J.G., Calabresi, P. and Frackelton, A.J. (1990) Actions of TNF and IFN-gamma on angiognesis *in vitro*. *J. Invest. Dermatol.* 95: 85S–89S.

Sweeney, T.M., Kibbey, M.C., Zain, M., Fridman, R. and Kleinman, H.K. (1991) Basement membrane and the SIKVAV laminin-derived peptide promote tumor growth and metastases. *Cancer Metastasis Rev.* 10: 245–254.

Timpl, R., Wiedemann, H., Delden, V.V., Furthmayr, H. and Kuhn, K. (1981) A network model for the organization of type IV collagen molecules in basement membrane. *Euro. J. Biochem.* 120: 203–211.

Timpl, R., Dziadek, M., Fujiwara, S., Nowack, H. and Wick, G. (1983) Nidogen: a new, self-aggregating basement membrane protein. *Euro. J. Biochem.* 137: 455–465.

Tsilibary, E.C. and Charonis, A.S. (1986) The role of the main noncollagenous domain (NCI) in type IV collagen self-assembly. *J. Cell Biol.* 103: 2467–2473.

Vukicevic, S., Kleinman, H.K., Luyten, F.P., Roberts, A.B., Roche, N.S. and Reddi, A.H. (1992) Identification of multiple active growth factors in basement membrane Matrigel suggests caution in interpretation of cellular activity related to extracellular matrix components. *Exp. Cell. Res.* 202: 1–8.

Yurchenco, P.D., Tsilibary, E.C., Charonis, A.S. and Furthmayr, H. (1986) Models of the self-assembly of basement membrane. *J. Histochem. Cytochem.* 34: 93–102.

Epithelial–Mesenchymal Interactions in Cancer
ed. by I.D. Goldberg & E.M. Rosen
© 1995 Birkhäuser Verlag Basel/Switzerland

Mammary tumor fibroblasts are phenotypically distinct from non-tumor fibroblasts

A.E. Hornby and K.J. Cullen

Vincent T. Lombardi Cancer Research Center, Georgetown University, Washington, D.C. 20007, USA

Introduction

A tumor contains cellular and extracellular elements. Although breast tumors themselves are primarily of epithelial origin, these tumor cells do not exist in isolation, but require interaction with their surrounding environment including fibroblasts, fat cells, extracellular matrix and vascular elements. Traditionally, studies of tumor growth have focused on the malignant epithelial cells themselves. There is increasing evidence that there is a role for interaction between the tumor epithelium and stroma in the development and progression of breast cancer. This chapter focuses on the fibroblast component of mammary tumors: the signals they produce and their potential influence on the growth of the tumor itself.

Stromal–epithelial interactions

Stromal–epithelial interactions are important for all stages of mammary development from the early embryo to lactation in the adult.

The influence of stroma on epithelium has been observed for many years. Grobstein (1953, 1966) demonstrated the need for mesenchymal interaction in salivary gland development. Since then, experiments using other organs such as teeth (Slavkin, 1978; Jowett et al., 1993), prostate (Cunha et al., 1980), pancreas (Wessells and Cohen, 1967), and breast (Sakakura et al., 1976, 1987) have demonstrated that the mesenchyme determines the morphology and function of organs.

The mouse has provided a model for studies on mammary stromal–epithelial interactions both in normal development and in mammary carcinoma. During embryogensis a dense mesenchyme made up of fibroblast cells surrounds the epithelial components of the mammary gland and can be characterized by the presence of the androgen receptor (Durnberger and Kratochwil, 1980) and tenascin (Chiquet-Ehrismann et al., 1986; Inaguma et al., 1988). It appears that tissue development is initiated by the adjacent mesenchyme. Propper (1967) suggested the presence of a specialized mammary mesenchyme within the dermal cells of the developing embryo that determines the fate of mammary epithelium. In this experiment Propper isolated the epidermis from prospective mammary and non-mammary regions of 12-day rabbit embryos and cultured them with mesenchymal tissue from mammary or non-mammary regions. He found that both mammary and non-mammary epithelia could make a mammary bud when associated with mammary stroma, but not when associated with non-mammary stroma. This implied that determination of a neutral epithelium was directed by the stroma. Another example is tooth morphogenesis in which tooth mesenchyme from rat embryos was shown to induce crown formation in avian oral epithelium (Koller and Fisher, 1980) suggesting that the absence of teeth in birds is not due to the loss of genes necessary for the development of teeth, but because of a loss of the mesenchyme required for teeth induction.

Several groups have studied embryonic stromal–epithelial interactions in mice (Sakakura, 1991). Fetal mammary epithelium when transplanted into the mammary fat pads of juvenile or adult mice will undergo glandular differentiation, an effect which is specific to mammary epithelium since lung, pancreas or salivary gland epithelium will not differentiate when transplanted into the mammary fat pad (Sakakura, 1979). Conversely, when embryonic mesenchyme from the mammary fat pad or salivary gland was transplanted into the mammary gland of virgin adult mice the glandular epithelium both proliferated and differentiated, whereas, mesenchyme from lung, pancreas or kidney did not have this effect on mammary epithelium, indicating that there are organ-specific interactions between stroma and epithelium (Sakakura, 1979).

Stromal–epithelial interactions are required for normal development and maturation of the mammary gland. In hormone-mediated pubertal development it has been shown in the mouse that stromal cells acquire estrogen receptors earlier than epithelial cells, and that both stromal and epithelial cells have a mitogenic response to estrogen, (Haslam, 1989) although the functional significance of stromal responsiveness to estrogen is not yet understood. Stromal–epithelial interactions may also be important in the regulation of lactation. Reichmann et al. (1989) performed co-culture experiments using mouse epithelial and fibroblast cell lines. When cultured together, the epithelial cells synthesized significant quantities of the milk protein β-casein in response to stimulation with prolactin and insulin and formed three-dimensional duct-like structures. However, when grown separately, the epithelial cells were unable to form the duct-like structures or produce β-casein.

Estrogens are also important in the development of breast cancer in that breast cancer cells, in response to estrogen, produce several factors which influence not only their own growth, but that of the surrounding stromal cells, which in turn produce factors that influence the growth of the epithelial cells (McGrath, 1983; Haslem and Levely, 1985). Mammary gland epithelial cells which are responsive to estrogen stimulation *in vivo* do not respond to estrogen stimulation when isolated and cultured *in vitro* (Iguchi et al., 1983). However, this responsiveness is restored *in vitro* when the epithelial cells are co-cultured with live mammary fibroblasts (Haslam, 1986). These experiments suggest that some estrogen actions are indirect and require interaction with stromal tissues.

Several observations suggest a labile secreted fibroblast product or products is required for estrogen-induced epithelial growth. Normal mouse mammary epithelial cells require the presence of mammary fibroblasts to respond to the mitogenic effects of estrogen (McGrath, 1983). Progesterone receptors can be induced by estrogen by glutaraldehyde-killed fibroblasts, type I collagen or conditioned medium from fibroblasts. However, estrogen-dependent DNA synthesis requires the presence of live fibroblasts implying that the fibroblasts secrete a labile factor required for epithelial response to estrogen (Haslam, 1986). Conditioned media from short term cultures of human fibroblasts has been demonstrated to be mitogenic for a number of breast cancer cell lines, increasing both DNA synthesis in short term culture and cell number in a long term proliferation assay (Ryan et al., 1993) lending further support to the concept that stromal cells may play an important role in the regulation of human breast tumor growth, through the production of factors that modulate tumor epithelial growth (Lippman et al., 1986a,b; Yee et al., 1988, 1989; Cullen et al., 1991).

Further evidence of a paracrine role for fibroblasts in the development of breast cancer was presented by Ryan et al. (1993) in which conditioned media from 14 short-term fibroblast cell lines, including fibroblasts from fibroadenomas, normal breast, malignant breast and skin, were mitogenic for human breast cancer epithelial cell growth in serum-free culture as measured by tritiated thymidine incorporation and the MTT colorimetric assay. Conditioned media from endothelial cells were also stimulatory; however, conditioned media from benign and malignant epithelial cells were not stimulatory for breast cancer cells (Ryan et al., 1993), thus demonstrating, once again, the specificity of stromal–epithelial interaction.

Adams et al. (1988) reported that conditioned medium from fibroblasts cultured from benign and malignant tumors had a stimulatory effect on MCF-7 cells, wherease conditioned medium from fibroblasts cultured from normal tissue had an inhibitory effect on the growth of MCF-7 cells suggesting that the growth modulatory effects of fibroblasts are due to soluble factor(s). On the other hand Mukaida et al. (1991) found that fibroblasts from mammary tissue were stimulatory while fibroblasts from other tissues had no effect on MCF-7 growth or were inhibitory.

Both *in vivo* and *in vitro* studies have demonstrated an effect of fibroblasts on tumor formation and growth rate (Van den Hooff, 1988; Zipori, 1990). When transformed rat prostatic fibroblasts were co-inoculated with breast or bladder tumor cell lines into nude mice the latency period was shortened and the growth rate of the tumors was increased (Camps et al., 1990). In a similar experiment human breast and skin fibroblasts were co-inoculated with MCF-7 cells into nude mice. Tumors formed in 94% of co-inoculated mice compared with 49% of mice inoculated with MCF-7 cells alone. No difference was seen in the rate of tumor development or the size of tumors whether the fibroblasts were derived from tumor or normal breast tissue (Horgan et al., 1987, 1988).

It would appear then, that the interaction of hormones and growth factors is required for growth of both the normal mammary gland and breast tumors and that the stroma provides, in a paracrine manner, many of the mediators required for epithelial growth. Several lines of evidence suggest that fibroblasts derived from breast tumors provide different signals than those provided by fibroblasts derived from normal breast tissue, although the reports are somewhat conflicting at this point. Horgan et al. (1987) demonstrated that human breast cancer cell growth *in vivo* was stimulated by fibroblasts independent of tissue source, however Mukaida et al. (1991) reported that only fibroblasts from breast or lung stimulated MCF-7 cell growth in soft agar and that fibroblasts from other tissues were inhibitory or had no effect. Also, Adams et al. (1988) reported that fibroblasts from breast tumors stimulated MCF-7 growth *in vitro*, but that fibroblasts from reduction mammoplasties were inhibitory and Van Roozendaal et al. (1992) reported that all fibroblasts, regardless of source, stimulated MCF-7 growth *in vitro*. The data reported by Ryan et al. (1993) suggests that the extent of proliferation response depends on the source of the fibroblasts, with tumor and skin fibroblasts inducing a greater mitogenic response than those derived from benign breast tissue.

Although there are conflicting reports regarding the effects of stroma from a variety of sources on mammary epithelial cell growth under experimental conditions, this in itself is consistent with the specificity observed for mesenchymal induction of epithelium during development. The source of the cultured fibroblasts, whether they are cell lines or short term cultures that senesce, and *in vitro* or *in vivo* culture conditions will influence the factors secreted and therefore their effect on adjacent epithelium. Also, fibroblasts are a heterogeneous population of cells, and therefore in any experimental situation one subpopulation of cells with a specific set of characteristics may predominate over the others.

Stromal cell variation

Are normal breast fibroblasts phenotypically and/or genotypically different from breast tumor fibroblasts? There is evidence that tumor fibroblasts, although not transformed themselves in breast tumors, are different from normal fibroblasts in several ways. A subset of fibroblasts within breast lesions and in many primary cultures derived from both non-malignant and malignant breast tissues demonstrates elements of smooth muscle differentiation and have been designated myofibroblasts (Rinnov-Jessen et al., 1990). They are distinguished from myoepithelial cells by the presence of vimentin and the absence of keratin. Likewise, they are distinguished from smooth muscle cells immunohistochemically by the absence of desmin (Ahmed, 1988). Immunostaining for α-smooth muscle actin showed actin expression in the stromal tissue derived from malignant breast tissue and was barely detectable in normal breast tissue (Sappino et al., 1990; Brouty-Boye and Magnien, 1994). Also, most primary cultures containing myofibroblasts produced a fibronectin isoform (ED-B) that is normally found in fetal and tumor tissue, but not in normal adult tissue (Oyama, 1990; Brouty-Boye and Magnien, 1994). Although the functional significance of this difference is not understood, there does appear to be a stromal population within breast tumors that is functionally different from normal breast stroma. Several groups have demonstrated phenotypic differences in skin and tumor fibroblasts from patients with cancer, such as familial polyposis of the colon (Pfeffer et al., 1976) and breast cancer (Azzarone et al., 1984; Maciera-Coelho and Azzarone, 1987) compared with normal controls. Schor et al. (1985) have observed that fetal and breast tumor fibroblasts migrate into a collagen gel matrix to a greater extent than their normal counterparts. They have purified a soluble factor called 'migration stimulating factor' (MSF) (Grey et al., 1989) produced by fetal and cancer patient fibroblasts, but not by normal adult fibroblasts (Schor et al., 1988). They have demonstrated that adult fibroblasts, when exposed to MSF, contained within the serum of breast cancer patients, show increased motility into collagen gels (Picardo et al., 1991). While the significance of these findings has not been defined in breast cancer models, it is possible that soluble factors such as MSF may alter the biological behavior of breast cancers by altering the motility of cells within the tumor, which may then affect tumor invasion and metastasis.

Stromal matrix proteins and their role in epithelial differentiation

Extracellular matrix

Fibroblasts produce a wide range of products that influence adjacent

Table 1. Comparison of biological activities of type I collagen, laminin, fibronectin, and tenascin

Matrix component	Biological function	Reference
Collagen	Cell shape, growth, polarity adhesion, migration and differentiation	Hay, 1981
	Mammary epithelial morphogenesis	Hall et al., 1982 Lee et al., 1985
Laminin	Cell adhesion, growth, differentiation, migration and matrix architecture	Kleinman et al., 1985
	Milk protein production by mammary epithelial cells	Streuli et al., 1991
Fibronectin	Cell adhesion, migration, morphology, growth Differentiation	Yamada et al., 1985
	positive – erythroblast	Patel and Lodish, 1987
	negative – keratinocyte	Adams and Watt, 1989
	– chondrocyte	West et al., 1979
	– adipocyte	Speigelman and Ginty, 1983
Tenascin	Cell attachment	Spring et al., 1989
	Cell proliferation	Chiquet-Ehrismann et al., 1986
	Cell migration – promotion	Halfter et al., 1989
	– inhibition	Tan et al., 1987
	Cell differentiation	Mackie et al., 1987b
	Morphogenesis	Inaguma et al., 1988
	Induction by diffusible factors	Hiraiwa et al., 1993

epithelial cells such as collagens, proteoglycans (heparan sulfate), and glycoproteins (fibronectin, laminin and tenascin). These extracellular matrix proteins have biological roles in cell attachment, differentiation, growth, morphogenesis and migration. The biological functions of these molecules are summarized in Table 1.

Stromal cells produce several extracellular matrix proteins and growth factors required by breast epithelium. The extracellular matrix provides mechanical support and influences orientation and differentiation of the epithelium while growth factors may act directly on the breast epithelium influencing its growth and differentiation, both in normal development and in carcinogenesis.

Similar studies in which mouse mammary epithelial cells grown on matrigel (a reconstituted basement membrane preparation) showed an increase in β-casein production as well as formation of structures resembling ducts, lumina, and alveoli. Ultrastructural examination of these epithelial cells revealed that they resumed a polar orientation which was lost during growth on plastic and that there was apical secretion of milk proteins that was not seen with cells grown on type IV collagen or fibronectin alone (Li et al., 1987). Bissell et al. (1989)

demonstrated that while the isolated matrix proteins heparan sulfate proteoglycan and laminin induced small increases in milk protein mRNA, they did not induce the full functional and histological differentiation seen when cells were grown on matrigel. The results of both these studies demonstrate that important interactions between epithelial cells and stromal matrix can have a profound impact on epithelial cell gene expression and morphology.

The same group has observed that mammary epithelial cells cultured on floating type I collagen gels are able to synthesize and deposit new basement membrane, while the same cells cultured on plastic are not. Interestingly, mRNA levels of basement membrane components (laminin, fibronectin, and type IV collagen) were highest in cells grown on plastic, therefore they hypothesized that mRNA expression is not sufficient for the production of a functional basement membrane, but that a specific environment that permits polar differentiation of epithelium is necessary for basement membrane formation (Streuli and Bissell, 1990). Stromal regulation of the functional differentiation of mammary epithelium, therefore appears to require both the induction of specific gene expression and a specific physical and chemical microenvironment.

Induction of specific gene expression in fibroblasts may be the result of signals from adjacent epithelial cells. Tenascin is an extracellular matrix glycoprotein that, unlike other extracellular matrix proteins, is expressed in a tissue-specific and developmentally-specific manner such as in the stroma surrounding growing epithelia, fetal epithelia, and malignant tissues (Bourdon et al., 1983; Chiquet-Ehrismann et al., 1986; Inaguma et al., 1988; Mackie et al., 1989; Natali et al., 1990) and in the mesenchyme of developing villi of the intestine (Beaulieu et al., 1991; 1993). It interferes with cell adhesion and may participate in epithelial migration in normal development or in the process by which tumor cells lose their normal adhesion characteristics and metastasize to other sites. Chiquet-Ehrismann et al. (1988) demonstrated that conditioned media from MCF-7 cells induced tenascin secretion by chick embryo fibroblasts, an effect that could be blocked by an antibody to TGF-β1, a product of breast tumor epithelial cells (Knabbe et al., 1987). One possible hypothesis, illustrated in Figure 1, is that TGF-β1 induces stromal production of tenascin, which in turn can promote epithelial cell dissociation and metastasis (Chiquet-Ehrismann et al., 1989).

Tenascin induction in epithelial cancer cells has also been observed. In another set of experiments, tenascin (TN) expression by human mammary epithelial cells was found to be inducible by embryonic mesenchyme. Human cell lines that do not produce TN *in vitro* were found to produce TN when injected into nude mice or co-cultured with embryonic mesenchyme. Also, conditioned media from these embryonic mesenchyme could induce TN expression suggesting that TN induction

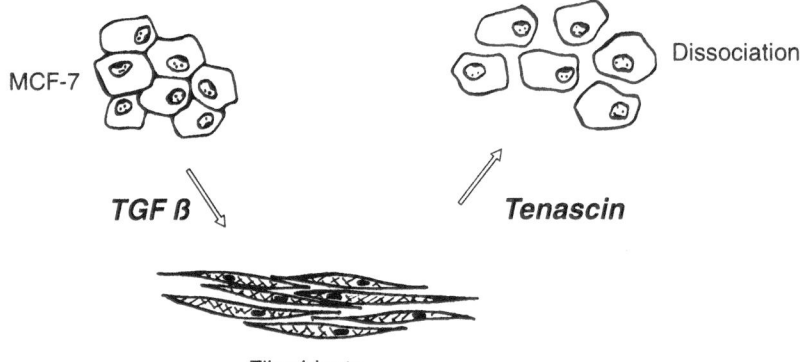

Figure 1. Illustration of a paracrine model by which tumor epithelial cells may lose their normal adhesion characteristics proposed by Chiquet-Ehrismann et al. (1989). The tumor epithelial cells produce TGF-β which then induces stromal production of tenascin. Tenascin, then promotes dissociation of the epithelial cells.

in tumor epithelia likely depends on interaction with the surrounding environment and that these interactions may be mediated by soluble factors(s) derived from the surrounding mesenchyme (Hiraiwa et al., 1993) suggesting a paracrine role for fibroblasts in the simulation of epithelial cell growth. It may be that the function of tenascin from epithelial cancers is different from mesenchymal tenascin. However, in both cases tenascin is deposited at sites of epithelial–stromal interactions and may, therefore, be involved in tissue remodeling.

Proteases

For a tumor to be able to grow and metastasize it must be able to degrade the surrounding extracellular matrix. While most proteases that will degrade extracellular matrix such as Type IV collagenase (Barsky et al., 1983) and plasminogen activator (Laug et al., 1975; Butler et al., 1979) are produced by the tumor cells themselves, some such as interstitial collagenase (Ellis et al., 1989) whose secretion by fibroblasts can be stimulated by the epithelial breast cancer cell line MDA-MB-231 (Noel et al., 1993), gelatinase A (Clavel et al., 1992; Pyke et al., 1992; Poulson et al., 1992; Noel et al., 1994) and stromelysin-3 (ST3)(Bassett et al., 1993) are produced by the stromal cells.

Immunohistochemical studies of breast tumor samples revealed the presence of gelatinase A on tumor epithelial cells (Monteagudo et al., 1990) and on the fibroblasts surrounding the tumor (Clavel et al., 1992). *In situ* hybridization studies have demonstrated that mRNA expression of gelatinase A was confined to the peritumoral fibroblasts, but not the fibroblasts distant from the tumor cells (Noel et al., 1994). Stromal expression of gelatinase A has now been demonstrated in human skin cancer (Pyke et al., 1992) and colorectal cancers (Poulson et al., 1992).

Using *in situ* hybridization studies, Basset et al. (1993; 1994) showed that the metalloprotease, stromelysin-3, is also secreted by the surrounding stromal cells, but not in the tumor epithelial cells themselves. ST3 was found to be expressed in 95% of invasive breast carcinomas, many metastases and *in situ* carcinomas in a manner that correlated with their invasive potential. Other groups have also demonstrated that the frequency of ST3 gene expression in a variety of carcinoma subgroups correlated well with the known risk of each of these carcinomas to become invasive (Wolf et al., 1993; Harris et al., 1992; Page et al., 1991). Growth factors such as PDGF and basic FGF which are known to be secreted by breast tumor epithelial cells caused an increase in ST3 expression in human fetal fibroblasts (Basset et al., 1990). Given the specificity of ST3 expression and its implication in the early stages of tumor invasion, ST3 is a potential prognostic marker for aggresive cancer and may be a target for cancer treatment.

These studies convincingly show that peritumoral fibroblasts actively synthesize basement membrane-degrading proteases that may participate in the invasive process by participating in degradation of the basement membrane. This is strong evidence that stromal cells can interact with tumor cells and influence cancer progression.

Growth factors

Table 2 summarizes some of the growth factors and receptors thought to be involved in stromal–epithelial interaction in breast cancer.

Several growth factors are produced by tumor epithelium and exert an autocrine effect in stimulation of tumor growth. Some growth factors produced by the tumor exert a paracrine effect on the surrounding stromal cells, which in turn, produce factors that stimulate tumor growth or metastasis. Other growth factors appear to be produced by the fibroblasts themselves and act in a paracrine fashion to promote tumor growth.

There is evidence for direct interaction between mesenchyme and epithelia mediated by small molecules. Rat mammary epithelial cells grown on purified acellular matrix derived from host mammary gland demonstrated significant differentiation, as measured by an increase in the production of α-lactalbumin and microvilli formation. These effects were not seen when cells were grown on simple collagen gels (Wicha et al., 1982).

Peptide growth factors have been shown to be an important means of communication between different cell types. The secretion of PDGF by tumor epithelial cells has been shown to be a stimulus for the growth of

Table 2. Polypeptide growth factors in breast cancer. (Adapted from Rosen et al., 1991)

Ligand	Receptor	Effects of receptor stimulation on growth	Role in breast cancer
TGF-α (Clarke et al., 1989; Ennis et al., 1989)	EGF-receptor; tyrosine kinase; expressed in mammary cells	Stimulates growth; transforms 3T3 cells	Autocrine growth factor
Heregulin (Slamon et al., 1989; Hudziak et al., 1987; Lupu et al., 1992b)	HER-2/HER-4; tyrosine kinase; amplified in 30% of breast cancers	HER-2 transforms fibroblasts	Associated with poor prognosis
FGF family (Wellstein et al., 1990)	Family of tyrosine kinases	Stimulates some tumor cells	May paracrine regulate stroma; regulates angiogenesis
PDGF (Bronzert et al., 1987)	Not expressed on breast cancer cells	?	Expressed in some tumor epithelia; stimulates stroma
TGF-β (Knabbe et al., 1987)	Expressed in mammary cells; uncharacterized	Inhibits breast epithelial cell growth	?
IGF-I (Yee et al., 1989)	IGF-I receptor; tyrosine kinase; expressed in mammary cells	Stimulates growth	Expressed in normal mammary stroma; paracrine regulator
IGF-II (Yee et al., 1988)	IGF-I receptor	Stimulates growth	Expressed in stroma and some tumor epithelial cells; potential paracrine and autocrine regulator of stroma
	IGF-II receptor	?	?

stromal cells (Peres et al., 1987) and the production of fibroblast growth factors by tumor epithelium appears to promote the development of vascular tissues required by the tumor for a blood supply (Gospo-darowicz et al., 1987; Folkman et al., 1989). Little is known, however, about how growth factor production by tumor stromal elements modulates the growth of transformed tumor epithelial cells.

TGF-α, found in many normal and malignant proliferating tissues, is a member of a family of growth factors related to epidermal growth factor (EGF) which act through the EGF receptor (Carpenter et al., 1983). TGF-α is produced by the malignant epithelial cells themselves, and is thought to act in an autocrine fashion. It has been suggested, however, that TGF-α may be produced by some stromal cells and, therefore act on the tumor in a paracrine fashion. Using *in situ* hybridization, Liscia et al. (1990) demonstrated the presence of TGF-α mRNA in a small subpopulation of mammary stromal cells in human and rat mammary glands during pregnancy. However, no human mammary fibroblasts have been shown to express TGF-α mRNA (Cullen et al., 1991).

Fibroblast growth factors (FGFs) are mitogenic for both fibroblasts (Brouty-Boye and Magnien, 1994) and epithelial cells (Smith et al., 1984; Takahashi et al., 1989). Basic FGF, secreted by breast tumor epithelial cells, has been shown to cause an increase in ST3 expression by human fetal fibroblasts (Bassett et al., 1990), illustrating how growth factor secretion by epithelial cells causes changes in gene expression in stromal cells which, in turn, may affect the growth of the tumor.

Heregulin, the ligand for *erbB2* (Lupu et al., 1990) and/or *erbB4* (Plowman et al., 1993), has been reported to have many isoforms (Wen et al., 1992; Holmes et al., 1992). Heregulin induces differentiation in mammary epithelial cells inducing growth arrest, production of milk proteins and ICAM-l induction (Bacus et al., 1992). This group also found a correlation with the presence of a prominent *in situ* component and a statistically significant reduction in lymph node metastases. They hypothesize that ICAM-l induction by heregulin is associated with a restricted ability of the tumor to metastasize (Bacus et al., 1993). Fibroblast cell lines derived from tumor and normal peripheral breast tissue have been examined for heregulin expression by our group and it was found that the majority of fibroblasts derived from breast cancers expressed higher levels of heregulin, while fibroblasts derived from normal peripheral or benign breast tissue expressed lower levels of the ligand (Hornby et al., 1994). Studies are currently underway using *in situ* hybridization to examine the possibility that stromal cells may be active in heregulin synthesis and interact with the tumor cells in a paracrine manner.

These studies demonstrate that growth factors made by both stroma and epithelia may facilitate cross-talk between stroma and epithelia of both normal and tumor tissue. Therefore, an understanding of the role the stroma plays in the development and maintenance of a tumor will lead to new possibilities for breast cancer treatment.

Current studies

Our group is currently studying two families of growth factors that appear to have a role in breast cancer: the insulin-like growth factors, IGF-I and IGF-II, and the ligands for tyrosine kinase receptors *erbB2*, *erbB3* and *erbB4*.

Insulin-like growth factors

Primary cultures of fibroblasts derived from benign and malignant breast lesions were examined by RNase protection for the expression of several growth factors to identify stromal growth factors that may be important in the growth regulation of malignant breast cancers. Most of the growth factors examined showed similar patterns of mRNA expression regardless of whether the fibroblasts originated from benign or malignant breast tissue. All fibroblasts examined expressed PDGF A chain, TGF-β1, FGF and FGF-5 mRNA, while none expressed PDGF B chain or TGF-α. Only the insulin-like growth factors showed differences in expression that correlated with whether the fibroblasts studied were derived from a malignant or benign lesion. The majority (7 out of 8) of fibroblast lines derived from benign lesions, but only 1 of 9 lines derived from malignant tumors expressed IGF-I. The inverse was seen for IGF-II where only 1 of 9 lines derived from benign lesions, but 5 of 9 lines derived from malignant tumors expressed IGF-II (Cullen et al., 1991; Cullen et al., 1992; Singer et al., 1992).

The insulin-like growth factors, IGF-I and II, are closely related peptide hormones having structural homology to insulin that have been implicated in the pathogenesis of both mesenchymal and epithelial tumors (Cohick and Clemmons, 1993). Both hormones are present in a variety of fetal and adult tissues and have complex and distinct roles in normal growth and development (Baker et al., 1993). Both hormones have also been shown to be potent mitogens for breast cancer epithelial cells (Furlanetto and DiCarlo, 1984; Myal et al., 1984; Yee et al., 1988; Karey and Sirbasku, 1988). The majority of breast tumors examined express both IGF-I and IGF-II mRNA, however no cultured breast epithelial cells were found to express IGF-I mRNA (Yee et al., 1989) and only a single epithelial cell line, T47D, was found to express IGF-II mRNA (Yee et al., 1988).

Normal and malignant breast tissue was then analyzed for IGF-I and IGF-II mRNA expression by *in situ* hybridization. In these studies, stromal cells were the source of IGF expression. *In situ* hybridization studies revealed IGF-I expression in normal stromal tissue at a distance from the tumor, but not in the stromal or epithelial cells of the tumor itself. Stromal cells in the vicinity of invasive tumors had reduced IGF-I expression and the tumor epithelial cells were always negative for IGF-I mRNA expression (Yee et al., 1989). IGF-II mRNA, however, was detected in the stroma of invasive breast cancers, some benign breast cancers and, in one case, the malignant epithelial component of the tumor (Paik, 1992).

Taken together these results suggest that the stroma is the source of IGF production in breast tissue and that normal production of IGF-I is reduced while IGF-II mRNA expression is induced in breast cancers.

IGFs as a model for paracrine interaction

Both IGF-I and IGF-II exert their mitogenic function through the IGF-I receptor (type I), however IGF-II binds this receptor with lower affinity than IGF-I. IGF-II binds the IGF-II/M6P receptor (type II) with much greater affinity and IGF-I has no significant affinity for this type II receptor.

These results, discussed previously, raise the possibility that IGF expression is primarily stromal in origin and that IGFs may be acting in a paracrine fashion to stimulate breast cancer epithelial cells. In order to postulate a paracrine stimulatory role for stromal secretion of IGF-I or IGF-II in breast cancer, breast cancer epithelial cells must have receptors for these ligands. When various cell lines have been examined for IGF-I, IGF-II and insulin receptors by binding competition and affinity crosslinking studies both IGF-I and IGF-II receptors were found to be present on the surface of a number of breast cancer cell lines (DeLeon et al., 1988). The presence of type I and type II IGF receptors has also been demonstrated in breast tumor tissues (Pollak et al., 1987; Peyrat et al., 1988b; Cullen et al., 1990) and type I receptors have also been demonstrated in benign breast lesions (Peyrat et al., 1988a).

There are two possible explanations for the pattern of IGF expression observed by Cullen et al. (1991). It may be that the tumor epithelial cells themselves can recruit and promote expansion of a preexisting subpopulation of stromal cells that already express IGF-II. On the other hand, the tumor epithelium may induce IGF-II expression in the fibroblast population.

The fibroblasts used in the studies by Cullen et al. (1991) were not transformed cells in that they show normal phenotypic behavior in tissue culture, with normal contact inhibition, monolayer growth, and senescence after approximately 15–20 passages. The cells were not monoclonal as each line was pooled from all of the fibroblasts that grew from the original tumor. Therefore, the differences in IGF-I and II expression seen betweeen fibroblasts derived from malignant and benign lesions are not the result of a transformation event within those cells, but represent as closely as possible, the *in vivo* phenotype.

An interesting observation is that the IGF phenotype of the fibroblasts persists in serial culture in the absence of tumor epithelial cells, therefore, the differences in gene expression between normal and tumor fibroblasts are stable and once initiated do not appear to require epithelial interaction to be maintained.

Having established a pattern of growth factor expression by fibroblasts and epithelial cells, Cullen et al. (1991) then asked whether any of these growth factors were involved in paracrine loops between the stroma and tumor epithelium. Breast tumor epithelia produce both PDGF A and B chains, but have not been found to have PDGF receptors and therefore, do not respond to PDGF in an autocrine fashion. Fibroblasts, however, have receptors for PDGF B chain. In an experiment in which fibroblasts grown in serum-free media treated with recombinant PDGF BB homodimer the cells were harvested at various time points. RNase protection was used to quantitate the expression of IGF-II. PDGF was found to increase IGF-II production three-fold. A complementary experiment was performed to analyze the effect of IGF-I and IGF-II treatment on the expression of PDGF by breast cancer epithelial cells. A two-fold increase in PDGF A chain and a slight increase in PDGF B chain was observed.

In the experiments described above, epithelial production of PDGF stimulated stromal production of IGF, which in turn stimulates greater epithelial production of PDGF, etc. The importance of such a mechanism in promoting actual tumor growth is still speculative. These experiments do demonstrate, however, the possibility of paracrine mediated communication between the different cell types within a breast tumor and illustrate how communication via soluble growth factors could contribute to tumor growth.

The biological effects of IGFs in breast cancer

Several experiments have identified the biological effects of IGF-I and II expression in breast cancer. The IGFs are potent mitogens for fibroblasts in both normal and maligant breast epithelium as well as a wide range of other cell types, including osteogenic sarcoma (Pollak et al., 1990), pancreatic cancer (Ohmura et al., 1990), squamous cell carcinomas of the skin (Neely et al., 1991) and others. The mitogenic response to IGF-I can be blocked by a monoclonal antibody (αIR3) directed against the IGF-I receptor indicating that the mitogenic signal of IGF-I is mediated by the IGF-I receptor (Flier et al., 1986). The mitogenic effect of IGF-II can also be blocked by αIR3, suggesting that its mitogenic effects are also mediated by the IGF-I receptor (Cullen et al., 1990; Osborne et al., 1989). IGF-II is two to five times less potent than IGF-I in inducing mitogenic effects, probably explained by the lower binding affinity of IGF-II for the IGF-I receptor (Fradkin et al., 1989).

Insulin and insulin-like growth factors are capable of acting as chemotactic agents in several cell culture models (Stracke et al., 1988). This motility response was inhibited by blockade of the IGF-I receptor with the monoclonal antibody αIR-3, indicating that the IGF-I receptor also mediates the motility response (Stracke et al., 1989; Kohn et al., 1990).

In order to identify an *in vivo* role for IGF-I and IGF-II these genes were disrupted by homologous recombination in murine embryonic stem cells. *IGF-I*-null mice showed reduced embryonic and post-natal growth. These mice have a homozygous deletion of the IGF-I gene, have hypoplastic reproductive organs and are sterile (Lui et al., 1993). Mice with a homozygous deletion of the IGF-II gene show reduced placental growth, however, they have normal reproductive organs and are fertile (DeChiara et al., 1990). These studies indicate that IGF-I plays a role in differentiation, whereas IGF-II may have a function as a growth promotor.

Transfection studies provide further evidence for a possible role of IGF-II in breast cancer. When MCF-7 cells, a well differentiated breast cancer cell line, were transfected with a retroviral expression vector containing IGF-II, positive clones showed loss of estrogen dependence which is characteristic of poorly differentiated, more aggressive breast tumor cell lines. The IGF-II-transfected cells showed phenotypic changes in colony morphology as demonstrated by enhanced growth in soft agar as well as enhanced anchorage-independent growth in suspension culture (Cullen et al., 1992a; Daly et al., 1991). Although IGF-II expression is more commonly found in the stromal cells, autocrine overexpression of IGF-II significantly enhanced the malignant characteristics of a well-differentiated breast epithelial cell line.

It would appear from the studies discussed above that IGF-I and IGF-II are mitogenic growth factors for epithelial cells produced primarily by the stroma. IGF-I appears to act on normal epithelial tissue at different stages of growth and differentiation, whereas IGF-II is associated with malignant growth. The role of IGF-II in normal postnatal growth and physiology is not yet understood.

The family of erbB receptors and their ligands

We have begun to study the possibility that the *erbB* receptors (*erbB2, erbB3,* and *erbB4*) and their ligands, heregulin and p75, interact in a paracrine fashion. c-*erbB2* specifies a transmembrane receptor phosphoprotein that is structurally related to the epidermal growth factor receptor (Yamamoto et al., 1986). It is over-expressed in approximately 30% of breast cancers and has been associated with a poor prognosis (Slamon et al., 1987; Paik et al., 1990). *In vitro* studies suggest that over-expression of *erbB2* may play a role in malignant progression (Seggato et al., 1988). In order to understand the function of *erbB2* and related receptors it is necessary to study interactions with specific ligands.

It has been demonstrated that purified heregulin, one of the ligands identified for *erbB2* activation (Lupu et al., 1992a), stimulates the phosphorylation of *erbB2* in a dose dependent manner in cells that over-express *erbB2* (Lupu et al., 1990). We have examined receptor expression in breast fibroblasts and epithelial cells by reverse transcription PCR and shown that most cells express *erbB2* at the mRNA level and that different cell types express different amounts (RNase protection) (Hornby et al., 1994). Expression of the ligand, heregulin, was similar to that of IGF-II using the RNase protection assay in which the majority of fibroblasts derived from breast cancer expressed higher levels of heregulin (8 out of 15), while fibroblasts derived from surrounding normal breast tissue or from benign lesions expressed lower levels or no heregulin mRNA. By PCR analysis, mRNA expression of the different isoforms of heregulin that we have examined, appear to be differentially expressed. When heregulin expressing fibroblasts were co-cultured with epithelial cells that overexpress c-*erbB2*, both an increase in growth and a change in morphology of the epithelial cells was observed (Hornby et al., 1994). We conclude from these experiments that heregulin produced by fibroblasts exerts a paracrine effect on breast tumor epithelial cells. It remains to be determined what the effect of heregulin is on the other *erbB* receptors. Recent evidence suggests that activation of *erbB2* requires either homodimerization or heterodimerization with other *erbB* receptors such as *erbB3* and *erbB4* (Carraway III and Cantley, 1994; Samanta et al., 1994; Sliwkowski et al., 1994).

Conclusions

It is clear that stromal–epithelial interaction is essential for both normal and neoplastic growth. The work described here also strongly supports the concept that tumor fibroblasts are phenotypically different in several respects from normal fibroblasts and that soluble factors produced by tumor fibroblasts may contribute to tumor growth.

Future directions

A picture of interaction, balance and redundancy is beginning to emerge as more is learned about stroma–epithelial interaction. It seems clear that factors between different cell types interact in both normal and tumor cell populations and that identification of these factors and how their pathways interact will broaden our understanding of cell growth and differentiation. An understanding of the balance of these molecules that promote or inhibit cell growth or differentiation, or degrade surrounding matrix should lead to more specific means of controlling the growth of cancers. The answers, however, will not be straight-forward as many of these factors are probably redundant and can be compensated for by alternative factors involved in similar pathways. An example of this was presented by Saga et al. (1992), in which nude mice that were homozygous null mutants for tenascin were born live with no phenoytypic abnormalities. More sensitive *in vitro* systems that take into account the architecture of cell–cell interaction will have to be developed in order to study these complex interactions in a relevant manner.

References

Adams, E.F., Newton, C.J., Braunsberg, H., Shaikh, N., Ghilchik, M. and James, V.H.T. (1988) Effects of human breast fibroblasts on growth and 17β-estradiol dehydrogenase activity of MCF-7 cells in culture. *Breast Canc. Res. Treat.* 11: 165–172.

Adams, J.C. and Watt, F.M. (1989) Fibronectin inhibits the terminal differentiation of human keratinocytes. *Nature* 340: 307–309.

Ahmed, A. (1990) The myofibroblasts in breast disease. *Pathol. Annu.* 25: 237–286.

Azzarone, B., Mareel, M., Beiilard, C., Scemama, P., Chaponnier, C. and Maciera-Coelho, A. (1984) Abnormal properties of skin fibroblasts form patients with breast cancer. *Int. J. Cancer* 33: 759–764.

Azzarone, B. and Maciera-Coelho, A. (1987) Further characterization of the defects of skin fibroblasts from cancer patients. *J. Cell Sci.* 87: 155–162.

Bacus, S.S., Stancovski, I., Huberman, E., Chin, D., Hurwitz, E., Mills, G.B., Ullrich, A., Sela, M. and Yarden, Y. (1992) Tumor inhibitory monoclonal antibodies to the HER2/neu receptor induce differentiation of human breast cancer cells. *Cancer Res.* 52: 2580–2589.

Bacus, S.S., Gudkov, A.V., Zelnick, C.R., Chin, D., Stern, R., Stancovski, I., Peles, E., Ben-Baruch, N., Farbstein, H., Lupu, R., Wen, D., Sela, M. and Yarden, Y. (1993) Neu differentiation factor (heregulin) induces expression of intercellular adhesion molecule 1: implications for mammary tumors. *Cancer Res.* 53: 5251–5261.

Baker, J., Liu, J.P., Robertson, E.J. and Efstratiadis, A. (1993) Role of insulin-like growth factors in embryonic and postnatal growth. *Cell* 75: 73–82.

Barsky, S.H., Togo, S., Speidione, G. and Liotta, L. (1983) Type IV collagenase immunoreactivity in invasive breast carcinoma. *Lancet* 1: 296–297.

Basset, P., Bellocq, J.P., Wolf, C., Stoll, I., Hutin, P., Limacher, J.M., Padhajcer, O.L., Chenard, M.P., Rio, M.C. and Chambon, P. (1990) A novel metalloproteinase gene specifically expressed in stromal cells of breast carcinomas. *Nature* 348: 699–704.

Basset, P., Wolf, C. and Chambon, P. (1993) Expression of the stromelysin-3 gene in fibroblastic cells of invasive carcinomas of the breast and other human tissues: a review. *Breast Canc. Res. Treat.* 24: 185–193.

Basset, P., Wolf, C., Rouyer, N., Bellocq, J.-P., Rio, M.-C. and Chambon, P. (1994) Stromelysin-3 in stromal tissue as a control factor in breast cancer behaviour. *Cancer* 74: 1054–1059.

Beaulieu, J.-F., Vachon, P.H. and Chartrand, S. (1991) Immunolocalization of extracellular matrix components during organogenesis in the human small intestine. *Anat. Embryol.* 183: 363–369.

Beaulieu, J.-F., Jutrs, S., Kusakabe, M. and Perreauult, N. (1993) Expression of tenascin in the developing human small intestine. *Biochem. Biophys. Res. Commun.* 192: 1086–1092.

Bissell, M.J. and Ram. T.J. (1989) Regulation of functional cytodifferentiation and histogenesis in mammary eptithelial cells: Role of the extracellular matrix. *Environ. Health Perspect.* 80: 61–70.

Bourdon, M.A., Wikstrand, C.J., Furthmaryr, H., Matthew, T.J. and Bigner, D.D. (1983) Human glioma–mesenchymal extracellular matrix antigen defined by monoclonal antibody. *Cancer Res.* 43: 2796–2805.

Bronzert, D.A., Pantazis, P., Antoniades, H.N., Kasid, A., Davidson, N., Dickson, R.B. and Lippman, M.E. (1987) Synthesis and secretion of platelet-derived growth factor by human breast caner cell lines. *Proc. Natl. Acad. Sci. USA.* 84: 5763–5767.

Brouty-Boye, D. and Magnien, V. (1994) Myofibroblasts and concurrent ED-B fibronectin phenotype in human stromal cells cultured from non-malignant and malignant breast tissue. *Eur. J. Cancer* 30A: 66–73.

Butler, W.B., Kirkland, W.L. and Jorgenson, T.L. (1979) Induction of plasminogen activator by estrogen in a human breast cancer cell line MCF-7. *BBRC* 90: 1328–1334.

Camps, J.L., Chang, S.-M., Hsu, T.C., Freeman, M.R., Hong, S.-J., Zhau, H.E., von Eschenbach, A.C. and Chung, W.L. (1990) Fibroblast-mediated acceleration of human epithelial tumor growth *in vivo*. *Proc. Natl. Acad. Sci. USA* 87: 75–79.

Carpenter, G., Stoscheck, C.M., Preston, Y.A. and DeLarco, J.E. (1983) Antibodies to epidermal growth factor receptor block the biological activities of sarcoma growth factor. *Proc. Natl. Acad. Sci. USA* 80: 5627–5630.

Carraway III, K.L. and Cantley, L.C. (1994) A new acquaintance for *erbB3* and *erbB4*: a role for receptor heterodimerization in growth signaling. *Cell* 78: 5–8.

Chiquet-Ehrismann, R., Mackie, E.J., Pearson, C.A. and Skakakura, T. (1986) Tenascin: an extracellular matrix protein involved in tissue interactions during fetal development and oncogenesis. *Cell* 47: 131–139.

Chiquet-Ehrismann, R., Kalla, P., Pearson, C.A., Beck, K. and Chiquet, M. (1988) Tenascin interferes with fibronectin action. *Cell* 53: 383–390.

Chiquet-Ehrismann, R., Kalla, P. and Pearson, C.A. (1989) Participation of tenascin and transforming growth factor-B in reciprocal epithelial–mesenchymal interactions of MCF-7 cells and fibroblasts. *Cancer Res.* 49: 4322–4325.

Clarke, R., Brunner, N., Katz, D., Glenz, P., Dickson, R.B., Lippman, M.E. and Kern, F. (1989) The effects of a constitutive production of TGF-a on the growth of MCF-7 human breast cancer cells *in vitro* and *in vivo*. *Mol. Endocrinol.* 3: 372–380.

Clavel, C., Polatte, M., Doco, M., Binniger, I. and Birembaut, P. (1992) Immunolocalization of matrix metalloproteases and their tissue inhibitor in a mammary pathology. *Bull. Cancer* 79: 261–270.

Cohick, W.S. and Clemmons, D.W. (1993) The insulin-like growth factors. *Ann. Rev. Physiol.* 55: 131–153.

Cullen, K.J., Yee, D., Sly, W.S., Perdue, J., Hampton, B., Lippman, M.E. and Rosen, N. (1990) Insulin-like growth factor receptor expression and function in human breast cancer. *Cancer Res.* 50: 48–53.

Cullen, K.J., Smith, H.S., Hill, S., Rosen, N. and Lippman, M.E. (1991) Growth factor messenger RNA expression by human breast fibroblasts from benign and malignant lesions. *Cancer Res.* 51: 4798–4985.

Cullen, K.J., Lippman, M.E., Chow, D., Hill, S., Rosen, N. and Zwiebel, J. (1992a) IGF-II overexpression in MCF-7 cells induces phenotypic changes associated with malignant progression. *Mol. Endocrinol.* 6: 91 100.

Cullen, K.J., Singer, C., Allison, A.A., Martire, I., Lippman, M.E. and Smith, H.S. (1992b) Stromal growth factor expression and function in human breast cancer. *J. Cell. Biochem. Supp.* 16D: 90.

Cunha, G.R., Chung, L.W.K., Shannon, J.M. and Reese, B.A. (1980) Stromal–epithelial interactions in sex differentiation. *Biol. Reprod.* 22: 19–42.

Daly, R.J., Haris, W.H., Wang, D.Y. and Darbre, P.D. (1991) Autocrine production of insulin-like growth factor II using an inducible expression system results in reduced estrogen sensitivity of MCF-7 human breast cancer cell. *Cell Growth & Diff.* 2: 457–464.

DeChiara, T., Efstratiadis, A. and Robertson, E. (1990) A growth deficiency phenotype in heterozygous mice carrying an insulin-like growth factor II gene disrupted by targeting. *Nature* 345: 78–80.

DeLeon, D.D., Bakker, B., Wilson, D.M. and Resenfeld, R.G. (1988) Demonstration of insulin-like growth factor (IGF-I and -II) receptors and binding protein in human breast cancer cell lines. *Biochem. Biophys. Res. Comm.* 152: 398–405.

Durnberger, H. and Kratochwil, K. (1980) Specificity of tissue interaction and origin of mesenchymal cells in the androgen-responsive of the embryonic mammary gland. *Cell* 19: 465–471.

Ellis, S., Nabeshima, K. and Biswas, C. (1989) Monoclonal antibody preparation and purification of a tumor-cell collagenase-stimulatory factor. *Cancer Res.* 49: 3385–3391.

Ennis, B.W., Valverius, E.M., Lippman, M.E., Bellot, F., Kris, R., Schlessinger, J., Masui, H., Goldberg, A., Mendelsohn, J. and Dickson, R.B. (1989) Anti-EGF receptor antibodies inhibit the autocrine stimulated growth of MDA-MB-468 breast cancer cells. *Mol. Endocrinol.* 3: 1830–1838.

Flier, J.S., Usher, P. and Moses, A.C. (1986) Monoclonal antibody to the type I insulin-like growth factor receptor blocks IGF-I receptor mediated DNA synthesis: Clarification of the mitogenic mechanisms of IGF-I and insulin in human skin fibroblasts. *Proc. Natl. Acad. Sci. USA* 83: 664–668.

Folkman, J., Watson, K., Ingber, D. and Hananhan, D. (1989) Induction of angiogenesis during the transition from hyperplasia to neoplasia. *Nature* 339: 58–61.

Fradkin, J.E., Easman, R.C., Lesniak, M.A. and Roth, J. (1989) Specificity spillover at the hormone receptor – exploring its role in human disease. *N. Engl. J. Med.* 320: 640–645.

Furlanetto, R.W. and DiCarlo, J.N. (1984) Somatomedin-C receptors and growth effects in human breast cells maintained in long term tissue culture. *Cancer Res.* 44: 2122–2128.

Gospodarowica, D., Ferrara, N., Schweigerer, L. and Neufeld, G. (1987) Structural characterization and biological functions of fibroblast growth factor. *Endocrine Rev.* 8: 95–114.

Grey, A.M., Schor, A.M., Rushton, G., Ellis, I. and Schor, S.L. (1988) Production of migration stimulating factor produced by fetal and breast cancer patients fibroblasts. *Proc. Natl. Acad. Sci. USA* 86: 2438–2442.

Grobstein, C. (1953) Morphogenic interaction between embryonic mouse tissues separated by membrane filter. *Nature* 172: 869–871.

Grobstein, C. (1967) Mechanisms of organogenetic tissue interaction. *Natl. Cancer Inst. Monogr.* 26: 279–299.

Halfter, W., Chiquet-Ehrismann, R. and Tucker, R.P. (1989) The effect of tenascin and embryonic basal lamina on the behaviour and morphology of neural crest cells *in vitro*. *Dev. Biol.* 132: 14–25.

Hall, H.G., Farson, D.A. and Bissell, M.J. (1982) Lumen formation by epithelial cell lines in response to collagen overlay: a morphogenetic model in culture. *Proc. Natl. Acad. Sci. USA* 79: 4672–4678.

Harris, J.R., Lippman, M.E., Veronesi, U. and Willett, W. (1992) Breast Cancer. *N. Engl. J. Med.* 327: 319–328, 390–398, 473–480.

Haslam, S.Z. and Lively, M.L. (1985) Estrogen responsiveness of normal mouse mammary cells in primary cell culture: association of mammary fibroblasts with estrogenic regulation of progesterone receptors. *Endocrinology* 116: 1835–1844.

Haslam, S.Z. (1986) Mammary fibroblasts influence on normal mouse epithelial cell responses to estrogen *in vitro*. *Cancer Res.* 46: 310–316.

Haslam, S.Z. (1989) The ontogeny of mouse mammary gland responsiveness to ovarian steroid hormones. *Endocrinology* 125: 2766–2772.

Hay, E.D. (1981) *In*: E.D. Hay (ed.): *Cell Biology of Extracellular Matrix*, Plenum Press, New York, pp 379–409.

Hiraiwa, N., Kida, H., Sakakura, T. and Kusakabe, M. (1993) Induction of tenascin in cancer cells by interactions with embryonic mesenchyme mediated by a diffusible factor. *J. Cell Sci.* 104: 289–296.

Holmes, W.E., Sliwkowski, M.X., Akita, R.W., Henzel, W.J., Lee, J., Park, J.W., Yansura, D., Abadi, N., Raab, H., Lewis, G.D., Shepard, M., Wood, W.I., Goeddel, D.V. and Bandlea, R.L. (1992) Identification of heregulin, a specific activator of p185erbB2. *Science* 256: 1205–1210.

Horgan, K., Jones, D.L. and Mansel, R.E. (1987) Mitogenicity of human fibroblasts *in vivo* for human breast cancer cells. *Br. J. Surg.* 74: 227–229.

Horgan, K., Jones, D.L. and Mansel, R.E. (1988) Stromal stimulation of human breast cancer growth and development. *Progr. Cancer Rese. Ther.* 35 (Hormones Cancer 3): 179–182.

Hornby, A., Singer, C., Rassmussen, A., Cho, C., Smith, H., Lupu, R. and Cullen, K. (1994) Stromal expression of growth factors in breast cancer. *J. Cell. Biochem.* Supp: 18D: Y113, 1994.

Hudziak, R.M., Schlessinger, J. and Ullrich, A. (1987) Increased expression of the putative growth factor receptor p185[HER2] causes transformation and tumorigenesis of NIH3T3 cells. *Proc. Natl. Acad. Sci. USA* 84: 7159–7162.

Iguchi, T., Uchima, F.-D.A., Ostrander, P.L. and Bern, H.A. (1983) Growth of normal mouse vaginal cells in and on collagen gels. *Proc. Natl. Acad. Sci. USA* 80: 3743–3747.

Inaguma, Y., Kusakabe, M., Mackie, E.J., Pearson, C.A., Chiquet-Ehrismann, R. and Sakakura, T. (1988) Epithelial induction of stromal tenascin in the mouse mammary gland: from embryogenesis to carcinogenesis. *Dev. Biol.* 128: 245–255.

Jowett, A.K., Vanino, S., Ferguson, M.W.J., Sharpe, P.T. and Thesleff, I. (1993) Epithelial-mesenchyme interactions are required for *msx1* and *msx2* gene expression in the developing murine molar tooth. *Development* 117: 461–470.

Karey, K.P. and Sirbasku, D.A. (1988) Differential responsiveness of human breast cancer cell lines MCF-7 and T47D to growth factors and 17β estradiol. *Cancer Res.* 48: 4083–4092.

Kleinman, H.K., Cannon, F.B., Laurie, G.W., Hassell, J.R., Aumailley, M., Terranova, V.P., Marin, G.R. and DuBois-Dalcq, M. (1985) Biological activities of laminin. *J. Cell Biochem.* 27: 317–325.

Knabbe, C., Lippman, M.E., Wakefield, L.M., Flanders, K.C., Kasid, A., Derynck, R. and Dickson, R.B. (1987) Evidence that transforming growth factor beta is a hormonally regulated negative growth factor in human breast cancer cells. *Cell* 48: 417–428.

Kohn, E.C., Francis, E.A., Liotta, L.A. and Schiffman, E. (1990) Heterogenity of the motility responses in malignant tumor cells: a biological basis for the diversity and homing of metastatic cells. *Int. J. Cancer* 46: 287–292.

Kollar, E.J. and Fisher, C. (1990) Tooth induction in chick epithelium: expression of quiescent genes for enamel synthesis. *Science* 207: 993–995.

Laug, W.E., Jones, P.A. and Benedict, W.F. (1975) Relationship between fibrinolysis of cultured cells and malignancy. *J. Natl. Cancer Inst.* 54: 173–179.

Lee, E. Y.-H., Lee, W.-H., Kaetzel, C.S., Parry, G. and Bissell, M.J. (1985) Interaction of mouse mammary epithelial cells with collagen substrata: regulation of casein gene expression and secretion. *Proc. Natl. Acad. Sci. USA* 85: 1419–1423.

Li, M.L., Aggeler, J., Farson, D.A., Hatier, C. Hassell, J. and Bissell, M.J. (1987) Influence of a reconstituted basement membrane and its components on casein gene expression and secretion in mouse mammary epithelial cells. *Proc. Natl. Acad. Sci. USA* 84: 136–140.

Liscia, D.S., Merlo, G., Giardiello, F., Kim, N., Smith, G.H., Callahan, R.H. and Saloman, D.S. (1990) Transforming growth factor-α messenger RNA localization in the developing adult rat and human mammary gland by *in situ* hybridization. *Dev. Biol.* 140: 123–131.

Lippman, M.E., Dickson, R.B., Bates, S., Knabbe, C., Huff, K., Swain, S., McGManaway, M., Bronzert, D., Kasid, A. and Gelmann, E.P. (1986a) Autocrine and paracrine growth regulation of human breast cancer. *Breast Canc. Res. Treat.* 7: 59–70.

Lippman, M.E., Dickson, R.B., Kasid, A., Gelmann, E., Devidson, N., McManaway, M., Juff, K., Branzert, D., Bates, S., Swain, S. and Knabbe, C. (1986b) Autocrine and paracrine growth regulation of human breast cancer. *J. Steroid Biochem.* 2: 147–154.

Lui, J., Baker, J., Perkins, A.S., Robertson, E.J. and Efstratiadis, A. (1993) Mice carrying null mutations of the genes encoding insulin-like growth factor I (*igf-I*) and type I IGF receptor (*igf-Ir*). *Cell* 75: 59–72.

Lupu, R., Colomer, R., Zugmaier, G. and Lippman, M.E. (1990) A candidate ligand of the *erbB2* protooncogene interacts directly with both EGF receptor and *erbB2*. *Science* 249: 1552–1555.

Lupu, R., Dickson, R.B. and Lippman, M.E. (1992a) The role of *erbB2* and its ligands in growth control of malignant breast epithelium. *J. Steroid Biochem. Molec. Biol.* 43: 229–236.

Lupu, R., Wellstein, A., Sheridan, J., Ennis, B.W., Augmaier, G., Katz, D., Lippman, M.E. and Dickson, R.B. (1992b) Purification and characterization of a novel growth factor from human breast cancer cells. *Biochem.* 31: 610–616.

Mackie, E.J., Thelsleff, I. and Chiquet-Ehrismann, R. (1987) Tensascin is associated with chondrogenic and ostrogenic differentiation *in vitro* and promotes chondrogenesis *in vitro*. *J. Cell. Biol.* 105: 2569–2579.

Mackie, E.J., Chiquet-Ehrismann, R., Pearson, C.A., Inaguma, Y., Taya, K., Kawarada, Y. and Sakakura, T. (1989) Tenascin is a stromal marker for epithelial malignancy in the mammary gland. *Proc. Natl. Acad. Sci. USA* 84: 4621–4625.

McGrath, C.M. (1983) Augmentation of the response of normal mammary epithelial cells to estradiol by mammary stroma. *Cancer Res.* 31: 1355–1360.

Monteagudo, C., Merino, M.J., Sanjuan, J. and Liotta, L.A. (1990) Immunohistochemical distribution of type-IV collagen in normal, benign, and malignant breast tissue. *Amer. J. Pathol.* 136: 585–592.

Mukaida, H., Hiabayashi, N., Hirai, T., Iwata, T., Saeki, S. and Toge, T. (1991) Significance of freshly cultured fibroblasts from different tissues in promoting cancer cell growth. *Int. J. Cancer* 48: 423–427.

Myal, Y., Shiu, R.P.C. and Bhaumick, B. (1984) Receptor binding and growth-promoting activity of insulin-like growth factors in human breast cancer cells (T-47D) in culture. *Cancer Res.* 44: 5486–5490.

Natali, P.G., Nictra, M.R., Bortolazzi, A., Mottolese, M., Coscia, N., Bigotti, A. and Zardi, L. (1990) Expression and production of tenascin in benign and malignant lesions of melanocyte lineage. *Int. J. Cancer* 46: 586–590.

Neely, E.K., Morhenn, V.B., Hintz, R.L., Wilson, D.M. and Rosenfeld, R.G. (1991) Insulin-like growth factors are mitogenic for human keratinocytes and a squamous cell carcinoma. *J. Invest. Dermatol.* 96: 104–110.

Noel, A., Munaut, C., Nusgens, B., Lapiere, Ch.M. and Foidart, J.M. (1993) Different mechanisms of extracellular matrix remodeling by fibroblasts in response to human mammary neoplastic cells. *Invas. Metast.* 13: 72081.

Noel, A.C., Polette, M., Lewalle, J.-M., Munant, C., Emonard, H.P., Birembaut, P. and Foidart, J.-M. (1994) Coordinate enhancement of gelatinase A mRNA and activity levels in human fibroblasts in response to breast-adenocarcinoma cells. *Int. J. Cancer* 56: 331–336.

Ohmura, E., Okada, M., Onoda, N., Kamiya, Y., Murakami, H., Tsushima, T. and Shizume, K. (1990) Insulin-like growth factor I and transforming growth factor α as autocrine growth factors in human pancreatic cancer cell growth. *Cancer Res.* 50: 103–107.

Osborne, C.K., Coronado, E.B., Kitten, L.J., Arteata, C.I., Fuqua, S.A., Ramasharma, K., Marshall, M. and Li, C. (1989) Insulin-like growth factor II (IFG-II): A potential autocrine/paracrine growth factor for human breast cancer actin via the IGF-I receptor. *Mol. Endocrinol.* 3: 1701–1709.

Oyama, F., Hirohashi, S., Shimosato, Y., Titani, K. and Sekiguchi, K. (1990) Oncodevelopmental regulation of the alternative splicing of fibronectin pre-messenger RNA in human lung tissues. *Cancer Res.* 50: 1075–1078.

Page, D.L., Kidd, T.E., Dupont, W.D., Simpson, J.F. and Rogers, L.W. (1991) Lobular neoplasia of the breast: higher risk for subsequent invasive cancer predicted by more extensive disease. *Human Pathol.* 22: 1232–1239.

Paik, S., Hazan, R., Fisher, E.R., Sass, R.E., Fisher, B., Redmond, C., Schlessinger, J., Lippman, M.E. and King, C.R. (1990) Pathological findings from National Surgical Adjuvant Breast Cancer Project (protocol B-06), prognostic significance of erbB-2 protein overexpression in primary breast cancer. *J. Clin. Oncol.* 8: 103–112.

Patel, V.P. and Lodish, H.F. (1987) Fibronectin matrix is required for differentiation of murine erythroleukemia cells into reticulocytes. *J. Cell Biol.* 105: 3105–3118.

Peres, R., Betsholtz, C., Westermark, B. and Heldin, C. (1987) Frequent expression of growth factors for mesenchymal cells in human mammary carcinoma cell lines. *Cancer Res.* 47: 3425–3429.

Peyrat, J.-P., Bonneterre, J. and Laurent, J.C. (1988a) Presence and characterization of insulin-like growth factor 1 receptors in human benign breast disease. *Eur. J. Cancer Clin. Oncol.* 24: 1425–1431.

Peyrat, J.-P., Bonneterre, J., Beuscart, R., Dijane, J. and Demaille, A. (1988b) Insulin-like growth factor 1 receptors in human breast cancer and their relation to estradiol and progesterone receptors. *Cancer Res.* 48: 6429–6433.

Pfeffer, L., Lipkin, M., Stutman, O. and Kopelovich, L. (1976) Growth abnormalities of cultured human skin fibroblasts derived from individuals with hereditary adenomatosis of the colon and rectum. *J. Cell Physiol.* 80: 29–38.

Picardo, M., Schor, S.L., Grey, A.-M., Howell, A., Laidaw, I., Redford, J. and Schor, A.M. (1991) Migration stimulating activity in serum of breast cancer patients. *Lancet* 337: 130–133.

Plowman, G.D., Green, J.M., Culouscou, J.-M., Carlton, G.W., Rothwell, V.M. and Buckley, S. (1993) Heregulin induces tyrosin phosphorylation of HER4/p180erbB4. *Nature* 366: 473–475.

Pollak, M.N., Perdue, J.F., Margolese, R.G., Baer, K. and Richard, M. (1987) Presence of somatomedin receptors on human breast and colon carcinomas. *Cancer Lett.* 38: 223–230.

Pollak, M.N., Polychronakos, C. and Richard, M. (1990) Insulin-like growth factor I: a potent mitogen for human osteogenic sarcoma. *JNCI* 82: 301–305.

Poulsom. R., Pignatelli, M., Stetler-Stevenson, W.G., Liotta, L., Wright, P.A., Jeffrey, R.E., Longcroft, J.M., Rogers, L. and Stamp, G.W.H. (1992) Stromal expression of 72 kDa type-IV collagenase (MMP2) and TIMO-2 mRNAs in colorectal neoplasia. *Amer. J. Pathol.* 141: 389–396.

Propper, A.Y. (1967) Tissue interactions during organogenesis of the mammary gland in the rabbit embryo. *C. R. Acad. Sci. Hebd. Seances Acad. Sci. D.* 264: 2573–2575.

Pyke, C., Ralfkiaer, E., Huhtala, P., Hurskainen, T., Dano, K. and Tryggvason, K. (1992) Localization of messenger RNA for Mr 72,000 and 92,000 type-IV collagenases in human skin cancers by *in situ* hybridization. *Cancer Res.* 52: 1336–1341.

Reichmann, E., Ball, R., Groner, B. and Friis, R.R. (1989) New mammary epithelial and fibroblastic cell clones in coculture form structures competent to differentiate functionally. *J. Cell Biol.* 108: 1127–1138.

Rinnov-Jessen, L., van Deurs, B., Celis, J.E. and Peterson, O.W. (1990) Smooth muscle differentiation in cultured human breast gland stromal cells. *Lab Invest.* 63: 532–543.

Rosen, N., Yee, D., Lippman, M.E., Paik, S. and Cullen, K.J. (1991) Insulin-like growth factors in human breast cancer. *Breast Cancer Res. & Treat.* 18 Suppl. 1: S55–S62.

Ryan, M.C., Orr, D.J.A. and Horgan, K. (1993) Fibroblasts stimulation of breast cancer cell growth in a serum-free system. *Br. J. Cancer* 67(6): 1268–1273.

Sakakura, T., Nishizuka, Y. and Dawe, C.J. (1976) Mesenchyme-dependent morphogenesis and epithelium-specific cytodifferentiation in mouse mammary gland. *Science* 194: 1439–1441.

Sakakura, T., Nishizuka, Y. and Dawe, C.J. (1979a) Capacity of mammary fat pads of adult C3H/HeMs mice to interact morphogenetically with fetal mammary epithelium. *J. Natl. Cancer Inst.* 63: 733–736.

Sakakura, T., Sakagami, Y. and Nishizuka, Y. (1979b) Persistence of responsiveness to adult mammary gland to induction by embryonic mesenchyme. *Dev. Biol.* 72: 201–210.

Sakakura, T., Kusano, I., Kusakabe, M., Inaguma, Y. and Nishizuka, Y. (1987) Biology of mammary fat pad in fetal mouse: capacity to support development of various fetal epithelia *in vivo*. *Development* 100: 421–430.

Sakakura, R. (1991) New aspects of stroma-parenchyma relations in mammary gland differentiation. *Int. Rev. Cytol.* 125: 165–202.

Samanta, A., LeVea, C.M., Dougall, W.C., Qian, X. and Greene, M.I. (1994) Ligand and p185c-neu density govern receptor interactions and tyrosine kinase activation. *Proc. Natl. Acad. Sci. USA* 91: 1711–1715.

Sappino, A.P., Skalli, O., Jackson, B., Schurch, W. and Gabbiani, G. (1988) Smooth-muscle differentiation in stromal cells of malignant and non-malignant breast tissues. *Int. J. Cancer* 41: 707–712.

Sappino, A.P., Schurch, W. and Gabbiani, C. (1990) Differentiation repertoire of fibroblastic cells: expression of cytoskeletal proteins as marker of phenotypic modulations. *Lab. Invest.* 63: 144–161.

Schor, S.L., Schor, A.M., Durning, P. and Rushton, G. (1985) Skin fibroblasts obtained from cancer patients display fetal-like behaviour in collagen gels. *J. Cell. Sci.* 73: 235–244.

Schor, S.L., Schor, A.M., Grey, A.M. and Rushton, G. (1988) Fetal and cancer patient fibroblasts produce an autocrine migration stimulating factor not made by normal adult cells. *J. Cell Sci.* 90: 391–399.

Segatto, O., King, C.R., Peirce, J.H., Di Diore, P.P. and Aaronson, S.A. (1988) Different structural alterations upregulate *in vitro* tyrosine kinase activity and transforming potency of the *erbB-2* gene. *Mol. Cell Biol.* 8: 5570–5574.

Singer, C., Smith, H.S., Lippman, M.E. and Cullen, K.J. (1992) IGF-I and IGF-II expression in fibroblasts derived from tumor, normal breast and skin of breast cancer patients. *Proc. Am. Assoc. Cancer Res.* 33: 270.

Slamon, D.J., Clark, G.M., Wong, S.G., Levin, W.J., Ullrich, A. and McGuire, W.L. (1987) Human breast cancer: coorelation of relapse and survival with amplification of the HER-2/neu oncogene. *Science* 235: 177–182.

Slamon, D.J., Goldolphin, E., Jones, L.A., Holt, J.A., Wong, S.G., Keith, D.E., Levin, W.J., Stuart, S.G., Udove, J., Ullrich, A. and Press, M.J. (1989) Studies of the HER-2/neu protooncogene in human breast and ovarian cancer. *Science* 244: 621–624.

Slavkin, H.C. (1978) The nature and nurture of epithelial–mesenchymal interaction during tooth morphogenesis. *J. Biol. Bucca.* 6: 189–204.

Sliwkowski, M.X., Schaefer, G., Akita, R., Logfren, J.A., Fitzpatrick, V.D., Nuijens, A., Fendly, B.M., Cerione, R.A., Vandlen, R.L. and Carraway III, K.L. (1994) Coexpression of erbB2 and erbB3 proteins reconstitutes a high affinity receptor for heregulin. *J. Biol. Chem.* 269: 14661–14665.

Smith, J.A., Winslow, D.P. and Rudland, P.S. (1984) Different growth factors stimulate cell division of rat mammary epithelial myoepithelial, and stromal cell lines in culture. *J. Cell. Physiol.* 119: 320–326.

Speigelman, B.M. and Ginty, C.A. (1993) Fibronectin modulation of cell shape and lipogenic gene expression in 3T3-adipocytes. *Cell* 35: 657–666.

Spring, J., Beck, K. and Chiquet-Ehrismann, R. (1989) Two contrary functions of tenascin: dissection of the active sites by recombination tenascin fragments. *Cell* 59: 325–334.

Stracke, M.L., Kohn, E.C., Aznavoorian, S.A., Wilson, L.L., Salomon, D., Krutzsch, H.C., Liotta, L.A. and Schiffman, E. (1988) Insulin-like growth factors stimulate chemotaxis in human melanoma cells. *Biochem. Biophys. Res. Commun.* 153: 1076–1083.

Stracke, M.L., Engel, J.D., Wilson, L.W., Rechler, M.M., Liotta, L.A. and Schiffman, E. (1989) The type I insulin-like growth factor receptor is a motility receptor in human melanoma cells. *J. Bio. Chem.* 264: 21544–21549.

Streuli, C.J. and Bissell, M.J. (1990) Expression of extracellular matrix components is regulated by substratum. *J. Cell Biol.* 110: 1405–1415.

Streuli, C.H., Bailey, N. and Bissell, M.J. (1991) Control of mammary epithelial differentiation: basement membrane induces tissue-specific gene expression in the absence of cell–cell interaction and morphological polarity. *J. Cell Biol.* 115: 1385–1395.

Takahashi, K., Suzuki, K., Kawahara, S. and Ono, T. (1989) Growth stimulation of human breast epithelial cells by basic fibroblast growth factor in serum-free medium. *Int. J. Cancer* 43: 870–874.

Tan, S.-S., Crossin, K.L., Hoffman, S. and Edelman, G.M. (1987) Asymmetric expression in somites of cytotactin and its proteoglycan ligand is correlated with neural crest cell distribution. *Proc. Natl. Acad. Sci. USA* 84: 7977–7981.

Van den Hooff, A. (1988) Stromal involvement in malignant growth. *Adv. Cancer Res.* 50: 159–196.

Van Roozendaal, C.P., Van Oojen, B., Klijn, J.M.G., Classen, C., Eggermont, A.M.M., Hensen-Logmans, S.C. and Foekens, J.A. (1992) Stromal influence on breast cancer cell growth. *Br. J. Cancer* 65: 6691–6696.

Wellstein, A. and Lippman, M.E. (1990) Fibroblast growth factors and breast cancer. *In*: S. Broder (ed.): *Molecular Foundation of Oncology.* Williams and Wilkins, Baltimore.

Wen, D., Peles, E., Cupples, R., Suggs, S.V., Bacus, S.S., Luo, Y., Trail, G., Hu, S., Silbiger, S.M., Levy, R.B., Lus, Y. and Yarden, Y. (1992) Neu differentiation factor: a transmembrane glycoprotein containing an EGF domain and an immunoglobulin homology unit. *Cell* 69: 559–572.

Wessels, N.K. and Cohen, J.H. (1967) Early pancreas organogenesis: morphogenesis, tissue interactions, and mass effects. *Dev. Biol.* 15: 237–270.

West, C.M., Lanza, R., Rosenbloom, J., Lowe, M. and Holtzer, H. (1979) Fibronectin alters the phenotypic properties of cultured chick embryo chondroblasts. *Cell* 17: 491–501.

Wicha, M.S., Lowerie, G., Kohn, E., Bagavvandoss, P. and Mahn, T. (1982) Extracellular matrix promotes mammary epithelial growth and differentiaton *in vitro*. *Proc. Natl. Acad. Sci. USA* 79: 3213–3217.

Wolf, C., Rouyer, N., Lutz, Y., Adida, C., Loriot, M., Bellocq, J.P., Chambon, P. and Basset, P. (1993) Stromelysin-3 belongs to a subgroup of proteinases expressed in breast carcinoma fibroblastic cells and possibly implicated in tumor progression. *PNAS* 24: 185–193.

Yamada, K.M., Humphries, M.J., Hasegawa, T., Hasegawa, K., Olden, K., Chien, W.-T. and Akiyama, S.K. (1985) *In*: G.M. Edelman and J.-P. Thiery (eds): *The Cell in Contact.* Wiley, New York, pp 303–332.

Yamamoto, T., Ikawa, S., Akiyama, T., Semba, K., Nomura, N., Miyajima, N., Saito, T. and Toyoshima, K. (1986) Similarity of protein encoded by the human c-erbB-2 gene to epidermal growth factor receptor. *Nature* 312: 230–234.

Yee, D., Cullen, K.J., Paik, S., Perdue, J.F., Hampton, B., Schwartz, A., Lippman, M.E. and Rosen, N. (1988) Insulin-like growth factor II mRNA expression in human breast cancer. *Cancer Res.* 48: 6691–6696.

Yee, D., Paik, S., Lebovic, G.S., Marcus, R., Faboni, R., Cullen, K., Lippman, M.E. and Rosen, N. (1989) Analysis of IGF-I gene expression in malignance: evidence for a paracrine role in human breast cancer. *Mol. Endocrinol.* 3: 509–517.

Zipori, D. (1990) Stromal cells in tumor growth and regression. *Cancer J.* 3: 164–169.

Epithelial–Mesenchymal Interactions in Cancer
ed. by I.D. Goldberg & E.M. Rosen
© 1995 Birkhäuser Verlag Basel/Switzerland

Fibroblast subpopulations as accelerators of tumor progression: The role of migration stimulating factor

S.L. Schor

Department of Dental Surgery and Periodontology, The Dental School, University of Dundee, Dundee, DD1 4HR, Scotland, UK

Summary. Tumor progression is a relatively indolent process, with many years commonly intervening between the inception of an initiating genetic lesion and the development of overt malignant disease. We suggest that the perturbation of normal epithelial–mesenchymal interactions caused by the inappropriate presence of fibroblast subpopulations displaying various 'fetal-like' phenotypic characteristics may significantly alter the kinetics of tumor progression and hence enhance susceptibility to cancer development. In this communication, we review our own data indicating the presence of fetal-like fibroblasts in cancer patients and put these observations in the context of similar published reports. We then discuss our interpretation of these findings, emphasising the possible direct involvement of fetal-like fibroblasts in cancer pathogenesis and putting forward an epigenetic 'clonal modulation' model to account for their presence in cancer patients.

Introduction

Cancer pathogenesis is generally accepted to be a multistep process involving the occurrence of an initiating genetic lesion followed by a series of subsequent events collectively referred to as 'progression' (Foulds, 1975; Kundson, 1977). The gradual acquisition during tumor progression of more 'aggressive' phenotypic characteristics, such as diminished growth control (Helin and Harlow, 1993), local invasion (Birchmeier et al., 1991) and metastasis (Steeg, 1991), also appear to involve the accumulation of genetic lesions within the emerging population of (pre-) neoplastic cells. The primary role of genetic lesions in the process of cancer development is further suggested by the occurrence of well-defined hereditary syndromes characterized by increased susceptibility to specific types of cancer (eg. carcinoma of the breast) (Lynch et al., 1984). The identification and mapping of these 'cancer susceptibility' genes is currently being pursued with great vigor.

Although the acquisition of genetic damage clearly plays a fundamental role in cancer pathogenesis, other mechanisms operative at higher (tissue) levels of organization also contribute to the course of disease progression. These higher level mechanisms involve a complex network of reciprocal cell–cell interactions which modulate various aspects of tumor and stromal cell behavior. For example, interactions between tumor cells and fibroblasts affect such potentially relevant phenomena as the production of matrix degrading enzymes (Nakajima and Chop, 1991) and tumor cell motility (Zetter and Brightman, 1990). Such tissue level interactions are clearly dependent upon the nature of the participating stromal cells. Viewed in this context, it is of interest to note that several groups have documented the presence of aberrant stromal fibroblasts in patients with various types of both sporadic and hereditary cancers (previously reviewed in Schor et al., 1987). Our own work in this area has indicated that fibroblasts obtained from a majority of breast cancer patients resemble fetal cells in terms of their apparently persistent production of a 'migration stimulating factor' (MSF) which is not made by their normal adult counterparts. Studies concerned with the mode of action of MSF have suggested means whereby it may perturb epithelial–stromal interactions in a manner which might favor cancer progression.

With these various points in mind, the objectives of this communication are to (a) summarize our observations leading to the initial characterization of MSF, its mode of action and the identification of MSF-secreting fibroblasts in cancer patients, (b) review the broader literature documenting the presence of aberrant fibroblasts in cancer patients, (c) discuss the possibility that these fibroblasts make a significant contribution to cancer progression, and (d) present an epigenetic 'clonal modulation' model to account for the presence of such aberrant fibroblasts in cancer patients. We finish by speculating that the development of therapeutic modalities targeted at the aberrant stromal cell populations might provide novel means of impeding the course of cancer progression.

Assessment of fibroblast migratory phenotype and the persistence of 'fetal-like' fibroblasts in cancer patients

Fibroblasts are highly motile cells *in vitro* and it is possible to quantitate various aspects of their migratory phenotype. We have developed an assay for measuring their migration into 3D collagen gels (Schor, 1980) and have demonstrated that this parameter is affected by cell density (Schor et al., 1985b). In our assay, fibroblasts are plated on the surface of a collagen gel at standard subconfluent (10^3 cells cm^{-1}) and confluent (2×10^4 cells cm^{-1}) densities. After a 4 day incubation period, the percentage of plated cells which have migrated down into the gel matrix is ascertained by microscopic observation. These data are then used to calculate the *cell density migration index* (CDMI), defined as follows: CDMI = log(%SD/%CD), where %SD and %CD are the percentage of cells found in the gel matrix in subconfluent and confluent cultures, respectively (Schor et al., 1985b). In an initial survey of approximately

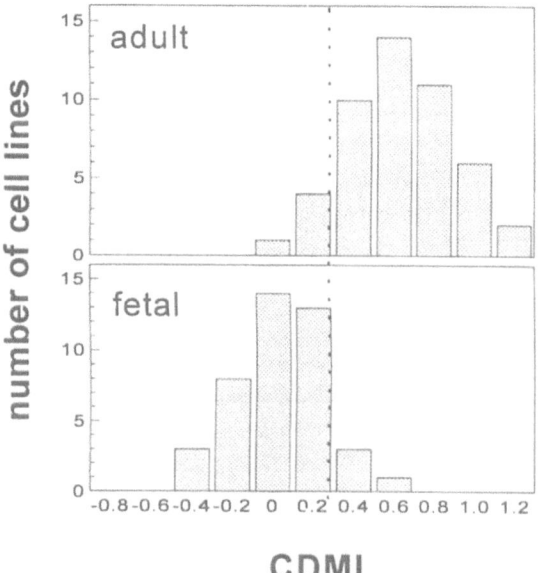

Figure 1. Distribution of CDMI values expressed by adult and fetal skin fibroblasts. Vertical dotted line indicates division between fetal and adult CDMI values.

100 distinct fetal and adult skin fibroblast lines, we found that approximately 90% of adult cells expressed CDMI values greater than +0.4, whilst approximately 90% of the fetal lines expressed CDMI values less than +0.4 (Fig. 1). On the basis of these observations, we empirically defined characteristic 'adult-like' and 'fetal-like' modes of migratory behavior based on whether the CDMI fell above or below +0.4, respectively.

The CDMI has proved to be a highly reproducible feature of a given fibroblast line's *in vitro* phenotype. For example, both uncloned and cloned populations of adult fibroblasts consistently expressed 'adult-like' CDMI values throughout their entire *in vitro* lifespan (Schor et al., 1985b); in contrast, fetal fibrobasts (both uncloned and cloned) consistently displayed CDMI values falling within the fetal range for the first 3/4 of their *in vitro* lifespan, whereupon the majority underwent a spontaneous transition to an 'adult-like' migratory phenotype which persisted until the cells became senescent.

Subsequent studies indicated that (a) tumor-derived fibroblasts obtained from approximately 50% of sporadic breast cancer patients expressed a fetal-like migratory phenotype (Schor et al., 1985a), (b) paired skin fibroblasts obtained from the same individuals were also fetal-like, thereby indicating the systemic nature of this stromal cell abnormality (Durning et al., 1984; Schor et al., 1985a), and (c) skin fibroblasts obtained from approximately 90% of patients with familial

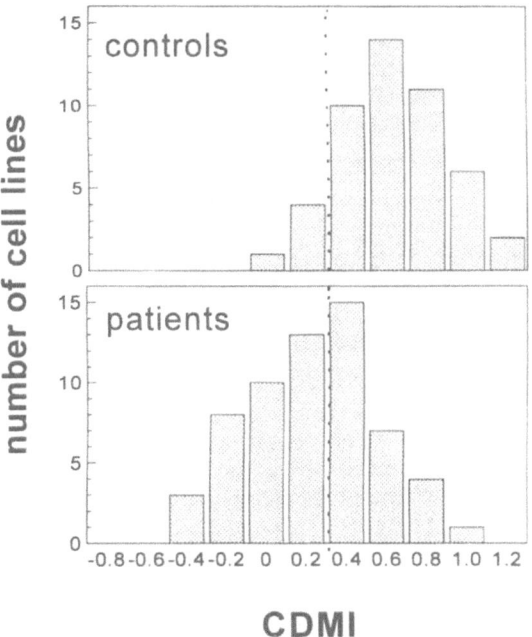

CDMI

Figure 2. Distribution of CDMI values expressed by control and cancer patient skin fibrob-
lsts. Summary data are presented concerning the CDMI values expressed by skin fibroblasts
obtained from patients with carcinoma of the breast (n = 32), carcinoma of the colon
(n = 12), carcinoma of the lung (n = 5), carcinoma of the prostate (n = 2), melanoma (n = 7)
and soft tissue sarcoma (n = 4). Vertical dotted line indicates division between fetal and adult
CDMI values.

breast cancer behaves in a similar fetal-like fashion, as did greater than
50% of their unaffected first-degree relatives (Schor et al., 1986; Haggie
et al., 1987). The high incidence of fibroblasts exhibiting a fetal-like
migratory phenotype in the unaffected first-degree relatives of the familial patients (ie. a population with a clearly documented elevated risk of
developing breast cancer) (Ottman et al., 1983) is of particular significance as it indicates that the systemic presence of these cells is likely to
precede the development of overt malignant disease in these individuals.
The presence of fetal-like skin fibroblasts is not unique to breast cancer;
ongoing related studies (Schor et al., 1985a; unpublished observations)
have indicated that skin fibroblasts obtained from approximately 50%
of patients with a variety of common carcinomas similarly display a
fetal-like migratory phenotype (Fig. 2). In all of these studies, there was
an approximate 10% incidence of fetal-like migratory behavior by skin
fibroblasts obtained from age- and sex-matched healthy controls.

A significant proportion (approximately 25%) of the fetal-like
fibrobasts obtained from breast cancer patients also differed from
normal adult fibroblasts with respect to their ability to form colonies in
suspension (unpublished observations); they were, however, indistin-

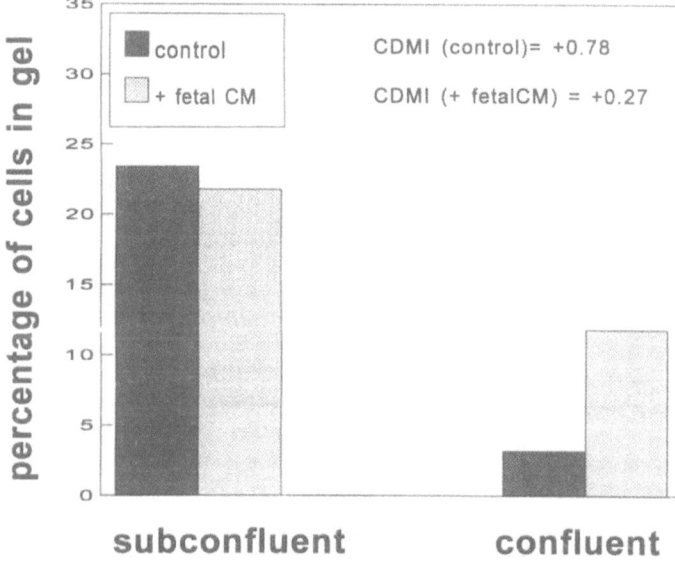

Figure 3. The effect of fetal conditioned medium on the migration of adult skin fibroblasts. Confluent adult skin fibroblats display a characteristic down regulation in migration compared to cells plated at subconfluence. Incubation in the presence of fetal fibroblast conditioned medium (CM) had no apparent effect on the migration of subconfluent cells, but significantly stimulated the migration of confluent ones.

guishable from their normal adult counterparts with respect to morphology, saturation cell density and serum dependence for growth. Finally, it should be noted that the cancer patient fetal-like fibroblasts differed from *bona fide* fetal cells in that they did not undergo a spontaneous transition in CDMI as a function of passage, but rather continued to display CDMI values falling within the fetal range for the entire duration of their *in vitro* lifespan (Schor et al., 1988a).

Identification of *migration stimulating factor*

The distinct distribution profiles of CDMI values expressed by fetal and adult fibroblasts appear to result from the differential effect of cell confluency on migration: ie. the migration of confluent adult fibroblasts is down regulated (Fig. 3), whilst that of fetal cells remains relatively unchanged (Schor et al., 1985b). Initial attempts to understand the biochemical basis of the different responses of fetal and adult fibroblasts to cell confluency indicated that the migration of confluent adult fibroblasts was stimulated by medium conditioned by fetal cells, this leading to the adult cells displaying a characteristic fetal-like CDMI (Fig. 3); the converse incubation of fetal fibroblasts with adult cell conditioned medium was without effect (data not shown). These and related obser-

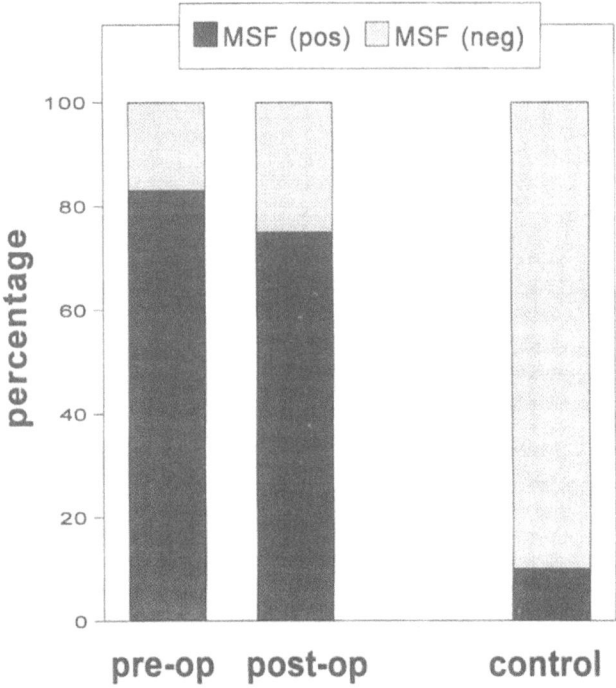

Figure 4. The presence of detectable levels of MSF in the serum of both treated and untreated breast cancer patients. Serum was collected from breast cancer patients 24 hr prior to surgery (pre-op) and 4 days after resection of their tumours (post-op). Serum was also collected from age- and sex-matched healthy controls. The presence of detectable MSF activity in these samples was ascertained using the adult fibroblast migration bioassay and confirmed by specific neutralization with the anti-MSF antibodies. The percentage of each patient group containing detectable MSF activity is indicated by the dark bar and the percentage devoid of activity by the stacked light bar. Details regarding patients and experimental procedures may be found in Picardo et al. (1991).

vations led to the conclusion that fetal skin fibroblasts produced a soluble 'migration stimulating factor' (MSF) which was not made by adult cells (Schor et al., 1988b). This putative soluble factor has subsequently been purified from fetal fibroblast conditioned medium (Grey et al., 1989; Schor et al., 1993). Our recent data indicate that it is a protein which easily degrades into biologically active fragments of approximately 120, 60 and 33 kDa. Amino acid sequence analyses indicate that the 120 and 60 kDa fragments contain apparently novel sequences, whilst the 33 kDa fragment contains an N-terminus exhibiting considerable homology with the gelatin-binding domain of fibronectin. Ongoing studies are concerned with the molecular cloning and complete sequencing of the gene coding for MSF.

Related studies indicated that fetal-like skin fibroblasts obtained from cancer patients also produce MSF (Schor et al., 1988a; unpublished observations). Interestingly, detectable levels of MSF are present in the serum of sporadic breast cancer patients (Picardo et al., 1991). In this study, serum was collected from newly diagnosed patients (n = 12) from whom serum was collected both 24 hours prior to surgical resection of the primary tumor and 4 days postoperatively. Serum samples were also collected from age-matched healthy controls with no family history of breast cancer (n = 20). Our data indicated that MSF was present in 10/12 (83.3%) of serum samples obtained from patients prior to surgery and 9/12 (75%) of these same individuals four days post-operatively (Fig. 4). In marked contrast, we detected MSF activity in only 2/20 (10%) of control sera. Biochemical characterisation of the serum-derived MSF indicated that it was indistinguishable from the fibroblast-produced molecule. The presence of MSF in post-operative serum suggests that it is not a marker of tumor burden, but may rather reflect the systemic and persistent presence of a population of fetal-like (MSF-producing) fibroblasts in these individuals.

Initial studies concerned with the mode of action of MSF indicated that it stimulates the synthesis of a high molecular weight species of hyaluronic acid (HA) (Ellis et al., 1992). HA is a linear glycosaminogly-can consisting of repeating disaccharide subunits of glucuronic acid and N-acetylglucosamine. It is a major biosynthetic product of fibroblasts and has been shown to promote the migration of a number of cell types both during embryonic development (Toole and Trelstad, 1971; Mark-wald et al., 1978) and *in vitro* (Turley and Torrance, 1984; Docherty et al., 1989). The involvement of HA in modulating cell migration *in vivo* appears to continue in the adult, with various lines of evidence implicating it in the regulation of cell movement in pathological processes, such as wound healing and tumor invasion (Knudson et al., 1989).

Our data further indicated that the stimulation of fibroblasts migration by MSF is in fact a secondary consequence of its effect upon HA synthesis. This was first suggested by the observation that the stimulation of both cell migration and HA synthesis by MSF follow parallel biphasic dose-response curves (Schor et al., 1989). The stimulatory effect of MSF on cell migration was also completely blocked by co-exposure of fibroblasts to *Streptomyces* hyaluronidase during the four day duration of the assay. Finally, the addition of exogenous high molecular mass HA to control adult fibroblasts was found to induce the stimulation of their migration in a biphasic dose-dependent fashion similar to that produced by MSF (Ellis et al., 1992). The stimulation of HA synthesis by MSF and the apparent subsequent effect of this HA on cell migration should caution us that the ascription of a biologial function in the naming of a cytokine (eg., *migration* stimulating factor) generally reflects its activity in the particular bioassay first used in its identification rather than being a necessarily accurate description of its principal physiological function *in vivo*. In this regard, many well-characterized

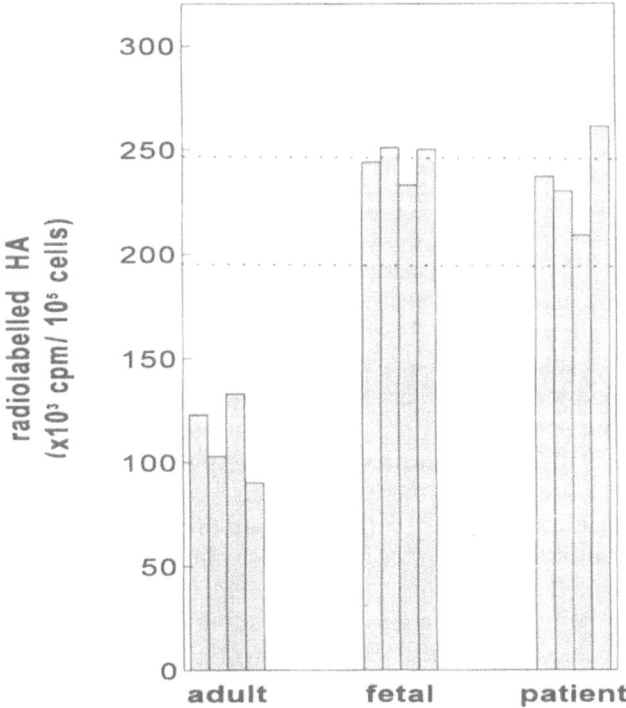

Figure 5. The synthesis of HA by confluent adult, fetal and breast cancer patient skin fibroblasts. Data are present for four different lines of adult, fetal and breast cancer patient skin fibroblasts. Synthesis of HA by subconfluent fibroblasts of each type were indistinguishable and fell within the range defined by the two dotted lines. The results obtained with confluent cells are indicated by bars. Confluent adult fibroblasts displayed a clear down regulation in HA synthesis at confluence, whilst the fetal and fetal-like cancer patient cells did not.

'growth factors' (such as PDGF and EGF) also affect a variety of aspects of cell behavior in addition to proliferation, these including cell migration (Rosen and Goldberg, 1989). With this multi-functionality of cytokine action in mind, it should be noted that the principal biological activity of MSF *in vivo* may not relate to cell mortility *per se*, but rather some other aspect of cell function, such as HA production (Schor et al., 1993).

In a related series of experiments, we noted that HA synthesis by fetal and adult fibroblasts displayed the same characteristic dependence on cell density as previously reported for cell migration (Chen et al., 1989): ie., adult fibroblasts exhibited a down-regulation in HA synthesis at confluence, whilst fetal cells continued to synthesize relatively elevated levels. Data collected in a recently completed study indicate that the fetal-like fibroblasts of breast cancer patients also fail to display a down-regulation in HA synthesis at confluence (Fig. 5).

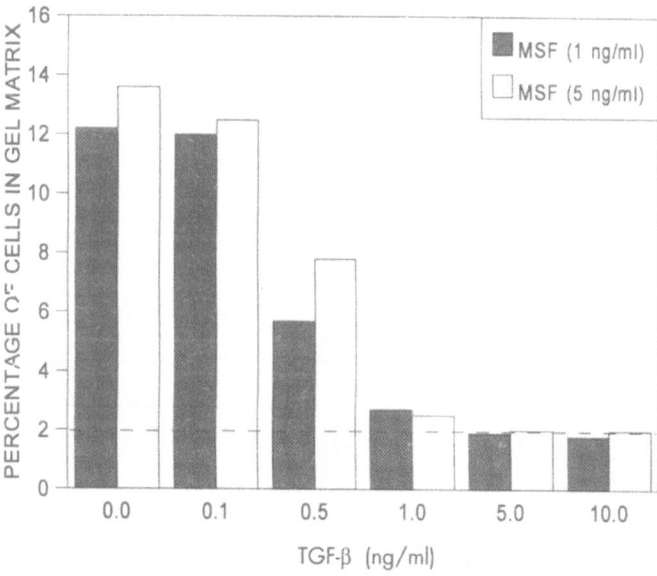

Figure 6. The effects of TGF-β on the stimulation of fibroblast migration by MSF. Confluent adult fibroblasts were incubated with 1.0 and 5.0 ng/ml MSF in the presence of various concentrations of TGF-β1 and the percentage of cells within the collagen matrix determined 4 days later. Control cells incubated in the absence of both MSF and TGF-β (dotted line) displayed a low level of migration (2.0 $\pm$ 0.2%).

The biological activity of a particular cytokine is influenced by the presence of other cytokines in the microenvironment. In view of this interdependence of cytokine function, we have been particularly interested in ascertaining the interaction of MSF with other potentially relevant molecules. In this regard, we have reported that the stimulatory effects of MSF on both cell migration (Fig. 6; Ellis et al., 1992) and HA biosynthesis (Ellis et al., 1992) are inhibited by TGF-β1. TGF-β2 and -β3 were equally effective inhibitors of MSF activity, whilst other cytokines examined (EGF, TGF-α, PDGF, IL-1, aFGF, bFGF) were without effect. The observed antagonistic interaction between MSF and TGF-β may reflect their 'balancing' role in the control of cell behavior during various physiological and pathological events, such as wound healing.

We have recently generated two polyclonal rabbit antibodies to MSF; the first (pAb-1) was raised against MSF purified from fetal fibroblast conditioned medium and the second (pAb-2) against a synthetic peptide based on the apparently novel N-terminal sequence of the 60 kDa fragment of MSF. Both antibodies exerted identical effects on the behavior of adult and fetal(-like) fibroblasts (Fig. 7). In summary: (a) the relatively elevated migration of all three cell types at subconfluent densities was unaffected by incubation with either antibody, (b) the

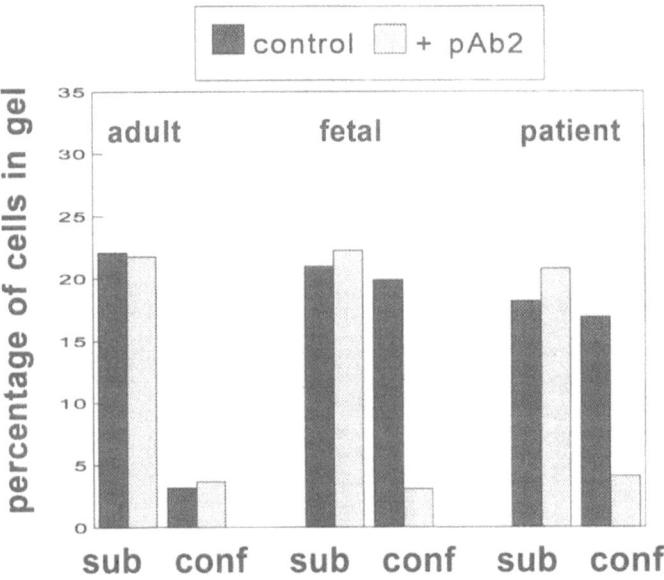

Figure 7. The effects of anti-MSF antibody on the migration of adult, fetal and cancer patient fibroblasts. Control adult fibroblasts displayed a characteristic down regulation of cell migration at confluence, whilst fetal and cancer patient fibroblasts did not (dark bars). Incubation of cells in the presence of either the pAb-1 (not shown) or pAb-2 (light bars) anti-MSF antibodies had no effect on the migration of subconfluent fibroblasts, but significantly inhibited the elevated migration displayed by confluent fetal and cancer patient cells. (sub = subconfluent; conf = confluent).

down-regulated migration of confluent adult fibroblast was similarly unaffected by antibody, and (c) the elevated migration of confluent fetal fibroblasts and the fetal-like fibroblasts of cancer patients was significantly inhibited by antibody. Interestingly, the effect of both antibodies on HA synthesis by these cell (at both subconfluent and confluent densities) exactly paralleled their effects on migration (manuscript in preparation). These observations are consistent with a mechanistic link between the effect of MSF on HA synthesis and cell migration, and further suggest that the observed differences between fetal(-like) and adult fibroblasts with respect to cell migration and HA synthesis are primarily due to the autocrine production of MSF by the former. As expected, both antibodies also completely neutralized the stimulation of adult fibroblast migration and HA synthesis by exogenus MSF (data not shown).

Other observations documenting the presence of aberrant fibroblasts in cancer patients

The detection of MSF-secreting fibroblasts in a significant proportion of cancer patients is consistent with numerous published reports documenting the presence of aberrant skin and tumor-derived fibroblasts in patients with a variety of both sporadic and hereditary cancers (reviewed earlier by Schor et al., 1987). More recently published data continue to expand our awareness of the presence of aberrant fibroblasts in cancer patients. For example, Thielmann et al. (1987) reported that dermal fibroblasts obtained from patients with squamous cell carcinoma (SCC) and (to a lesser extent) basal cell carcinoma exhibited impaired DNA repair synthesis. Danes et al. (1990) similarly reported that skin fibroblasts obtained from patients with oral SCC exhibited an increased tendency towards hyperdiploidy when cultured *in vitro*; they further demonstrated a correlation between the precise anatomical location of SCC within the oral cavity and the expression of this skin fibroblast abnormality. Svendsen et al. (1989) provided evidence that skin fibroblasts from patients with apparently sporadic colon cancer displayed a similar propensity to develop *in vitro* hyperdiploidy.

Tumor-derived fibroblasts have been reported to stain with antibodies specific for a cell surface glycoprotein (F19) previously shown to be expressed by sarcomas and proliferating cultured fibroblasts, but not by normal resting cells (Garin-Chesa et al., 1990). In this study, stromal fibroblasts associated with a number of apparently sporadic cancers (including colorectal, breast, ovarian, bladder and lung) were found to stain positively for F19. These workers further noted the co-distribution of F19 reactive fibroblasts with the fetal-associated protein tenascin.

Taken together with our own observations relating to the presence of fetal-like (MSF-secreting) fibroblasts in cancer patients, these findings raise two important questions, namely: (1) What, if any, is the role of these apparently aberrant fibroblasts in cancer pathogenesis?, and (2) What factors account for their presence in cancer patients? Our views regarding these key issues are presented in the following two sections.

Aberrant fibroblasts as accelerators of cancer progression

The vast majority of studies documenting the presence of aberrant fibroblasts in cancer patients have avoided ascribing any potential function to these cells in cancer pathogenesis. Indeed, where this question is explicitly addressed, it is generally assumed that such aberrant fibroblasts result from the inheritance of a germ line mutation (as will be discussed in the following section) which only contributes to cancer pathogenesis when also expressed in the relevant target epithelial cell population. According to this model, the presence of aberrant fibroblasts are viewed merely as convenient phenotypic markers of such putative genetic lesions and are not themselves considered to be relevant to disease progression.

We have taken a rather different view based on the well documented role of epithelial–stromal interactions in the control of epithelial growth and differentiation during normal development (Hay, 1993) and tissue homeostasis in the adult (Cunha et al., 1985). These tissue interactions play a significant role in the control of fundamental aspects of cell behavior such as proliferation, migration and gene expression. It has therefore been suggested that perturbation of epithelial–stromal interactions in the adult may contribute to the pathogenesis of various disease states (including cancer) characterized by abnormalities in epithelial cell proliferation and differentiation (Rubin, 1985). Several studies have lent experimental support to this view by indicating that interactions between tumor cells and fibroblasts enhance tumor cell growth (Camps et al., 1990) and metastasis *in vivo* (Picard et al., 1986; Tanaka et al., 1988; Gärtner et al., 1992). *In vitro* studies by Dabbous et al. (1987) revealed the existence of significant clonal heterogeneity in the production of proteases by tumor-associated fibroblasts in response to inductive signals from co-cultured carcinoma cells. Matsumoto et al. (1989) reported that fibroblasts promoted the invasion of oral squamous cell carcinoma cells into 3D collagen gels.

Such interactions between tumor cells and fibroblasts are complex; the precise effects of fibroblast populations have been reported to vary both as a function of disease progression (early *versus* late stages) and fibroblast site of origin (Cornil et al., 1991; Lu et al., 1992; Fabra et al., 1992). Perhaps the most dramatic manifestation of such tissue level mechanisms relates to the control of gene expression. In this regard, several studies have indicated that the appropriate host environment may actually suppress the expression of oncogenes ordinarily sufficient to result in tumor development (Mintz and Illmensee, 1975; Stoker et al., 1990).

Signalling between epithelial cells and fibroblasts involves the complex interplay of (soluble) cytokines and (insoluble) matrix macromolecules. The effects of cytokines and matrix macromolecules are mutually interdependent in the sense that (a) the specific response of cells to cytokines is modulated by the nature of the surrounding matrix, and (b) the deposition of this matrix is modulated by the action of cytokines (Schor and Schor, 1987; Nathan and Sporn, 1991; Schor, 1994). Interactions between tumor cells and fibroblasts have been reported to affect the production of both cytokines and matrix macromolecules in a bidirectional fashion (Edward et al., 1992). These interactions involve positive feedback loops which may result in an expansion in cell number and/or amplification of signal molecule synthesis. For example, Cullen et al. (1991) reported that mammary carcinoma cells synthesize PDGF which stimulates fibroblast proliferation and synthesis of IGF I and II; interestingly, the fibroblast-produced IGF in turn stimulates mammary carcinoma cell proliferation and synthesis of PDGF.

We have previously speculated that the apparently persistent fetal-like fibroblasts of cancer patients may contribute to the course of disease progression by creating a *milieu* which promotes the clonal expansion and invasive behavior of the emerging neoplastic cell population (Schor et al., 1987; Schor et al., 1993). Our data suggest that the MSF produced by these fibroblasts may function in this capacity, perhaps as a consequence of its effect on HA synthesis. In this regard, various studied have noted elevated levels of HA associated with the stroma of different types of tumors and demonstrated that this is correlated with more aggressive invasive behavior (Toole et al., 1979). HA has also been reported to modulate a number of processes of potential relevance to tumor progression, including the proliferation and differentiation of mammary epithelial cells (Elstad and Hosick, 1987) and angiogenesis (Feinberg and Beebe, 1983). According to our model, MSF functions as an autocrine regulator of fibroblast matrix synthesis, which in turn affects epithelial cell behavior.

A potential beneficial function of MSF has been suggested by our observation that it is present in 16/17 (94.1%) wound fluid samples (Picardo et al., 1992). The directed migration of fibroblasts into the wound site and the transient increase of HA in granulation tissue during the wound healing response are both consistent with the involvement of MSF. Although the source of MSF in wound fluid is not known, its absence from matched serum samples collected from the same patients suggests that it is not released from degranulating platelets nor derived from a plasma transudate. The postulated involvement of MSF in both cancer pathogenesis and wound healing may be relevant to the observation of Dvorak (1986) that the tumor-host interface resembles a 'wound that does not heal'. In this context, the same effects of MSF which contribute positively to wound healing when expressed *locally* and in a *transient* manner may facilitate cancer progression when they are *systematic* and/or *prolonged* in nature.

Tumor progression is an indolent process in which many decades may elapse between inception of the initiating genetic lesion and the emergence of a clinically recognizable malignancy. Information regarding the proportion of 'initiated' cells which eventually go on to develop into a tumor are not generally available, although recent data suggest that this figure may indeed be quite low. For example, Nielsen et al. (1987) have documented the presence of microfoci of carcinoma *in situ* in the breasts of apparently healthy women who died in road traffic accidents; this study revealed that more than 40% of women over the age of 40 had such histologically discernable lesions, although clearly only a relatively small proportion of these would presumably have proceeded to develop into an overt malignancy *during the normal lifespan of the individual*. These observations suggest that factors which may alter the kinetics of progression may actually play an important,

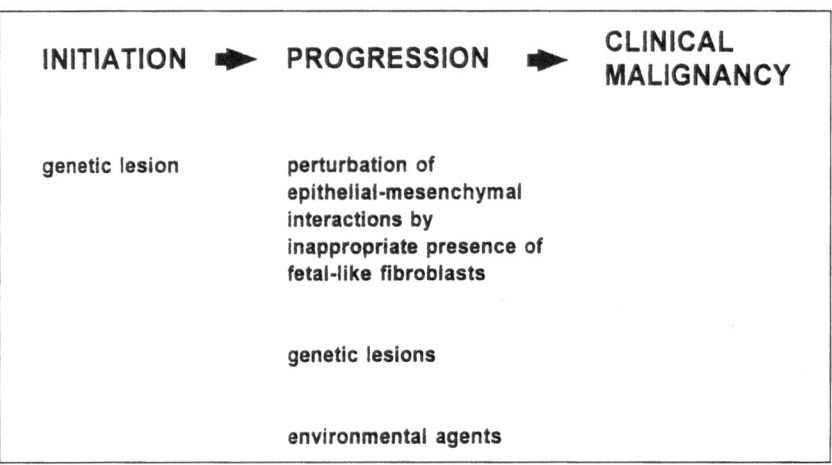

Figure 8. Multistep nature of cancer pathogenesis. We suggest that the perturbation in normal epithelial-mesenchymal interactions resulting from the inappropriate presence of fetal-like fibroblasts may alter the kinetics of tumor regression and thus effectively increase the risk of developing a clinically recognizable malignancy. Other factors (such as the accumulation of genetic lesions and host interactions with environmental agents) may also play a significant role.

and perhaps decisive, role in determining the probability of disease inception.

It is in this postulated role of an 'accelerator' of cancer progression that we view the potential contribution of fetal-like (MSF-secreting) fibroblasts to disease pathogenesis. We have intentionally avoided the word 'promoter' in describing the mode of action of fetal-like fibroblasts in view of the rigorous (and distinct) definition of this term in the carcinogenic literature (Pitot and Cambell, 1987). According to our model, we propose that fetal-like fibroblasts (by virtue of their perturbation of epithelial–mesenchymal interactions) contribute to the creation of an environment which is conducive to the clonal expansion of the evolving (pre-)neoplastic cell population (Fig. 8). Whether this involves permissive or inductive mechanisms, as well as other fundamental questions relating to mechanism, remain to be elucidated. Other factors (such as the continued accumulation of genetic lesions and host interactions with environmental agents) also contribute to the kinetics of tumor progression.

The origin of aberrant fibroblasts in cancer patients: The development of an epigenetic clonal modulation model

As mentioned above, the majority of previous studies documenting the presence of aberrant fibroblasts in cancer patients have considered them to result from the inheritance of a 'partially transforming' genetic lesion. Although this is certainly a plausible explanation for the presence of such fibroblasts in patients with hereditary cancer susceptibility syndromes, it is difficult to reconcile this model with the well-documented presence of aberrant fibroblasts in patients with apparently sporadic disease. In order to address this difficulty, we have suggested that the aberrant skin fibroblasts in cancer patients are not necessarily 'partially transformed', but may instead be expressing a spectrum of phenotypic characteristics displayed by normal fetal cells. In this regard, it should be noted that many of the transformation-associated characteristics displayed by the aberrant cancer patient fibroblasts are also commonly expressed by fetal cells, eg. colony formation in semi-solid medium (Nakano and Ts'O, 1981). The 'fetal-like' nature of tumor-associated fibroblasts has been more explicitly demonstrated by both Azzarone et al. (1984) and Wynford-Thomas et al. (1986) who independently reported that such cells exhibit an *extended* lifespan *in vitro*, characteristic of fetal fibroblasts (NB. not an *indefinite* lifespan characteristic of transformed cells). In a comprehensive study of the pathogenesis of benign prostatic hyperplasia and its subsequent malignant transformation, McNeal (1984) noted the early appearance of a 'fetal-like' stroma and postulated that its interaction with the emerging population of aberrant epithelial cells was a key factor in disease progression. Tumor-associated fibroblasts have also been reported to resemble fetal cells in terms of their expression of a fetal-specific cell surface antigen (Bartal et al., 1986) and production of stromelysin-3 (Basset et al., 1990). Nicolo et al. (1990) have reported that the oncofetal ED-B isoform of fibronectin is preferentially associated with the stroma of several types of malignant neoplasms.

This shift in emphasis from 'partially transformed' to 'fetal-like' is more than just a question of semantics. The designation 'partially transformed' carries with it the implication that this is an aberrant state resulting from an expressed genetic lesion. In contrast, the term fetal-like implies that the particular phenotypic attributes which define this state (eg., continued production of MSF) are inherently physiological, although their expression may be inappropriate in the adult.

Which factors, then, may lead to such a postulated inappropriate persistence of fetal phenotypic characteristics by fibroblasts in the adult? Our consideration of this question is based on an appreciation of the literature indicating that fibroblasts (in spite of their similar morphology) are actually a highly diverse cell population, exhibiting a significant degree of intersite, intrasite and developmental heterogeneity (Schor and Schor, 1987; Bayreuther et al., 1988; Sappino et al., 1990; Gleave et al., 1991; Mukaida et al., 1991). Indeed, on the basis of this well documented heterogeneity, we have proposed a 'clonal modulation' model to account for the presence of MSF-secreting fibroblasts in cancer patients (Schor and Schor, 1987). According to this model, we suggest that (a) distinct subpopulations of MSF-secreting fibroblasts exist in the healthy adult where they may function in processes such as wound healing, (b) there is a prolonged and systemic elevation in the relative number of these fibroblasts in cancer patients, and (c) the factors leading to this systemic effect may involve internal signals (eg. as might be produced by the emerging population of neoplastic cells), as well as host interactions with environmental agents. According to this epigenetic model, the MSF-secreting fibroblasts detected in cancer patients are not considered to be intrinsically aberrant, but rather an expanded subpopulation of cells also present in the healthy adult.

Our recent data clearly confirm that subpopulations of MSF-secreting fibroblasts are in fact present in the healthy adult and that these cells display specific inter- and intra-site heterogeneity with respect to their tissue distribution. For example, fibroblasts obtained from 100% (12/12) of the gingival biopsies examined produced detectable amounts of MSF, whereas none (0/9) of paired forearm skin fibroblasts obtained from the same individuals did so (Irwin et al., 1994). Interestingly, wound healing in the oral mucosa is clinically distinguished from dermal healing in terms of both its rapidity and lack of scar formation. It is possible that the presence of MSF-producing fibroblasts in the oral mucosa contributes to this regenerative and characteristically 'fetal-like' mode of wound healing. This study further revealed the existence of intra-site heterogeneity in the oral mucosa with respect to the tissue distribution of MSF producing fibroblasts. This involved the separation of gingival lamina propria (connective tissue) from its overlying epithelium by exposure to trypsin and the subsequent microdissection of the lamina propria to allow the selective culture of fibroblasts derived from the tips of the papillae and deeper reticular tissue. Only fibroblasts derived from the papillae produced MSF. Prolonged subculture of papillary fibroblasts resulted in their cessation in MSF production and their adoption of a reticular fibroblast phenotype. Staining of gingival tissue with purified anti-MSF antibody confirmed the preferential localisation of MSF in the papillae; interestingly, this particular pattern of MSF distribution is identical to that previously described for tenascin (Sloan, Schor and Lopes, 1990), another molecule preferentially produced by fetal fibroblasts (Chiquet-Ehrismann et al., 1986). We have already commented on the observed co-distribution of F19-positive fibroblasts and tenascin in the stroma of various common carcinoma (Garin-Chesa, Old and Retting, 1990).

We have also observed intra-site heterogeneity with respect to the distribution of MSF-secreting fibroblasts in the normal breast (Schor et al., 1994). Intra- and interlobular fibroblasts were isolated by controlled enzymatic digestion and differential sedimentation from *histologically normal breast tissue* from both non-cancer and cancer patients. Data are

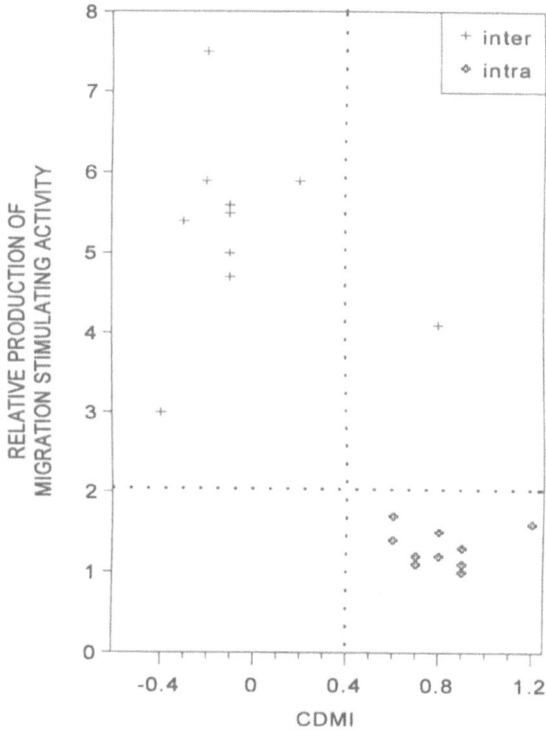

Figure 9. Correlation between CDMI and production of MSF by interlobular and intralobular fibroblasts obtained from non-cancer patients. Fibroblast lines displaying a cell density migration index (CDMI) equal to or greater than +0.4 (dotted line) have been classified as adult-like. The relative migration stimulating activity of the conditioned medium (CM) was assessed by calculating the ratio of the percentage migration of a target adult skin fibroblast cell line in the test CM to that achieved in control cultures in the absence of CM. A ratio equal to or greater than 2 (dotted line) represents a significant difference between test and control samples and has therefore been regarded as indicative of the presence of motility factors in the CM.

presented in Figure 9 regarding the migratory phenotype and production of MSF by interlobular and intralobular mammary fibroblast subpopulations derived from individuals undergoing reduction mammoplasty or resection of benign breast lesions (ie. non-cancer patients). Our data indicate that 91% (10/11) of the interlobular fibroblasts displayed a fetal-like migratory phenotype, compared to 0% (0/10) of the intralobular cells. Significant differences were also observed with respect to the production of MSF these cells, with 100% (11/11) of the interlobular lines and none (0/10) of the intralobular lines secreting this factor. As expected, there is a highly significant correlation between the expression of a fetal-like CDMI and production of MSF. These data therefore indicate the presence of a clearly defined intra-site heterogene-

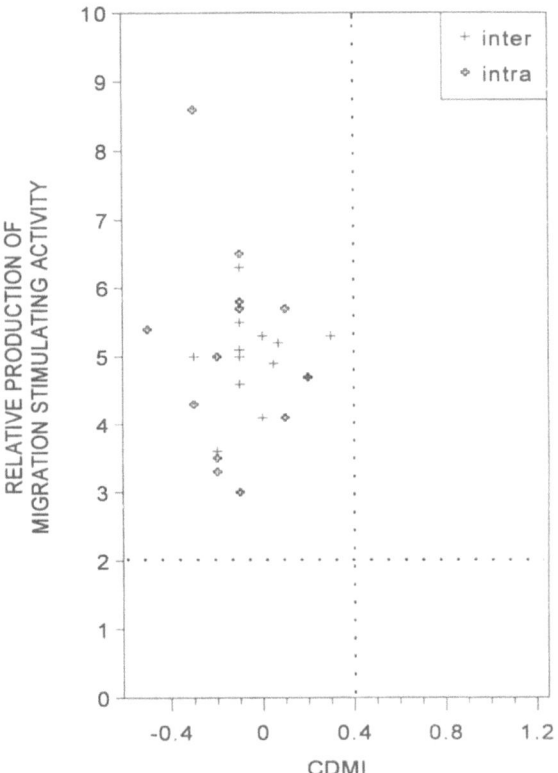

Figure 10. Correlation between CDMI and production of MSF interlobular and intralobular fibroblasts obtained from cancer patients. Fibroblast lines displaying a cell density migration index (CDMI) equal to or greater than +0.4 (dotted line) have been classified as adult-like, whilst cell lines displaying lower CDMI values have been classified as fetal-like. The relative migration stimulating activity of the conditioned medium (CM) was assessed by calculating the ratio of the percengate migration of a target adult skin fibroblast cell line in the test CM to that achieved in control cultures in the absence of CM. A ratio equal to or greater than 2 (dotted line) represents a significant difference between test and control samples and has therefore been regarded as indicative of the presence of motility factors in the CM.

ity with respect to MSF production by fibroblasts in the normal breast of non-cancer patients.

Comparable data are presented in Figure 10 for mammary fibroblasts obtained from histologically normal breast tissue adjacent to a carcinoma (ie. cancer patients). In these individuals, our results indicate that 100% of *both* interlobular (13/13) *and* intralobular (13/13) fibroblast lines displayed a fetal-like migratory phenotype and secreted detectable quantities of MSF. These observations therefore unambiguously indicate the presence of a functionally aberrant (ie., MSF-secreting) subpopulation of intralobular fibroblasts in histologically normal breast tissue adjacent to a carcinoma. Whether the presence of these fetal-like intralobular fibroblasts preceded carcinoma development or resulted from perturbation of the surrounding normal tissue by the carcinoma cannot be ascertained on the basis of the present data. In this regard, Slaughter et al. (1953) first postulated that a primary tumor may induce the 'cancerization' of adjacent normal tissue. This hypothesis was based on their observation that dysplastic epithelial foci are commonly present in apparently uninvolved tissue surrounding oral squamous cell carcinoma. Data supporting this 'field effect' theory has been obtained in a number of subsequent studies involving various types of primary malignancies (Neugut et al., 1992; Crissman et al., 1993; Ogden et al., 1993). In this context, it should be noted that cancer patient intralobular fibroblasts continued to express a fetal-like migratory phenotype and produce MSF *in vitro* (ie., *in the absence of carcinoma cells*), thereby suggesting that these 'aberrant' characteristics are not dependent upon continued proximity to the primary tumor, but are stable features of their *in vitro* phenotype.

Skin fibroblasts obtained from certain of these non-cancer and cancer patients were also examined. In keeping with our previous observations, we found that all skin fibroblast lines obtained from non-cancer patients exhibited an adult-like migratory phenotype (5/5) and did not secrete motility factors (4/4). In contrast, skin fibroblasts derived from 3/7 (43%) of cancer patients displayed a fetal-like migratory phenotype and secreted motility factors; the remaining 4/7 lines (57%) displayed an adult-like migratory phenotype and did not secrete detectable quantities of motility factors. The systemic presence of aberrant skin fibroblasts at uninvolved sites (eg. skin) has previously been attributed to an 'extended field effect' of the distant primary tumor (Nohammer et al., 1989), although the mechanisms by which this effect may be mediated were not discussed.

Little is known about the signals which lead to the postulated systemic expansion of MSF-producing fibroblasts. This may involve the proliferative expansion of pre-existent MSF-producing cells and/or the induction of MSF synthesis in facultative fibroblasts in response to soluble factors produced by the emerging population of pre-neoplastic cells (perhaps even before these are histologically recognizable). Alternatively, locally produced toxins (as in the gut) or host interactions with environmental agents may induce a ('wound healing') response characterized by the systemic expansion of MSF-producing fibroblasts (either by proliferation or induction of MSF synthesis).

The essential discriminating feature of our 'clonal modulation' model is that epigenetic, rather than genetic, mechanisms are postulated to account for the presence of fetal-like fibroblasts in cancer patients. These epigenetic mechanisms rely on continuing cell–cell interactions and environmental factors to regulate the postulated dynamic balance of phenotype expression within the fibroblast population.

Prospectives for future studies

In spite of the general acceptance of the multi-step nature of cancer pathogenesis, little information is currently available regarding either the frequency at which initiating genetic lesions occur or the proportion of the resulting initiated cells which proceed to develop into a *clinically recognizable* malignancy. As discussed above, our assessment of the literature suggests that initiating events may actually be relatively common occurrences and that only a small proportion of such initiated cells go on to produce clinical tumors. Seen in this context, the various events which contribute to tumor progression play a critical role in determining the kinetics of disease development and may thereby furnish a potential novel target for intervention in individuals believed to be at elevated risk. In this context, the observed antagonism between MSF and TGF-β1 may provide a useful starting point.

The presence of detectable levels of MSF in the serum of breast cancer patients suggests that its evaluation, presumably in concert with other relevant factors, may contribute to the identification of individuals at elevated risk of developing clinically identifiable malignancies. The postulated epigenetic nature of the perturbation in the relative abundance and/or tissue distribution of MSF-secreting fibroblast subpopulations would allow for the assessment of potentially relevant non-genetic insults and thereby complement screening protocols solely based on identifying genetic lesions.

Finally, we would like to emphasize our belief that cancer pathogenesis must be viewed within the context of the whole organism in which control mechanisms operate at various levels of complexity. It is therefore important to guard against adopting an overly reductionist approach and consider the complex interplay between molecular and tissue level control mechanisms.

Acknowledgements
This work was supported by grants from the Cancer Research Campaign and Medical Research Council.

References

Azzarone, B., Marell, M., Billard, C., Scemana, P., Chaponnier, C. and Marciera-Coelho, M. (1984) Abnormal properties of skin fibroblasts from patients with breast cancer. *Int. J. Cancer* 33: 759–764.

Bartal, A.H., Lichtig, C., Cardo, C.C., Feit, C., Robinson, E. and Hirshaut, Y. (1986) Monoclonal antibody defining fibroblasts appearing in fetal and neoplastic tissue. *J. Natl. Cancer Inst.* 76: 415–419.

Basset, P., Bellocq, J.P., Wolf, C., Stoll, I., Hutin, P., Limacher, J.M., Podhajcer, O.L., Chenard, M.P., Rio, M.C. and Chambon, P. (1990) A novel metalloproteinase gene specifically expresed in stromal cells of breast carcinomas. *Nature* 348: 699–704.

Bayreuther, K., Rodemann, H.P., Hommel, R., Dittman, K., Albiez, M. and Francz, P.I. (1988) Human skin fibroblasts *in vitro* differentiate along a terminal cell lineage. *Proc. Natl. Acad. Sci. USA* 85: 5112–5116.

Birchmeier, W., Behrens, J., Weidner, K.M., Frixen, U.H. and Schipper, J. (1991) Dominant and recessive genes involved in tumor cell invasion. *Curr. Opin. Cell Biol.* 3: 832–840.

Camps, J.L., Chang, S.M., Hsu, T.C., Freeman, M.R., Hong, S.J., Zhau, H.E., von Eschenbach, A.C. and Chung, L.W. (1990) Fibroblast-mediated acceleration of human epithelial tumor-growth in vitro. *Proc. Natl. Acad. Sci USA* 87: 75–79.

Chen, J., Grant, M.E., Schor, A.M. and Schor, S.L. (1989) Differences between adult and foetal fibroblasts in the regulation of hyaluronate synthesis: correlation with migratory activity. *J. Cell Sci.* 94: 577–589.

Chiquet-Ehrismann, R., Mackie, E.J., Pearson, C.A. and Sakaura, T. (1986) Tenascin: an extracellular matrix protein involved in tissue interactions during fetal development and oncogenesis. *Cell* 47: 131–139.

Cornil, I., Theodorescu, D., Man, S., Herlyn, M., Jambrosic, J. and Kerbel, R. S. (1991) Fibroblast cell interactions with human melanoma cells affect tumor cell growth as a function of tumor progression. *Proc. Natl. Acad. Sci. USA* 88: 6028–6032.

Crissman, J.D., Sakr, W.A., Hussein, M.E. and Pontes, J.E. (1993) DNA quantification of intraepithelial neoplasia and invasive carcinoma of the prostate. *Prostate* 22: 155–162.

Cullen, K.J., Smith, H.S., Hill, S., Rosen, N. and Lippman, M.E. (1991) Growth factor messenger RNA expression by human fibroblasts from benign and malignant lesions. *Cancer Res.* 51: 4978–4985.

Cunha, G.R., Bigsby, R.M., Cooke, P.S. and Sugimara, Y. (1985) Stromal–epithelial interactions in adult organs. *Cell Diff.* 17: 137–148.

Dabbous, M.K., Haney, L., Carter, L.M., Paul, A.K. and Reger, J. (1987) Heterogeneity of fibroblast response in host-tumor cell–cell interactions in metastatic tumors. *J. Cell. Biochem.* 35: 333–344.

Danes, B.S., De Angeles, P., Traganos, F., Ringborg, U., Nielsen, L.H. and Melamed, M.R. (1990) Comparison of anatomical location of squamous cell carcinoma within the oral cavity and oropharynx with the incidence of in vitro hyperdiploidly. *Clin. Genet.* 37: 188–193.

Docherty, R., Forrester, J.V., Lackie, J.M. and Gregory, D.W. (1989) Glycosaminoglycans facilitate the movement of fibroblasts through 3D collagen matrices. *J. Cell Sci.* 92: 263–267.

Durning, P., Schor, S.L. and Sellwood, R.A.S. (1984) Fibroblasts from patients with breast cancer show abnormal migratory behavior in vitro. *Lancet* ii: 890–892.

Dvorak, H.F. (1986) Tumors: wounds that do not heal. *New Eng. J. Med.* 315: 1650–1659.

Edward, M., Grant, A.W. and Mackie, R.M. (1992) Human melanoma cell-derived factor(s) stimulate fibroblast glycosaminoglycan synthesis. *Int. J. Cancer* 52: 499–503.

Ellis, I., Grey, A.M., Schor, A.M. and Schor, S.L. (1992) Antagonistic effects of transforming growth factor beta and MSF on fibroblast migration and hyaluronic acid synthesis: possible implications for dermal wound healing. *J. Cell. Sci.* 102: 447–456.

Elstad, C.A. and Hosick, H.L. (1987) Contribution of the extracellular matrix to the growth properties of cells from a preneoplastic outgrowth: possible role of hyaluronic acid. *Exp. Cell Res.* 55: 313–321.

Fabra, A., Nakajima, M., Bucana, C.D. and Fidler, I.J. (1992) Modulation of the invasive phenotype of human colon carcinoma cells by organ specific fibroblasts of nude mice. *Differentiation* 52: 101–110.

Feinberg, R.N. and Beebe, D.C. (1983) Hyaluronate in vasculogenesis. *Science* 220: 1177–1179.

Foulds, L. (1975) *Neoplastic Development*, Vol. 2. Academic Press, London.

Garin-Chesa, P., Old., L.J. and Retting, W.J. (1990) Cell surface glycoprotein of reactive stromal fibroblasts as a potential target in human epithelial cancers. *Proc. Natl. Acad. Sci. USA* 87: 7235–7239.

Gärtner, M.F.R.M., Wilson, E.L. and Dowdle, E.B. (1992) Fibroblast-dependent tumorigenicity of melanoma xenografts in athymic mice. *Int. J. Cancer* 51: 788–791.

Gleave, M., Hsieh, J.-T., Gao, C., von Eschenbach, A.C. and Chung, W.K. (1991) Acceleration of human prostate cancer growth in vivo by factors produced by prostate and bone fibroblasts. *Cancer Res.* 51: 3573–3761.

Grey, A.M., Schor, A.M., Rushton, G., Ellis, I. and Schor, S.L. (1989) Purification of the migration stimulating factor produced by fetal and breast cancer patient fibroblasts. *Proc. Natl. Acad. Sci. USA* 86: 2438–2442.

Haggie, J., Schor, S.L., Howell, A., Birch, J.M. and Sellwood, R.A.S. (1987) Fibroblasts from relatives of hereditary breast cancer display foetal-like behaviour in vitro. *Lancet* i: 1455–1457.

Hay, E. (1993) Extracellular matrix alters epithelial differentiation. *Curr. Op. Cell Biol.* 5: 1029–1035.

Helin, K. and Harlow, E. (1993) The retinoblastoma protein as a transcriptional repressor. *Trends Cell Biol.* 3: 43–46.

Irwin, C.R., Picardo, M., Ellis, I., Sloan, P., Grey, A.M., McGurk, M. and Schor, S.L. (1994) Inter- and intrasite heterogeneity in the expression of fetal-like phenotypic characteristics by gingival fibroblasts: potential significance for wound healing. *J. Cell Sci.* 107: 1333–1346.

Knudson, A.G. (1977) Genetics and etiology of human cancer. *Adv. Hum. Genet.* 8: 1–66.

Knudson, W., Biswas, C., Li, X.-Q., Nemec, R.E. and Toole, B.P. (1989) The role and regulation of tumor-associated hyaluronanh. In: *The Biology of Hyaluronan*. Ciba Foundation Symposium 143. John Wiley and Sons, Chichester, pp 150–159.

Lu, C., Vickers, M.F. and Kerbel, R.S. (1992) Interleukin 6: a fibroblast-derived growth inhibitor of human melanoma cells from early but not advanced stages of tumor progression. *Proc. Natl. Acad. Sic. USA* 89: 9215–9219.

Lynch, H.T., Albano, W.A., Heieck, J.J., Mulcahy, G.M., Lynch, J.F., Layton, M.A. and Danes, B.S. (1984) Genetics, biomarkers and control of breast cancer: a review. *Cancer Genet. Cytogenet.* 13: 43–92.

Markwald, R.R., Fitzharris, T.P., Bank, H. and Bernanke, D.H. (1978) Structural analysis on the material organization of glycosaminoglycans in developing endocardial cushions. *Dev. Biol.* 62: 292–316.

Matsumoto, K., Horikoshi, M., Rikimaru, K. and Enomoto, S. (1989) A study of an *in vitro* model of squamous cell carcinoma. *J. Oral. Pathol. Med.* 18: 498–501.

McNeal, J.E. (1984) Anatomy of the prostate and morphogenesis of BPH: *In*: A.B. Kimbal, K. Buhl and H.C. Carter (eds): *New Approaches to the Study of Benign Prostatic Hyperplasia*. A.R. Liss, Inc., New York, pp 27–53.

Mintz, B. and Illmensee, K. (1975) Normal genetically mosaic mice produced from malignant teratorcarcinoma cells. *Proc. Natl. Acad. Sci. USA* 72: 3585–3589.

Mukaida, H., Hirabayashi, N., Hirai, T., Iwata, T., Saeki, S. and Toge, T. (1991) Significance of freshly cultured fibroblasts from different tissues in promoting cancer cell growth. *Int. J. Cancer* 48: 423–427.

Nakajima, M. and Chop, A. (1991) Tumor invasion and extracellular marix degradative enzymes: regulation of activity of organ factors. *Seminars Cancer Biol.* 2: 115–127.

Nakano, S. and Ts'O, P.O. (1981) Cellular differentiation and neoplasia: characterization of subpopulations of cells that have neoplasia related growth properties in Syrian hamster embryo cell cultures. *Proc. Natl. Acad. Sci. USA* 78: 4995–4999.

Nathan, C. and Sporn, M. (1991) Cytokines in context. *J. Cell Biol.* 113: 981–986.

Neugut, A.I., Sherr, D., Robinson, E., Murray, T. and Nieves, J. (1992) Differences between first and second primary lung cancer. *Cancer Epidemiol. Biomarkers* 1: 109–112.

Nielsen, M., Thomsen, J.L., Primdahl, S., Dyreborg, U. and Andersen, J.A. (1987) Breast flcancer and atypia among young and middle-aged women. *Br. J. Cancer* 56: 814–819.

Nicolo, G., Salvi, S., Oliveri, G., Borsi, L., Castellani, P. and Zardi, L. (1990) Expression of tenascin and the ED-B containing oncofetal fibronectin isoform in human cancer. *Cell. Differ. Dev.* 32: 401–408.

Nohammer, G., Bajardi, F., Benedetto, C., Kresbach, H., Rojanapo, W. and Schauenstein, E. (1989) Histophotometric quantification of the field effect and the extended field effect of tumors. *Free Rad. Res. Commun.* 7: 129–137.

Ogden, G.R., Lane, E.B., Hopwood, D.V. and Chisholm, D.M. (1993) Evidence for field change in oral cancer based on cytokeratin expression. *Br. J. Cancer* 67: 1324–1330.

Ottman, R., King, M.C., Pike, M.C. and Henderson, B.E. (1983) Practical guide for estimating risk for familial breast cancer. *Lancet* ii: 556–558.

Picard, O., Rolland, Y. and Poupon, M.F. (1986) Fibroblast-dependent tumorigenicity of cells in nude mice: implication for implantation and metastasis. *Cancer Res.* 46: 3290–3294.

Picardo, M., Schor, S.L., Grey, A.M., Howell, A., Laidlow, I., Redford, J. and Schor, A.M. (1991) Migration stimulating activity in serum of breast cancer patients. *Lancet* 337: 130–133.

Picardo, M., Grey, A.M., McGurk, M., Ellis, I. and Schor, S.L. (1992) Identification of migration stimulating factor in wound fluid. *Exp. Mol. Path.* 57: 8–21.

Pitot, H.C. and Cambell, H.A. (1987) An approach to the determination of the relative potencies of chemical agents during the stages of initiation and promotion in multistage hepatocarcinogenesis in the rat. *Environ. Health Perspect.* 76: 49–56.

Rosen, E.M. and Goldberg, I.D. (1989) Protein factors which regulate cell motility. *In Vitro* 25: 1079–1087.

Rubin, H. (1985) Cancer as a dynamic developmental disorder. *Cancer Res.* 45: 2935–2942.

Sappino, A.P., Schurch, W. and Gabbiani, G. (1990) Differentiation repertoire of fibroblastic cells: expression of cytoskeletal proteins as marker of phenotypic modulations. *Lab. Invest.* 63: 144–161.

Schor, S.L. (1980) Cell proliferation and migration within three-dimensional collagen gels. *J. Cell Sci.* 41: 159–175.

Schor, S.L., Schor, A.M., Durning, P. and Rushton, G. (1985a) Skin fibroblasts obtained from cancer patients display foetal-like migratory behaviour on collagen gels. *J. Cell Sci.* 73: 235–244.

Schor, S.L., Schor, A.M., Rushton, G. and Smith, L. (1985b) Adult, foetal, and transformed fibroblasts display different migratory phenotypes on collagen gels: Evidence for an isoformic transition during foetal development. *J. Cell Sci.* 73: 221–234.

Schor, S.L., Haggie, J., Durning, P., Howell, A., Sellwood, R.A.S. and Crowther, D. (1986) The occurrence of a foetal fibroblast phenotype in breast cancer. *Int. J. Cancer* 37: 831–836.

Schor, S.L. and Schor, A.M. (1987) Clonal heterogeneity in fibroblasts: implications for the control of epithelial-mesenchymal interactions. *BioEssays* 7: 200–204.

Schor, S.L., Schor, A.M., Howell, A. and Crowther, D. (1987) Hypothesis: persistent expression of fetal phenotypic characteristics by fibroblasts is associated with an increased susceptibility to neoplastic disease. *Exp. Cell Biol.* 55: 11–17.

Schor, S.L., Schor, A.M. and Rushton, G. (1988a) Fibroblasts from cancer patients display a mixture of both foetal and adult-like phenotypic characteristics. *J. Cell Sci.* 90: 401–407.

Schor, S.L., Schor, A.M., Grey, A.M. and Rushton, G. (1988b) Fetal and cancer patient fibroblasts produce an autocrine migration stimulating factor not made by normal adult cells. *J. Cell Sci.* 90: 391–399.

Schor, S.L., Schor, A.M., Grey, A.M., Chen, J., Rushton, G. and Ellis, I. (1989) Mechanism of action of the migration stimulating factor produced by fetal and cancer patient fibroblasts: effect on hyaluronic acid synthesis. *In Vitro* 25: 737–746.

Schor, S.L., Grey, A.M., Ellis, I., Schor, A.M., Coles, B. and Murphy, R. (1993) Migration stimulating factor: its structural homology to the gelatin-binding domain of fibronectin, mode of action and possible function in health and disease. *In*: G. Evans, C. Wigley and R. Warn (eds): *Cell Behaviour: Adhesion and Motility*. S.E.B. Symposium, No. 47, pp. 235–251.

Schor, S.L. (1994) Cytokine control of cell motility: modulation and mediation by the extracellular matrix. *Prog. Growth Factor Res.* 5: 223–248.

Schor, A.M., Rushton, G., Ferguson, J.E., Howell, A., Redford, J. and Schor, S.L. (1994) Phenotypic heterogeneity in breast fibroblasts: functional anomaly in fibroblasts from histologically normal tissue adjacent to carcinoma. *Int. J. Cancer.* 59: 25–32.

Slaughter, D.P., Southwick, H.W. and Smejkal, W. (1953) Field cancerization in oral stratified squamous epithelium. *Cancer* 6: 963–968.

Sloan, P., Schor, S.L. and Lopes, V. (1990) Immunohistochemical study of the heterogeneity of tenascin distribution within the oral mucosa of the mouse. *Archs. oral Biol.* 35: 67–70.

Steeg, P. (1991) Genetic control of the metastatic phenotype. *Seminars Cancer Cell Biol.* 2: 105–110.

Stoker, A.W., Hatier, C. and Bissell, M.J. (1990) The embryonic environment strongly attenuates v-*src* oncogenesis in mesenchymal and epithelial tissues, but not in endothelia. *J. Cell Biol.* 111: 217–228.

Svendsen, L.B., Thorup, J., Larson, J.K., Norgard, T., Willumsen, H. and Hansen, O.H. (1989) Association between tumor DNA aneuploidy and *in vitro* tetraploidy of skin fibroblasts in patients with colorectal neoplasms. *Scand. J. Gastroenterol.* 24: 755–760.

Tanaka, H., Mori, Y., Ishii, H. and Akedo, H. (1988) Enhancement of metastatic capacity of fibroblast-tumor cell interaction in mice. *Cancer Res.* 48: 1456–1459.

Thielmann, H.W., Edler, L., Burkhardt, M.R. and Jung, E.G. (1987) DNA repair synthesis in fibroblast strains from patients with actinic keratosis, squamous cell carcinoma, basal cell carcinoma, or malignant melanoma after treatment with ultraviolet light, N-acetoxy-2-acetyl-aminofluorene, methyl methanesulfonate and N-methyl-N-nitrosurea. *J. Cancer Res. Clin. Oncol.* 113: 171–186.

Toole, B.P. and Trelstad, R.L. (1971) Hyaluronate production and removal during corneal development in the chick. *Dev. Biol.* 26: 28–35.

Toole, B.P., Biswas, C. and Gross, J. (1979) Hyaluronate and invasiveness of the rabbit V2 carcinoma. *Proc. Natl. Acad. Sci. USA* 76: 6299–6303.

Turley, E.A. and Torrance, J. (1984) Localization of hyaluronate and hyaluronate-binding protein on motile and non-motile fibroblasts. *Expl. Cell Res.* 161: 17–28.

Wynford-Thomas, D., Smith, P. and Williams, E.D. (1986) Prolongation of fibroblast lifespan associated with epithelial rat tumor development. *Cancer Res.* 46: 3125–3127.

Zetter, B.R. and Brightman, S.E. (1990) Cell motility and the extracellular matrix. *Curr. Opin. Cell Biol.* 2: 850–856.

Subject index

(The page number refers to the first page of the chapter in which the keyword occurs)

P. Jollès, *University of Paris V and C.R.N.S., Paris, France (Ed.)*

Proteoglycans

1994. 280 pages. Hardcover
ISBN 3-7643-2957-2 (EXS 70)

Proteoglycans are glycoconjugates which constitute a topic of current interest in biochemistry, molecular biology, cell biology, and medicine. They can no longer be considered, as they were a few years ago, as simple inert building blocks.

Proteoglycans belong to a versatile protein family whose potential functions arise either from their glycosaminoglycan chains or from specific regions of their protein cores. They help maintain the essential microenvironment for cell adhesion, migration, and proliferation through their ability to function as links between cells and the extracellular matrix, and as growth-factor binders.

Their importance is gaining recognition in a wide spectrum of research areas such as the brain and reproduction. In this book, various new research aspects of proteoglycans are reviewed in a series of chapters written by well-known scientists involved in this exciting field of research.

Birkhäuser Verlag • Basel • Boston • Berlin

S. Papa, *Institute of Medical Biochemistry and Chemistry, University of Bari, Italy*
J.M. Tager, *E.C. Slater Institute, University of Amsterdam, The Netherlands (Eds)*

Biochemistry
of Cell Membranes
A Compendium of Selected Topics

1995. 376 pages. Hardcover. ISBN 3-7643-5056-3 (MCBU)

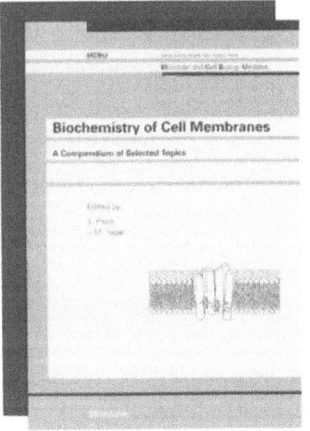

This book consists of a series of reviews on selected topics within the rapidly and vastly expanding field of membrane biology. Its aim is to highlight the most significant and important advances that have been made in recent years in understanding the structure, dynamics and functions of cell membranes.

Areas covered in this monograph include

- Signal Transduction
- Membrane Traffic: Protein and Lipids
- Bioenergetics: Energy Transfer and Membrane Transport
- Cellular Ion Homeostasis
- Growth Factors and Adhesion Molecules
- Structural Analysis of Membrane Proteins
- Membranes and Disease

Biochemistry of Cell Membranes should serve as a benchmark for indicating the most important lines for future research in these areas.

Birkhäuser Verlag • Basel • Boston • Berlin